LOGIC AND GENERAL THEORY OF SCIENCE
LECTURES 1917/18
WITH SUPPLEMENTARY TEXTS FROM THE
FIRST VERSION OF 1910/11

EDMUND HUSSERL
COLLECTED WORKS

EDITOR:
JULIA JANSEN

VOLUME XV

LOGIC AND GENERAL THEORY OF SCIENCE
LECTURES 1917/18
WITH SUPPLEMENTARY TEXTS FROM
THE FIRST VERSION OF 1910/11

TRANSLATIONS
PREPARED UNDER THE AUSPICES OF
THE HUSSERL-ARCHIVES (LEUVEN)

More information about this series at
http://www.springer.com/series/6059

EDMUND HUSSERL

LOGIC AND GENERAL THEORY OF SCIENCE

Lectures 1917/18
with Supplementary Texts from the First Version of 1910/11

TRANSLATED BY

CLAIRE ORTIZ HILL
PARIS, FRANCE

Springer

Edmund Husserl

Translated by
Claire Ortiz Hill
Paris, France

Husserliana: Edmund Husserl – Collected Works
ISBN 978-3-030-14528-6 ISBN 978-3-030-14529-3 (eBook)
https://doi.org/10.1007/978-3-030-14529-3

This Springer imprint is published by the registered company Springer Nature Switzerland AG.
The registered company address is: Gewerbestrasse 11, 6330 Cham, Switzerland

This translation is dedicated
to the memory of Dallas Willard,
my precursor

ACKNOWLEDGMENTS

As the years fly by, I grow increasingly grateful for the truly generous, steadfast support I have received from senior colleagues, who for decades, year after year, always made themselves available to help me. So once again, I must thank, in alphabetical order, Paul Gochet (†2011), Ivor Grattan-Guinness (†2014), Jaakko Hintikka (†2015), Ruth Barcan Marcus (†2012), and Dallas Willard (†2013). To the above list, I must add my spiritual director of 24 years Jacques Sommet s.j. (†2012), Barry Smith, my first philosophy teacher Roger Schmidt (†2018), my last philosophy teacher Maurice Clavelin, and Bernd Magnus (†2014), who in 1970 agreed to direct my Senior Honors Thesis on Husserl and in 1977 oriented me toward work on Husserl and contemporary logic. Then one thing led to another. I am profoundly grateful to you.

This translation was made possible by a fellowship from the National Endowment for the Humanities in Washington, D.C. I am very grateful to it for its support. It is dedicated to the memory of my precursor Dallas Willard.

CONTENTS

LOGIC AND GENERAL THEORY OF SCIENCE, LECTURES 1917/18, WITH SUPPLEMENTARY TEXTS FROM THE FIRST VERSION OF 1910/11

SECTION I

FUNDAMENTAL CONSIDERATIONS FOR THE DEMARCATION AND CHARACTERIZATION OF FORMAL LOGIC

TRANSLATOR'S INTRODUCTION

Logic and General Theory of Science is the translation of Edmund Husserl's *Logik und Allgemeine Wissenschaftstheorie* edited by Ursula Panzer under the auspices of the Husserl Archives in Cologne and published for the first time in 1996 as volume XXX of the Husserliana series. The main text of the volume is the final version of lecture courses on the philosophy of logic and science given by Husserl at the universities of Göttingen and Freiburg between 1910 and 1918. The course was initially entitled *Logik als Theorie der Erkenntnis* (Logic as Theory of Knowledge) and was taught in Göttingen for the first time during the Winter semester of 1910/11. According to Panzer's introduction (p. XIX), looking back, Husserl characterized it as the final version of his Göttingen lectures on formal logic. He would teach it a second and third time during the Winter semesters of 1912/13 and 1914/15 under the title *Logik und Einleitung in die Wissenschaftstlehre* (Logic and Introduction to the Theory of Science) and a fourth time at Freiburg during the Winter semester of 1917/18 under the present title.

Panzer explains that these lectures were considerably reworked over the years, a fact considered to be of particular significance because those changes were made during a decisive period in the evolution of Husserl's thought. She notes that Husserl wrote to Georg Misch on November 16, 1930, that, by the time his *Ideas* was published in 1913, he had already lost all the interest that formal logic and all real (*reale*) ontology had held for him in the face of a systematic grounding of a theory of transcendental subjectivity.[1] And, of the very subjects making up much of *Logic and General Theory of Science*, he wrote to Hermann Weyl on October 10, 1918, that for 20 years, a central part of his lectures had been the theory of functional judgments, of judgments with empty places, and the distinguishing of the different modes of this empty some-

[1] Cited Hua XXX, p. XXIII, n. 4.

thing and, in addition, the implementation of the fundamental distinctions between factual and formal ways of judgment, between proposition form and proposition (or judgment), proof- and theory-form and theory, as well as of the objective correlates associated with them. His concept of complete manifolds already acquired in the beginning of the 1890s had also, as had all the distinctions he had dealt with extensively in lessons in Göttingen, proved especially fruitful. But he added that, in spite of all that, and of all the work he had devoted to it, he had not found the time and serenity to carry that train of thought completely to the end, because it had had to be more important to him to develop his ideas about transcendental phenomenology.[1]

Prepared for oral delivery in the classroom, these lectures are refreshingly lively and spontaneous, clearer, more explicit, and readable than the books Husserl published during his lifetime. Appendices II and III convey the enthusiasm for logic and philosophy of science that he was intent upon injecting into his students by means of them. There he explains that through his lectures, he was seeking to provide something entirely different from what could be learned from books, because lectures do not in general exist for the purpose of replacing books or to be spoken books or excerpts from books. He warned students that the usual expositions of logic are dangerous for beginners, because they only too easily deaden their sense of genuine scientificity and cover up "the difficulties from them, like chasms covered over with greenery." It was his desire instead to introduce them to the inmost essence of work in logic and critique of knowledge, to the nature of their problems and methods, and prepare them to be able to derive benefit and excitement from reading important writings in the field, something which is achieved by laying bare the essential meaning of centuries of efforts as they relate to the state of things at the time and its insights, by imparting an understanding of the problems, goals, methods, and a sense of the deeper meaning of the theoretical efforts "in which the greatness, grandeur, and force of philosophical science lies," by teaching them to sense the inmost spirit of the intentions of logic and theory of knowledge and to

[1] Cited Hua XXX, p. XXIII n. 1.

feel profoundly concerned by the fact that the problems to be surmounted were of grave import for anyone interested in ultimate truth and philosophy.

The only justified task of the teacher, he declares, is to train beginners to philosophize. If "philosophy" is the word for the highest aims of knowledge and the sciences directly oriented toward them, he suggests, then the discipline we want to devote ourselves to is in fact first philosophy, because it is a matter of investigations that must precede and be attended to before further philosophizing can be contemplated in earnest. He defines his mission as being one of raising students above the naïve standpoint and showing them "the way to the mysterious solitudes in which one day the sphinx of knowledge must unveil its riddle." He says that he could also call it the way to the "mothers" of knowledge, to the essence-principles of knowledge in terms of its ultimate origins. Regarding these mothers, he says that he loves to recall Mephistopheles' words in Goethe's *Faust*: "Enthroned sublime in solitude are goddesses. Around them is no place, still less any time." Mephistopheles, he reminds students, had described that solitude in a horrifying manner and tried to deter aspirants from venturing into those untrodden, and not to be trodden, places where they would see nothing "in interminably empty farness," not hear their own footsteps, not find anything firm where they rested. Husserl, however, counseled his students to be unafraid and to respond with Faust, "Just keep on, we want to fathom it. In your nothing, I hope to find the universe."[1]

These lectures find Husserl enthusing that even though formal logic belongs to one of the oldest sciences of mankind, everything in it is evolving; the "ossified concepts and formulas of the tradition are coming alive again, moving, evolving" (§19b). In *Alte und Neue Logik*, taught in 1908/09, he speaks of what a delight it is to be alive and to share in striving after the greatness coming into being in those days, which are not, as often said, times of decadence, but rather the beginning of a truly great philosophical era in which age-old goals will finally be met at the cost of truly heroic strain from toil and new, higher, goals will everywhere be held out.

---

[1] Johann Wolfgang von Goethe, *Faust*, Section II, Act I.

We in modern philosophy, he said, are no less than visionaries (*Phantasten*). We have the courage and determination of the highest goals, but we strive after them on the most reliable paths, those of patient, constant work.[1]

When trying to answer questions that Husserl's philosophy has raised, philosophers of logic, science, and mathematics often find that he seemed to assume his readers to be in possession of facts about the development of his thought that are not found in the writings published during his lifetime. By making available clear, explicit, and revealing discussions of important topics less thoroughly and clearly treated in those writings, the publication of his lecture courses is gradually providing really significant material for putting together a complete picture of his teachings on crucial matters. By providing material necessary for assuring that his ideas will one day have the impact on the history, philosophy, invention, and pedagogy of modern logic, set theory, and the foundations of mathematics that they should have had from the beginning, these lecture courses afford needed insight into how the father of phenomenology managed to exercise the profound philosophical impact he did on the times that gave birth to twentieth-century philosophy.

This particular volume is a very important link in the chain of ideas connecting Husserl's *Logical Investigations* of the turn of the century to his *Formal and Transcendental Logic* of 1929 and helps to subvert many *idées reçues* about the development of his thought. For instance, it routs the widely held view that he lapsed back into psychologism after trouncing it in the earlier work. Indeed, these lectures show him still pursuing the course set out in that work well into the second decade of the century and displaying utter consistency with stands he had begun taking toward, for example, categoriality, analyticity, axiomatization, completeness, Platonism, idealism, empiricism, logicism, manifolds, mathematics, anti-psychologism, objective and subjective presentation, pure logic, and dependent and independent meanings, during the 1890s, and which were still pursued in his latest work, topics, moreover, that were of

[1] Husserl, Edmund, *Alte und Neue Logik, Vorlesung 1908/09*, Dordrecht, Kluwer, 2003, p. 6.

fundamental importance to the makers of twentieth-century philosophy in English-speaking countries and still are to their followers (e.g., ones enchanted by Michael Dummett's interpretation of Gottlob Frege's writings).

In addition to the topics just mentioned, philosophers trained in the Analytic tradition are presently especially well-primed to appreciate a wealth of interesting insights published here, for example, those into (in alphabetical order) the word "all," demonstratives, existential generalization, extensionality, formal logic, the foundations of arithmetic, functions and arguments, identity/equivalence/equality, imaginary numbers, logicism, mathematizing logic, modality, philosophical grammar, philosophy of language, quantification, relations, sense and meaning, set theory, states of affairs, theory of inference, theory of judgment, theory of probability, truth, and wholes and parts. This is of particular significance because these subjects have long been the bailiwick of philosophers who, loath to inquire into their own history and to acquire the linguistic skills needed to study works unavailable in English, have long viewed Husserl's work through a glass darkly.

In contrast, Husserl rarely mentions phenomenology here, to which there are only 11 references beginning in the last third of the course. Four of them appear in the appendices. Only in §69, the very last section of the work, where Husserl argues that a systematic phenomenology of consciousness and its consciousness-correlates is needed to solve the problems of reason, is phenomenology considered somewhat closely. The word "transcendental" is mentioned a mere 15 times and is so only in connection with transcendental numbers, Kantian philosophy, or to defend himself against charges that he was trying to penetrate into a mystical transcendental world by means of his own intellect.

The main text of this last version of the lecture course is divided into three sections. The first is devoted to defining and characterizing formal logic, the second to a systematic theory of meaning and judgment, and the third to the theory of science. It is followed by 18 appendices of related material which Husserl himself had designated for insertion into specific places within the text or individual pages which he had added to the bundle of papers making up the text of the course. All the texts published in the appendices are from *Logik als Theorie der Erkenntnis* 1910/11. Kant is the philosopher most cited here, followed closely behind by Brentano. Only Aristotle, Bolzano, and Euclid merit more than a few mentions.

Husserl states the subject and purpose of these lectures in the first few minutes of the course. He says that it can be designated by one word: "understanding," a synonym of "reason" signifying the many mental activities and achievements which span all areas of the life of the mind and are familiar to everyone from experiences prior to any logic. Understanding, he tells students, governs all the sciences, technical arts, and spheres of extra-theoretical life. Sciences are only truly sciences to the extent that the understanding governs their findings and gives them form. It is what sets norms for them, inquires into what is right, and demonstrates error. It creates the unity of theoretical knowledge out of unconnected experiences, isolated convictions, presumptions, and inferences. In all domains of nature, it unveils the inviolable system of laws that are called the theory of the particular field, and it is owing to it that nature figures before the mind's eye as ruled by laws. It casts light on the darkness of the inner life and the life of the will and judges beauty and ugliness, goodness and badness, and appropriateness and inappropriateness.

An essentially new line of inquiry opens up, Husserl teaches, when we turn our gaze away from the things we know through the understanding to consider the ways in which it operates and inquire into the forms in which this takes place, which forms they imprint upon the contents thought by means of it, which norms they are bound to in so doing, and how reaching or missing the goal of truth essentially depends upon the observance or nonobservance of these forms and norms. He cites John Locke who, in *An Essay Concerning Human Understanding*, compared the understanding to the eye, which, while making all things knowable for us, does not perceive itself and only with great difficulty can make itself its own object. He maintains that it is only in logic – whose goal it is to acquire systematically exhaustive, purely theoretical self-knowledge of the understanding's ways of experiencing and of all the kinds and forms of the thought-contents specific to it – that the understanding reflectively investigates itself. So it is that the purpose of this course is to provide a glimpse of the vast science that this awareness opens up and of the great importance the attainment of this aim has for knowledge. In so doing, he seeks to introduce beginners to great and interesting areas of logic to the extent that they can lay claim to scientificity. The imperfect state of knowledge of the understanding, he declares in Appendix II, explains the imperfect development of philosophy and why dogmatic-didactic presentations of philosophy prove fruitless. There, he cites Kant's words in the *Critique of Pure Reason* that one cannot learn philosophy but only to philosophize.[1]

Much of this work is devoted to an exploration of the realm of meaning, which Husserl credits Bolzano with having sighted. Meaning, he suggests, displays an inner structure which is amazingly regular in form and can be compared to that of a crystal. Just as crystals have their crystal form and conform to a crystal system, he theorizes, it is part of the essence of meanings that "they form fixed configurations into concrete meanings such that all meaning is bound, so to speak, to fixed crystal-configurations and only so crystallized can have concrete being." According to him, every judgment is itself a crystal, a crystal configuration, a crystal structure which is the formal structure of its components. He judges the metaphor to be apt except for the fact that in the sphere of meaning, every crystal must be from the fixed, coherent system; there is no amorphousness that the same matter can assume (§27b).

In §19a, Husserl importantly says that the Fourth Logical Investigation represented his first attempt to do justice to the theory of forms of meanings as a highly important underlying level of formal logic. In Note 3 found at the end of the Fourth Logical Investigation of the second edition of the *Logical Investigations*, he had lamented the fact that logic still lacked a prime foundation, a scientifically rigorous, phenomenologically elucidated distinction between primitive meaning elements and structures and essence-laws germane to knowledge, because logicians had never scientifically formulated a purely logical theory of forms. He maintained that that was why the many theories of concept or of judg-

[1] Immanuel Kant, *Critique of Pure Reason*, II. Transcendental Doctrine of Method, Chap. 3, The Architectonic of Pure Reason.

ment had yielded so few tenable results. He had concluded §13 of the Fourth Logical Investigation of that edition writing that he hoped that the much improved study of the theory of forms of meanings that he had hinted at in a note at the end of that section of the first edition, and that he had expounded in his courses at Göttingen since 1901, would be made available to a wider public. According to Panzer's introduction (p. XXVII), the more thorough presentation of the theory of meaning forms sketched in the Fourth Logical Investigation is published for the first time in this volume.

In these lectures, Husserl indeed further develops ideas set forth in the Fourth Logical Investigation. For example, §§20–21 find him teaching students that the difference between independent and dependent meanings is ideal, therefore a priori, and that pertaining to it is the wealth of a priori laws which he had called "purely" logical-grammatical in the *Logical Investigations*. These ideas about the distinction between independent and dependent meanings make for interesting comparison with Frege's and Russell's thoughts on the same, something which I studied in "On Fundamental Differences between Dependent and Independent Meanings" and "Incomplete Symbols, Dependent Meanings, and Paradox."[1]

All philosophical thinking has its share of special terminology to which philosophers are obliged to adjust if they are ever to enter into the ideas propounded there. For example, Husserl taught that there are forms of subjective discourse which make no reference to the contingent empirical persons investigating and substantiating it or for which reference to empirical-psychological facts is extra-essential. In the sciences, he held, it is not just a question of facts and meanings of statements, but it is also a question of insight into thinking itself with respect to its legitimacy.

---

[1]"On Fundamental Differences between Dependent and Independent Meanings", *Axiomathes, An International Journal in Ontology and Cognitive Systems* 20: 2–3, online since May 29, 2010, 313–32, (DOI https://doi.org/10.1007/s10516-010-9104-1). "Incomplete Symbols, Dependent Meanings, and Paradoxes", in *Husserl's Logical Investigations*, Daniel O. Dahlstrom (ed.), Dordrecht, Kluwer, 2003, 69–93. Both papers are anthologized in Claire Ortiz Hill and Jairo José da Silva, *The Road Not Taken, On Husserl's Philosophy of Logic and Mathematics*, London, College Publications, 2013.

In his discussions of these matters, he uses the term "noetics," which will be unfamiliar to some readers. He characterizes noetics as a systematically formal theory of justification of knowledge which stands opposite to formal logic, analytic ontology, and the theory of meaning but is attached to them in the most intimate way. It is theory of science in the highest sense and also the discipline that makes the ultimate and highest fulfillment of epistemological needs possible, because only absolute knowledge can provide ultimate epistemological satisfaction. Part of the essence of all knowledge, he maintains, is that the Idea of the absolute "hovers" above it as its "guiding star," so that if philosophy is to be the name for every kind of scientific investigation aspiring to serve the striving for absolute knowledge, then all logical disciplines, and noetics above all, deserve to be called philosophical disciplines (§§66–69).

It is well-known that Franz Brentano taught Husserl that thinking has what he calls here the "obvious, and therefore wonderful" characteristic of being intentional, of being thinking about something, which for him amounted to its having inherent meaning (§19d). He calls this the "miracle of consciousness" and considers that for the philosophically naïve, it seems most obvious that in subjective experiencing something can be intended which is not itself an experience but lies beyond the experience and that when it comes to such experiences, subjects can be certain of the objective validity of their intending. However, he objects, this obviousness actually proves to be the "enigma of all enigmas," the original one of which centers on the original fact that each experience called consciousness has intrinsic meaning (Appendix V).

Cognizant of the importance of the ambiguity, the two-sidedness, of the word "knowledge," i.e., the fact that it may signify either the knowing process or what is known, as it is known, in §7, Husserl very succinctly defines what he came to call the "noematic" perspective as the turning toward what is known as such – toward the noema – as the essential counterpart to noetic reflections on the forms of knowing in pure generality from the standpoint of correctness.

In *Ideas I*, he was much more forthcoming about exactly what noemata are. Explaining the need to distinguish between the parts and phases of the intentional experience and the fact that it is consciousness of something, he explained there that

> Every intentional experience… is noetic, it is its essential nature to harbour in itself a "meaning" of some sort… and on the ground of this gift of meaning, and in harmony therewith, to develop further phases which through it become themselves "meaningful." Such noetic phases include… the directing of the glance of the pure Ego upon the object "intended" by it in virtue of its gift of meaning, upon that which "it has in its mind as something meant"…. Corresponding… to the manifold data of the real (*reellen*) noetic content, there is a variety of data displayable in really pure (*wirklich reiner*)… intuition, and in a correlative "*noematic content*," or briefly "*noema*"…. Perception… has its noema, and at the base of this is its perceptual meaning… the *perceived as such*. Similarly, the recollection, when it occurs, has its own *remembered as such*, precisely as it is "meant" and "consciously known" in it; so again judging has its own the *judged as such*, pleasure, the pleasing as such…. We must everywhere take the noematic correlate, which (in a very extended meaning of the term) is here referred to as "meaning" (*Sinn*) *precisely* as it lies "immanent" in the experience of perception, of judgment, of liking…. (§88)

Thinking as thinking about something, Husserl teaches in these lectures, has an immanent a priori constitution and from the noematic perspective has necessary types and forms in which it alone can acquire a relationship to objectivity, in which it can intend – mean – something and do so prior to any question as to whether the intending is valid or not. The nature and form of the intentionality of thinking is reflected in the nature and form of logical meaning, so that understanding the basic composition of logical meaning affords insight into the a priori essence of thinking and vice versa. The theory of forms of meanings thereby gives a theory of forms of thinking as that of logical meaning. So the theory of forms of logical meanings would also correlatively be a theory of forms of thinking with respect to its meaning-like essence, with respect to the possible forms of its intentionality. He says that studying the wonders of the intentionality of thinking is certainly a matter of the greatest interest and that one thereby goes about this by investigating the basic constitution of the realm of meaning prior to all questions of validity (§19c–d).

So these lectures contain ample information about Husserl's theory of the essential parallelism obtaining between the various kinds of consciousness and the concept of linguistic meaning, thus supporting Dagfinn Føllesdal's thesis that with the theory of noemata, Husserl generalized the notion of linguistic meaning to the realm of all intentional acts.[1] However, the noemata must not be too closely identified with Fregean linguistic meanings, because, as I pointed out in *Word and Object in Husserl, Frege, and Russell*, for Husserl

> the logico-linguistic realm was but a stepping stone to an infinitely vaster realm of inquiry.... he undertook to do something that was radically different from anything Frege and his successors ever tried, or ever considered worth doing: he "plunged into the task of laying open the infinite field of transcendental experience" (CM, p. 31). Not just interested in the meaning of linguistic expressions, Husserl made it his life's work to investigate as thoroughly and painstakingly as humanly possible the meaning conferred upon objects by the intentional acts of consciousness. To explain how the human mind confers meaning on its objects, Husserl posited the presence of structures analogous to the intensional "meanings" used by his contemporaries and predecessors in their discussions of the meanings of words. These structures were the noemata.[2]

Importantly, it was precisely during the years in which Husserl gave the lectures published here that he fully developed the noematic notion of meaning, which did not figure in the *Logical Investigations*. In his 1913 Foreword to the second edition of that work,[3] he even signaled it as a "defect" of the First Logical Investigation that the essential ambiguity of "meaning" as an Idea had been not stressed, because attention had not been paid there to the fundamental role of the distinction between the "noetic" and "noematic" in all fields of consciousness and to the parallelism obtaining between them, which was only fully expounded for the first time in *Ideas I* published in 1913. He regretted the fact that the *Logical Investigations* had unduly, one-sidedly emphasized the noetic concept of meaning and that the problem had not been

---

[1] Dagfinn Føllesdal, "Husserl's Notion of Noema", *Journal of Philosophy* 66 (1969), 680–87.

[2] Claire Ortiz Hill, *Word and Object in Husserl, Frege, and Russell, the Roots of Twentieth Century Philosophy,* Athens, Ohio University Press, p. 30. The page reference in the text is to Husserl's *Cartesian Meditations*, The Hague, Martinus Nijhoff, 1973, p. 31.

[3] See his "Foreword II", *Logical Investigations* vol. I, London, Routledge and Kegan Paul, p. 48.

understood and remedied until the end of the work but considered that the distinction had nonetheless been implicit in many of the arguments of the Sixth Logical Investigation.[1] Along these lines, Panzer stresses in her introduction that one must above all bear in mind that Husserl had had compelling reasons to change the title of his lecture course from *Logik als Theorie der Erkenntnis* (1910/11) to *Logik und Einleitung in die Wissenschaftstlehre* (1912/13, 1914/15), because it had become chiefly a matter of a systematic theory of forms of meanings from the noematic perspective as a basic part of an a priori, and primarily formal, theory of science, whereby the noetic theory of justification was for the time being expressly left to the side (pp. XV–XVI).

Husserl further instructs students that they will encounter a pure logic in a new sense through investigating what is stated in the stating, what is known in the knowing, as it is known, meant, in knowledge. In particular, they will first encounter "apophantic logic," also a term which some readers might find mystifying and which Husserl variously defines as the logic of statements, the logic of asserting propositions, the pure logic of affirmative statements, and the logic of the affirmative predicative proposition, which he considered to be essentially the logic Aristotle dealt with under the heading of Analytics. Aristotle, he notes, called judgment, the statement as such, "apophansis," whose form is symbolized by "A is b." It is a mere predication in which something is stated about something. For Husserl, pure arithmetic, the whole of formal mathematics, the theory of manifolds were for essential reasons intertwined with the logic of affirmations and formed a higher tier of apophantics. He considered recognizing this and characterizing them so to be of the greatest importance philosophically (§§7, 9, 14, 15c, 19c).

---

[1] In this regard, it interesting to compare the theory of noetics expounded in these lectures with his ideas about the "subjective ideal conditions" making possible the operations of thinking in §§32, 64–65 of his "Prolegomena to Pure Logic", the first volume of his *Logical Investigations*, with his §§25–33 *Introduction to Logic and Theory of Knowledge, Lectures 1906/07*; and with §94 of *Ideas I*, where he speaks of noesis and noema in the sphere of judgment.

In §66, he reminds students that in seeking to obtain the first idea of formal logic, and of apophantic logic to begin with, he had told them that sciences live on objectively in the form of writings in which scientific theories are given expression and that this continued existence is one of interrelated statements having their meanings in judgments and relationships of judgments referring to the objectivities of the scientific domain, for example, to numbers in arithmetic, to geometrical forms in geometry, and to the things of the natural world and natural relationships in the natural sciences. He particularly examines, and criticizes, Kant's distinction between analytic and synthetic judgments. Kant, he notes in §46c, had been the first to see the difference between the analytic and synthetic a priori and had rightly called it a classic distinction for transcendental philosophy, but he neither had a genuine concept of the analytical as determined by the conceptual sphere of the formal categories nor had he understood the essence of formal logic. If, Husserl reasons, as Kant wished, we separate the logical a priori from everything empirical-methodological, we see that what is purely logical according to the tradition belongs exclusively in the apophantic sphere and that therefore formal logic to a certain extent sought to be apophantics. However, apart from Leibniz and some of those influenced by him, neither Kant nor anyone else had suspected that pure arithmetic and all the disciplines essentially related to it intrinsically belong together with the old formal logic. Kant's concept of the analytical did not extend beyond his concept of apophantic logic. His definition of analytic judgments is therefore limited to categorical judgments and implies that every categorical judgment is analytic whose predicate concept is contained in the subject concept. So Husserl concludes that, however right Kant's observation was, it blocked the way to the far more important realization of the fundamental separation of everything belonging to the realm of formal category from the sphere of the nonformal a priori, something which leads to a wholly unacceptable equation and equal treatment of arithmetic disciplines with other purely mathematical disciplines and the severing of both from what Kant called pure natural science.

Husserl explains here that by analytic truths in the broadest sense, he himself understands "analytic concept-truths, therefore all pure categorial truths, therefore, the entire pure mathesis, pure logic, then however, also their a priori and empirical individuations, therefore, the analytic necessities." For him, the pure categorial concept-truths essentially belong together as a whole and form a single system of scientific disciplines to be dealt with under the broadest heading of formal logic, or analytics, or the *mathesis universalis* in Leibniz' sense (§§45b, 47e, 50). In Appendix XV, he underscores the purely categorial nature of formal logic. Instead of formal logic, he tells his students, we can also say analytics or science of what is analytically knowable in general, the science that establishes and systematically substantiates analytic, categorial laws.

Among the stated goals of Husserl's late work *Formal and Transcendental Logic* was to redraw the boundary line between logic and mathematics in light of the new investigations into the foundations of mathematics and to examine the logical and epistemological issues such developments raise. This is something that the former student and assistant of Karl Weierstrass in Berlin, long-term friend of Georg Cantor in Halle, and member David Hilbert's circle in Göttingen undertook to do in these lectures. In particular, he argues here that those who had mixed the roles of the philosopher and the mathematician had only succeeded in creating their own closed worlds. Arithmetic, algebra, and analysis, he teaches, developed independently from philosophy and must remain independent. The theory of the analytic and traditional syllogistic logic, he maintains, is a piece of the pure mathematics of propositions and of predicates of possible subjects in general and as such is not the job of philosophers, but of mathematicians, who are the only competent engineers of deductive structures. If, in the nineteenth century, mathematicians had also adopted the deductive theories of traditional syllogistic logic and gradually developed a mathesis of propositional, conceptual, and relational meanings in the spirit of the solely proper mathematical method, they had only laid hold of a field that was their rightful possession (§50).

So, according to him here, all arguing against mathematizing logic was symptomatic of a lack of understanding. With formal mathematics, he asserts, we do not actually enter into an essentially new domain but are dealing with a field of pure concept-truths whose conceptual matter is inseparably linked to the original matter of the logic of meaning. Owing to the work of mathematizing logicians, he stresses, the disciplines of logical validity have reached a higher level of technical perfection in certain ways. They have seen the essential kinship between formal mathematics and formal logic and so have expanded the sphere of the exact mathematical disciplines to include the new ones of formal logic by carrying over to formal logic the same algebraic methods entirely suited to it. However, they have lacked a scientific understanding of thinking and so remain confused about the nature, meaning, and basic concepts of formal logic, and the idea of a theory of forms of meanings as a discipline that by its very nature is a comprehensive, difficult discipline prior to the disciplines of logical validity has also remained completely beyond their ken. The philosopher's task is therefore one of providing a complementary reflection on the essence and meaning of the governing basic concepts and basic laws of deductive theories (§§50, 19a, d).

In these lectures, Husserl additionally spells out for his students how they might convince themselves that there is something in logic that is akin to what mathematicians have in mind when they speak of functions. In particular, he discusses a theory of functional judgments, which he specifically links to what mathematicians call a function, something which he considers that, without arriving at the full descriptive analysis of the kinds of judgment concerned, the sharp-witted Frege, had had the merit of recognizing in his article "Function and Concept." The empty places, Husserl explains, are what mathematicians call arguments. Many empty places can occur in a judgment, so that the same judgment can have several places, or terms, of universality and several of particularity. These places can be pure empty places as, for example, "Something or other is red," in which a nominal something figures in the subject position and is the bearer of a particular function, or universal as, for example, "Everything is red." However, as a rule, the "something" is specified by a letter of the alphabet, as in the arithmetic example, $a + b = b + a$, where two terms function universally and are determined as the numbers a and b. Argument places are specifically nominal forms. He states as a principle that every nominal position in a judgment form can become an argument place and take on the generality forms in relation to it. Arguments or generality positions, terms of universality and particularity also occur in predication since nouns can occur there in many different ways. Everything formal, he stresses, is exclusively composed of terms of this kind. He considers the empty something to be of the greatest importance for the theory of meaning and all of formal logic. As what is specifically mathematical in the mathesis, specifically formal in formal logic, absolute emptiness of content was for him the hallmark of the formal logical. These ideas about functional propositions are expounded in §§26a–b, 32c, 40b–45a, and Appendix XIII.[1]

This is a good place to note that contrary to what is widely believed, Husserl did have a formal language. In §19d of these lectures, he argues that the rigorous carrying out of purely deductive theory requires one and the same method everywhere, namely, what is called the algebraic method, and that mathematicians are everywhere the technical experts qualified in deductive theory, whose development to technical perfection is everywhere the requirement of exact science. In his *Logik, Vorlesung 1896* and his *Logik, Vorlesung 1902/03*, he set out the axioms, notation, and rules of inference for the conceptual and propositional calculus he advocated.[2] In *Alte und Neue Logik*, for example, he resorted to it to show that a propositional form has blank terms, empty terms, or variables, so that one can then say: each system of values satisfying the propositional form or function F(αβ…) also satisfies the function F′ (αβ…) and that that satisfying signifies transforming $\prod_{\alpha\beta\ldots}$ F(αβ…) € F′ (αβ…) into a valid proposition, where "€" is the sign for implication and "′" marks the step from premise to conclusion.[3] Like his contemporaries in Germany, Husserl adopted C. S. Peirce's

[1] See also Claire Ortiz Hill, "Husserl and Frege on Functions", *Husserl and Analytic Philosophy*, Guillermo Rosado Haddock (ed.), Berlin, de Gruyter, 2016, pp. 89–117.

[2] Edmund Husserl, *Logik, Vorlesung 1896*, Dordrecht, Kluwer, 2001, pp. 272–73 and his *Logik, Vorlesung 1902/03*, Dordrecht, Kluwer, 2001, pp. 231, 239–49.

[3] *Op. cit.*, Husserl, *Alte und Neue Logik, Vorlesung 1908/09*, p. 213.

symbols for the universal and existential quantifiers $\prod$, $\sum$ which, unlike Frege's, were widely used.

In §39, Husserl discusses the traditional talk of extensions of concepts according to which the totality of objects that are to be subsumed under each valid universal concept is to belong to it as its extension. He taught that no *pure* concept has anything like an extension and that it is nonsense to say that for *every* concept, a distinction is to be made between intension and extension. Indeed as shown by the remarks he wrote on his copy of Frege's "Function and Concept" available at the Husserl Archives in Leuven, Husserl believed that there are extensionless concepts, that impossible, imaginary, absurd concepts are also concepts. It was in fact precisely the search to justify the use of apparently meaningless signs in calculations or deductive thought that had led him to develop his theory of manifolds as the third and highest level of pure logic. In the *Prolegomena* §70, he called his theory of complete manifolds the key to the only possible solution to how in the realm of numbers impossible, nonexistent, meaningless concepts might be dealt with as real ones. In §72 of *Ideas I*, he wrote that his chief purpose in developing his theory of manifolds – which he likened to Hilbert's axiom systems – was to find a theoretical solution to the problem of imaginary quantities.

Chapter 11 of this lecture course is devoted to a discussion of that theory of manifolds, which nicely complements the discussions of the same in *Introduction to Logic and Theory of Knowledge* (§19) and *Formal and Transcendental Logic* (Chap. 3). Here he teaches that since the procedure used is purely formal, and since not a single concept is used that does not arise out of the analytic sphere, his theory of manifolds, for him the highest level of mathematics, is the supreme consummation of analytics, the ultimate consummation of all purely categorial knowledge (§§57–59).

As is evident from the foregoing, *Logic and General Theory of Science* is replete with insights into matters that many philosophers have now been primed to appreciate out of enthusiasm for Frege's ideas. It in fact takes readers back to the place where two main logical roads diverged during the early part of the twentieth century and affords a look down the one less traveled by. It invites phenomenologists and analytic philosophers alike to overcome pride and prejudice. Indeed, had not so many twentieth-century philosophers resolved to barricade themselves behind the walls of ideological prejudices, the kinds of insights expounded here could have altered philosophical landscape in English-speaking countries, so much so that history will eventually show that logic and philosophy would have followed a different, and better, course in the twentieth century had Husserl's thoughts on these matters found their rightful place alongside the works of Frege, Russell, Carnap, Hilbert, and Gödel, for example.

However, although these lectures are laced with pertinent lessons about matters that lived on to become the stuff of twentieth-century logic and philosophy of science, they at the same time draw clear epistemological and metaphysical lines between Husserl's theories and those that dominated in English-speaking countries. This is particularly evident when one looks at the place Husserl accords to ideal entities here. Indeed, he instructs his students to adopt ontological views about the reality of essences, universals, Ideas, senses, meanings, concepts, attributes/essential properties, modalities, propositions, and intensions – anything hinting of the a priori – that are fundamentally antithetical to those that analytic philosophers have wanted to have. He plainly says that he completely means in absolute earnest that the recognition of ideal objects, or Ideas, as new kinds of atemporal, supra-empirical objects is the pivotal point of all theory of knowledge and that he considers the proper grasp of them to be decisive for all further considerations. He contends that it is imperative to force people to concede once and for all that Ideas are genuine, actual objects by contrasting them with the empirical objects that they see in the natural attitude and alone are inclined to recognize as objects.

He suggests that people who want to know what Ideas are need only to point to self-evident givens like the cardinal number series or to absolutely self-evident statements about members of the number series, which he claims everyone knows in a certain naïve way since they talk of numbers and do so in ideal ways. Only philosophers, he charges, shun Ideas. He contends that once one recognizes givens like the series of natural numbers as objectivities, one can only describe them in the way Plato did in his theory of

Ideas, as eternal, selfsame, nontemporal and nonspatial, unmoved, unchangeable, etc.

He fully realized that what he was claiming was "very hotly combated as being mysticism and scholasticism" that people trained in traditional philosophy instantly think Platonic Ideas and then such Platonic realism becomes associated with mysticism, Neo-Platonism, and a magical view of nature, something as far removed as possible from genuine natural science. Philosophers then recall how this merged into scholastic realism during the Middle Ages. Anyone advocating giving ideal objects their due faces charges of being a reactionary, mystic, and scholastic, the latter two being the strongest scientific terms of abuse of the time, in which formal logic is vilified as being empty scholasticism and espousing idealism for a pure logic is left undefended (§§4, 5a, 8, 19a).

Indeed, on December 29, 1916, Göttingen philosopher Leonard Nelson wrote to David Hilbert that Husserl

> admittedly also originally came from the mathematical school, but... bit by bit turned more and more away from it and turned towards a school of mystical vision, whereby he also deadened the feel in his school for the demands and value of a specifically scientific method. He even goes so far, after his own lack of success with it, as to see a danger in methodological thinking and thinks that it would ruin philosophers for whom the truth only reveals itself in mystical vision. Even though Husserl himself remains protected by certain inhibitions from mystical degeneracy by virtue of strong ties to mathematics that he has not been able to cast off, one must unfortunately nonetheless note with horror that after the school as such had torn down the bridges to mathematics behind them, how unrestrainedly his students lapsed into every excess of Neo-platonic mysticism....[1]

In the lectures published here, Husserl defends himself against such charges by explaining that he was not adopting Ideas and classes of idealities out of some desire to penetrate into a mystical transcendental world by means of his own intellectual intuition but rather for the same banal reason that he embraced things: because he saw them and in looking at them grasped them himself. He even compares ideal objects to ordinary stones found

---

[1]Translated in my book with Jairo da Silva, *The Road Not Taken, On Husserl's Philosophy of Logic and Mathematics* cited above, pp. 390–91.

lying on the road. In response to charges that he was espousing scholasticism, he insisted that he was only asking for the intellectual integrity to allow what is prior to any theory, because it is the most evident of evident facts, to count as being exactly what it proclaims itself to be. If this is enough to have him called a scholastic, he protests, then that is all fine and good, and he asks whether it was not better to have integrity and be called a scholastic or to lack integrity and be a modern empiricist! He says that he advocates integrity and does not fall flat on his face when labeled as a scholastic, because integrity stands the test of time (§§4, 8).

For those influenced by the Analytic tradition in philosophy who feel queasy about such talk of ideal entities, it is worthwhile to note here that in Russell's article entitled "The Philosophical Implications of Mathematical Logic," which is found translated in Husserl's notes on set theory,[1] Russell affirmed "that there is *a priori* and universal knowledge," that "all knowledge which is obtained by reasoning, needs logical principles which are *a priori* and universal." He further wrote that "it is necessary that there should be self-evident logical truths" and that these were "the truths which are the premises of pure mathematics as well as of the deductive elements in every demonstration on any subject whatever." "Logic and mathematics force us, then," Russell wrote, "to admit a kind of realism in the scholastic sense, that is to say, to admit that there is a world of universals and of truths which do not bear directly on such and such a particular existence. … We have immediate knowledge of an indefinite number of propositions about universals: this is an ultimate fact…."[2]

As is all too well-known, Husserlians and followers of the Frege-inspired Analytic school that dominated philosophy in English-speaking countries during the twentieth century have not in general spoken the same language. Fortunately, however, this translation appears at a time when the latter have heartily

---

[1] Published in part in German in G. E. Rosado Haddock (ed.), *Husserl and Analytic Philosophy*, Berlin, de Gruyter, 2016, pp. 289–319. Unfortunately, Russell's text is cited (p. 317), but not reproduced there.

[2] Bertrand Russell, "The Philosophical Implications of Mathematical Logic", *Essays in Analysis,* London, Allen & Unwin, 1973, pp. 292–93.

embraced Frege's thoughts and concerns, many of which he shared with Husserl.

Both Husserl and Frege faced the same terminological confusions, and they both fought their way through a terminological jungle to achieve conceptual clarity in spite of them.[1] Though Frege's choices are now more familiar to most English-speaking philosophers than Husserl's are, they were often eccentric. As Russell wrote when he introduced Frege to the English-speaking world in *The Principles of Mathematics*,

> Frege is compelled, as I have been, to employ common words in technical senses which depart more or less from usage. As his departures are frequently different from mine, a difficulty arises as regards the translation of his terms. Some of these, to avoid confusion, I shall leave untranslated, since every English equivalent that I can think of has already been employed by me in a slightly different sense.[2]

For example, of Frege's famous distinction between *Sinn* and *Bedeutung*, Russell wrote "The distinction between meaning (*Sinn*) and indication (*Bedeutung*) is roughly, though not exactly, equivalent to my distinction between a concept as such and what the concept denotes."[3] For his part, in these lectures, Husserl explains his use of the term "nucleus" in these terms,

> the word concept is so ambiguous and in particular also used so ambiguously in the field of formal logic itself that we cannot use it without thinking twice. In any event, it may be said that by means of my analysis, an extraordinarily important meaning of the word concept as "nucleus-content" has been scientifically defined. And at the same time, the general-nucleus narrows down the meaning of concept in a clear-cut way. It is indeed often said that generality is part of the essence of concepts. §25)

In the First Logical Investigation, he had written of how in the absence of fixed terminological landmarks concepts run confusedly together and fundamental confusions arise, and he went on to defend his decision to use "sense" and "meaning" as synonyms as follows,

---

[1]This is one of the main subjects of my *Word and Object in Husserl, Frege and Russell* cited above.

[2]Bertrand Russell, *The Principles of Mathematics*, London, Norton, 1903, p. 501.

[3]*Ibid*., p. 502.

> It is agreeable to have parallel, interchangeable terms in the case of this concept, particularly since the sense of the term 'meaning' is itself to be investigated. A further consideration is our ingrained tendency to use the two words as synonymous, a circumstance which makes it seem rather a dubious step if their meanings are differentiated, and if (as G. Frege has proposed) we use one for meaning in our sense, and the other for the objects expressed. To this we may add that both terms are exposed to the same equivocations, which we distinguished above in connection with the term 'expression', and to many more besides, and that this is so both in scientific and in ordinary speech. (§15)

In §42 of the Second Logical Investigation, he complained that the word "meaning" was equivocal so that people did not hesitate to call the object of a presentation a "meaning" and to say the same of its "intension," the sense of its name. He further noted that since a meaning is often called a concept, talk of concepts and objects of concepts is also ambiguous.

And this brings us to take a look at Husserl's use of the notoriously hard to translate word "*Vorstellung*," for which there is no satisfactory English equivalent. It has very frequently been translated as "idea," "imagination," or "representation," all words charged with philosophical connotations that are not his. He addresses the problem in §8 stating "One must, besides, also surely keep in mind the fact that the word idea (*Idee*) has taken on many meanings and that, especially in the parlance of English philosophy, it is an expression for subjective experiences, for presentations (*Vorstellungen*), that, however, Ideas in my Platonic sense are not presentations, but atemporal, supraempirical objectivities." In these lectures, he in fact uses "*Vorstellung*" ubiquitously in ways that are unusual in the English language. I have consistently translated "*Vorstellung*" as "presentation" and can only ask readers to try to enter into his thought and divine his meaning.

Readers should also be aware that Husserl's use of "*Vorstellung*" in these lectures differs from his use of it in other periods of his career. In his "Psychological Studies in the Elements of Logic" of 1894, he wrote that he thought it was a good principle to avoid using a word as equivocal as "*Vorstellung*" as much as possible.[1]

[1] Edmund Husserl, "Psychological Studies in the Elements of Logic", *Early Writings in the Philosophy of Logic and Mathematics*, Dordrecht, Kluwer, 1994, p. 146.

Students of his logic course of 1902/03 heard him complain that no psychological and logical term was laden with as many pernicious ambiguities as was "*Vorstellung*."[1] There he distinguished, as Frege had in *Foundations of Arithmetic*,[2] between subjective *Vorstellungen* as psychological experiences and objective logical *Vorstellungen*, which he regarded as the "completely lost" distinction that Bolzano had for the first time identified as a "cornerstone of all genuinely pure logic."[3] In §5 of the introduction to the second volume of the *Logical Investigations*, he told readers that he should "have to raise fundamental questions as to the acts, or alternatively, the ideal meanings, which in logic pass under the name of 'presentations' (*Vorstellungen*)." It was, he said, "important to clarify and prise apart the many concepts that the word 'presentation' has covered, concepts in which the psychological, the epistemological and the logical are utterly confused." After discussing 13 dangerous ambiguities associated with the word in §§44–45 of the Fifth Logical Investigation, he concluded that "However the notion of presentation is defined, it is universally seen as a pivotal concept, not only for psychology, but also for epistemology, and particularly for pure logic,"

In these lectures, Husserl speaks of how presentation may signify something psychological like the intuition or thought-presentation underlying the thought-act and having nothing to do with the theory of pure meanings, something like a mere neutral thought or something different and mixed in with this. As an example, he gives the fact that that nouns are said to be expressions of mere presentations, but not complete judgments. And he recalls that in previous lectures on logic, he had defined a different, more important, concept of presentation. What he was presently calling nominal "syntagma," he had called nominal presentation and had discerned as many kinds of presentations as there are syntagmas, thus making presentation the same as syntagma (§25).

---

[1] *Op. cit.*, Husserl, *Logik, Vorlesung 1902/03*, p. 82.

[2] Gottlob Frege, *Foundations of Arithmetic*, Blackwell, Oxford, 1884 (1986), §27n.

[3] *Op. cit.*, Husserl, *Logik, Vorlesung 1902/03*, p. 56.

Syntagmas in turn are defined as the syntactical "stuff," which is "always given in syntactical form that lends it the specific thought-function in the proposition unit and overall meaning in general." He says that he had found it necessary to introduce an artificial word that could be used when there was a danger of becoming ensnared in the myriad ambiguities surrounding the words "presentation" and "concept" and adds that if the Idea of presentation is to be appealed to in the realm of meaning, then the natural choice for the meaning of "presentational content" as distinct from the objectivity presented is the nucleus (§24).

Husserl takes care to distinguish the concept of syntagma from that of nucleus by pointing out that the former had been defined as concerning that which "is identical, which stands out as the same noun in a different predicative function, or as the same predicate–but in a different function–sometimes as actual predicate, sometimes as determining attribute." He explains that it was by contrasting presentations such as "similar" and "similarity" and "redness" and "red" that he had been led to the concept of the full nucleus. He had observed that primitive presentations of different syntactical categories which coincide in terms of a content have the same nucleus, which could be formed nominatively, non-nominatively, adjectivally, and relationally. In comparison, however, he saw that the syntagmatic category clings to the syntagma. Changing the function does not change the same noun but makes it stand out as such. There can be both something identical and something different in syntagmas of different category, and it is precisely the nucleus that remains the same. The syntagmas "similarity" and "similar" or "redness" and "red" enjoy the same nucleus. Comparing them, we find that syntagmas of a different category have an essence-nucleus. They differ in their nucleus-form, which is what forms the pure nucleus into the syntagma of the particular category (§25, Appendix VIII).

Every language also has its share of recalcitrant terms which for one reason or another frustrate translators' efforts to capture their precise meaning. In the case of the German language, translators must cope with the fact that it delights in inventing compound words. Husserl makes lavish use of his freedom to do so, and most of his creations are not found in dictionaries. I have very often had to resort to hyphenations which, though not elegant in English, are nevertheless easily understandable.

In reading these lectures devoted to the theory of science, it is naturally important to keep in mind that the English word "science" and the various words derived from it are narrower in meaning than the German word for science, "*Wissenschaft*," and the words that are derived from it are. It helps to remember that these words contain the noun "*Wissen*," meaning knowledge, or the verb "*wissen*," to know. And, it is good to keep in mind that the English word "science" has its roots in the Latin word *scientia*, meaning science, and *scire*, to know. In §60, Husserl sheds some light on his particular use of the term in these lectures. There he explains that taking up the Idea of the theory of science that had served as a guide from the beginning of the course, then

> as the formal theory of meaning and ontology described, logic is the first manifestation of this Idea. Knowing (*Wissen*) in the sense of science (*Wissenschaft*) is thinking or thought-state-of-mind that refers back to thinking. Corresponding to thinking is something thought, and so corresponding to every science is a system of judgments in my meaning-theoretical sense, a system of postulated truths and probabilities, and these refer to objects and states-of-affairs. The science of meanings in general, of truths, possibilities, probabilities in general, of objects in general in absolutely pure, formal universality, yields a system of absolute truths to which every science is obviously bound, and which are prior in terms of validity to every science in general–as already to every judgment in general.

Since Husserl here takes up the questions of logical grammar of the Fourth Logical Investigation, it is also important to note that, as he used them, the words "*Widersinn*" and "*widersinnig*" do not translate neatly into philosophical English. The German word "*wider*" means against, counter, contrary to, and in opposition to. So, some have chosen to translate "*Widersinn*" and "*widersinnig*" as "countersense" and "countersensical." Husserl himself used "*Absurdität*" and "*absurd*" as synonyms for "*Widersinn*" and "*Widersinnig*" (ex. *Logical Investigation* I, §19; *Logical Investigation* IV, Introduction, §12). These words may, however, also be understood in the sense of paradox or contradiction and paradoxical, contradictory, or illogical. In that case, they fall into the family of "*widersprechen*" (to contradict), "*Widerspruch*" (contradiction), and "*widersprechend*" and "*widerspruchsvoll*," two common German words meaning contradictory. Given the difficulties and the importance that Husserl accorded to these concepts, I have chosen to leave "*Widersinn*," "*Widersinnigkeit*," and "*widersinnig*" in German. "*Unsinn*" can be translated as "nonsense" and "*unsinnig*" as "nonsensical," but I have chosen to leave them in German where Husserl talked about *Widersinnigkeit.*

In relation to his theory of manifolds, in §57 of these lectures, Husserl points out that in the concrete spheres, the formal limits imposed by forbidding the use of any factual *Widersinn*, any *widersinnigen* imaginary concepts, act as a constraint and hindrance in deductively theorizing work but that the marvelous thing about manifolds is that they free us from such prohibitions and explain why by passing through the imaginary, what is meaningless, must lead, not to meaningless, but to true results. In speaking of sets in §39, he notes that even inferential thinking makes use of *widersinnigen* presentations to some degree. As an example, he points out that, although the set-presentation, the totality-presentation, is for the most part also realized by mathematicians to begin with, it is nevertheless of no use for argumentation because it involves a *Widersinn*. Holding fast to the meaning of totality, he explains, a totality of triangles, a totality of numbers, is not graspable intuitively and so cannot exist either. Here he was taking up an issue he had tackled at the end of Chap. 11 of *Philosophy of Arithmetic*, where he noted that despite the absurdity of the idea, analogies fostered a tendency to transpose the idea of constructing a collection for infinite sets, thereby creating what he called a kind of "imaginary" concept whose anti-logical nature was harmless in everyday contexts precisely because its inherent contradictoriness was never obvious in life. This was, he explained there, the case when "All S" was treated as a closed set. However, he warned, the situation changes when this imaginary construct is actually carried over into reasoning and influences judgments. It is clear, he concluded, that from a strictly logical point of view, we must not ascribe anything more to the concept of infinite sets than is actually logically permissible, and above all not the absurd idea of constructing the actual set.

The German word "*Evidenz*" is also without a good equivalent in English. In §30e of *Introduction to Logic and Theory of Knowledge, Lectures 1906/07*, Husserl called "*Evidenz*" "a word for the fact that, as noeticians affirm and prove, there is a difference between acts that not only think that something is thus and thus, but are fully certain and aware, in the manner of perspicacious seeing, of this being and being thus. Therefore, the thing, the state-of-affairs is given in insight." *Evidenz*, he maintained, was nothing other than the quality of givenness understood in a comprehensive way and not just limited to the being of individual real things. We face the problem of *Evidenz* when we come to understand the correlation of consciousness and object that concerns all consciousness – even dreaming, hallucinatory, and erring consciousness – and we then ask the closely interrelated questions as to how we come by the existence of any object in itself at all, how we know that any object at all exists in reality, where and when an object is truly given to us, how we know that an object is given, and what it means for an object to be given to us. In §67b of the lectures published here, he teaches that *Evidenz* is what necessarily assures us that we are in the possession of truth. He points out that we can judge without seeing, blindly, routinely, and so on, but only if we judge insightfully does our judging have objective cognitive value. If we were not capable of *Evidenz*, he maintains, then no talk of truth and science would make any rational sense.

"*Geist*" is another troublesome German word. "Mind" (the faculty of reasoning and understanding) is a proper, though imperfect, translation for it in philosophical contexts, and this is particularly so here, because Husserl is mainly talking about logic and philosophy of science and his lecture course is meant to be a critique of reason. I have used it and the corresponding adjective "mental" (of or pertaining to the mind) when it is a matter of the psychophysical reality of human beings. However, "*Geist*" also translates as "spirit," and in these lectures, it and its adjective "spiritual" are appropriate translations when speaking of supra-individual, immaterial, abstract realities, such as the communal spirit and the world of the spirit. I have also used it for animals where it is a question of "an animating or vital principle held to give life to physical organisms." My dictionary also defines "spirit" as meaning ghost, the third person of the Trinity, fairy, sprite, elf, angel, and demon.

It is likewise very difficult to know how to translate "*Gemüt*," which does not simply mean mind, but which refers to the emotive, affective dimension, in contrast to the *Verstand* (understanding) and *Wille* (will). Dictionaries typically give "mind," "soul," "temperament," "feeling," "spirit," and "heart" as translations for "*Gemüt*," but most of these words overlap in very complicated ways with other German philosophical terms such as "*Seele*" and "*Geist*." To convey the meaning of "*Gemüt*," I have usually appealed to expressions using the adjective "inner."

Paris, France Claire Ortiz Hill

<SECTION I

FUNDAMENTAL CONSIDERATIONS FOR THE DEMARCATION AND CHARACTERIZATION OF FORMAL LOGIC>[1]

[1] Regarding Sections I-III of the lecture, compare Appendix I: *For the 1910/11 Lecture. Concluding Remarks and Plans*. (Editor's note)

<Chapter 1

Acquiring the Idea of Pure Logic as Pure Theory of Norms 3
for Knowledge>

<§ 1. Logic as Purely Theoretical Self-Knowledge of the Understanding>

The field of scientific studies to which I wish to devote these lectures can be designated in a generally understandable way by a single word. It is the understanding. The name of our discipline already indicates this. It signifies: the science of the logos. Understanding–or as it is frequently also synonymously called, reason–is a word for manifold mental activities and mental achievements that are already familiar to us from occasional experiences prior to any logic. Ordinary language possesses a wealth of relevant terms, <such as> conceptualizing, judging, thinking, substantiating, deducing, proving, and so forth, and correlative expressions such as concept, judgment, thought, deduction, proof, theory. The activities and accomplishments so designated span all areas of the life of the mind. Understanding reigns in all sciences and technical arts. Sciences are sciences in the genuine sense precisely insofar as the understanding governs their findings, giving form to them, and setting norms for them. It is what creates unity of theoretical knowledge out of unconnected experiences, out of iso- 4
lated convictions, presumptions, inferences. Thanks to it, nature figures before our mind's eye as a cosmos governed by laws. In all domains of nature, it unveils the inviolable system of laws that we call the theory of the particular field.

But, by setting norms, inquiring into what is right or demonstrating error, the understanding also governs in the spheres of extra-theoretical life. With its light, it illuminates the darkness of

E. Husserl, *Logic and General Theory of Science*,
Husserliana: Edmund Husserl – Collected Works 15,
https://doi.org/10.1007/978-3-030-14529-3_1

the inner life and the life of the will. There it makes multiform appraisals of beauty and ugliness, of goodness and badness, of appropriateness and inappropriateness–appraisals that admittedly draw their content from the inner sphere, but beyond that precisely have an understanding-form that has to subordinate itself to the general norms of understanding.

Even though the understanding governs in all areas of mental activities in this manner, sets into play the manifold activities of conceptualizing, judging, inferring, and so forth proper to it in accordance with set norms in all of them, it is, however, not thereby its own object without further ado. For example, geometricians, physicists, grammarians act as such using the understanding. In this activity, they devote themselves to geometrical figures, physical processes, grammatical formations, not however either to their own mental acts, to the inner activity of their understanding, to the essential forms and norms with which, as being of the understanding, this activity is in general in conformity, or to the pure forms captured by the accomplishment of thinking as such.

An essentially new line of inquiry actually opens up when we turn our gaze from the things we know through the understanding back to the understanding's ways of operating and inquire into the forms in which they take place, or which forms they imprint upon the contents as such thought by means of the understanding, to which norms they are bound in so doing, and how reaching or missing the goal of truth essentially depends upon whether these forms and norms are observed or not. If it is to lead over to scientific fixing and theorization, this reflection, deflection of the gaze back from things to thinking about things, as well as to the con-
5 tents of thinking, with their forms and norms, is by no means an entirely easy matter. It certainly runs counter to our natural, habitual ways of thinking. To a degree, as mature people, already intellectually educated from our schooldays, we are all experienced, indeed quite experienced, in using the understanding in thinking itself. However, we are by no means trained in the reflective consideration, analysis and conceptual fixing of the operations of our thinking or understanding and the forms of their contents remote

from our dominant life interests. *Locke* already said in his epoch-making work *An Essay Concerning Human Understanding* published at the turn of the eighteenth century: like the eye, the understanding, while it makes all things knowable for us, does not perceive itself, and it requires no small amount of art and pains to confront it with itself and to make itself its own object. The difficulty does not generally lie in reflection, in the mere looking back upon the thinking experiences, but in the fixing observing and differentiating, as well as then further in making what has been differentiated into the object of scientific pursuits in every manner.

The fact that, in general, reflections upon what concerns the understanding prior to all logic are easily possible and not at all rare need only be mentioned. Rationally thinking about any scientific subjects, or even engaging in rational deliberations in the midst of practical life, often enough turns our gaze back to thinking and deliberating themselves. We are interested in ensuring their legitimacy, their rationality in the specific sense. Only by virtue of such reflections do we know in general that something such as thinking and rational thinking exists. Only by virtue of them do the various words of ordinary language referring to the intellectual sphere and to what is specifically rational originally acquire their meaning—words such as judging and judgment, conceptualizing and concept, truth and falsehood, presuming and probability, inferring and inference, proper and false inference, deducing and deduction, false deduction, and so forth. However, such reflections scarcely achieve any more for knowledge of the essence of the understanding than pre-scientific experience does for the knowledge of external nature.

The practice of the rigorous sciences certainly involves the fact that, for purposes of the reliability of their findings, reflection upon the soundness of the methods applied, the deductions, proofs etc. performed, is often, and must be, carried out in much more deeply 6
penetrating ways in them. In all sciences, therefore, logical reflections and reflections of a scientific nature occur from time to time. If reservations arise concerning the soundness of any methodological steps, concerning deductions made, concerning the conducting of proofs, assessments of probability, and so on, then the attention

paid to scientifically fixing those claimed accomplishments of the understanding shifts and they are then submitted to analysis in the reflective attitude. However, such analyses remain sporadic. They are centered on a particular deduction, on a particular proof that was to have been checked there. And, they are terminated as soon as clarity and insight into the soundness of that particular methodological step is reached. Only in logic does the understanding act reflectively, not in order to investigate any other objects–for instance, external nature or the human world–, but purely in order to investigate itself. It does not wish to reflect there upon its individual actions within the context of current scientific knowledge and upon its individual methodological structures, but the aim is now systematically exhaustive, purely theoretical self-knowledge of the understanding, of the understanding in general in terms of all its ways of experiencing and in terms of all kinds and forms of the thought-contents specific to it.

Beginners cannot on the basis of their occasional and fleeting reflections on logical facts even have an inkling of what a vast science opens up here and what a great thing for human knowledge depends upon its successful execution. Now, to provide you with an inkling of this and thereby succeed in showing you great and interesting chapters of logic, to the extent that they can presently lay claim to scientificity, is the purpose of my lecture course.[1]

<§ 2. The Historically First Concept of Logic: Logic as Theory of Norms for Rational Thinking>

And now! Let us get down to business. The word provisionally
7 used to describe the subject of the lecture for the general public may provide us with the link. It was understanding. However, this term "understanding" yields a field of activity for different scientific disciplines, and demarcating these in relation to one another and isolating with conceptual clarity and perspicacity those among them possessing such fundamental significance for all of philoso-

[1] Compare Appendix II, *Logic as Philosophical Science of Knowledge, as First Philosophy*. (Editor's note)

phy is an important and not entirely easy task. We shall have to devote a series of interesting and instructive reflections to it.

In ordinary language, understanding denotes a mental faculty, similar, for example, to will. By will, we certainly do not merely mean certain actual inner feelings, those experiences very familiar to us that we express with the words “I want,” “I wish,” “I desire,” possibly even in the form of imperative clauses or optative clauses, and the like, but also the mind’s general capacity to will, to wish, to desire, or to flee, to abhor.

So, the understanding as a mental faculty points to a class of mental experiences that are associated with it, to the experiences of what is called thinking. Now, we all know that there is a science that is called theory of the psyche, psychology, a science that does not, like somatology, deal with the physical things and events of what is called external nature, but rather whose factual domain is the inner sphere. In it belong, for example, perceptions (as distinct from perceived things), memories (as distinct from the objective facts that one remembers), free fantasizing, feelings, desires, wishes, and finally also thinking phenomena, for example, conceptual thinking in the use or understanding of a significant word, or judging in the use or understanding of a mathematical or other kind of proposition, whether otherwise true or false, or deducing, proving, theorizing, for example, when we conduct a proof in mathematics, construct or think through a theory, whether correctly or incorrectly. All those are mental experiences or mental phenomena, processes that we attribute to our, or to some, ego-consciousness. As psychic phenomena, they all belong *eo ipso* in psychology. And, all the psychic states corresponding to these phenomena also belong right in there, the innate abilities of perceiving, presenting, judging, etc., as well as abilities acquired through experience and 8
training such as, for example, mathematical or linguistic thinking skills and the like. In this sense, the understanding, faculty of the understanding, and acts of understanding belong in psychology.

Accordingly, it might appear that logic, the science of the understanding, is a mere branch of psychology, of the science of the psychic in general. However, this is not in tune with the fact that the research and theories that since Antiquity, since *Aristotle*’s days, have actually been dealt with under the heading “Logic,” or under an equivalent heading, were in no way meant to be considered psychological. We can thereby leave out of account the exceptions that first fell few and far between in the nineteenth century and will be discussed later on.

Historically, logic as a theory of norms and theory of the art of rational thinking has stood in contrast to psychology as the natural science of the psychic, and especially of thinking, and it meant thereby to establish itself as a science of the understanding or of reason in the specific sense. It is here a matter of a concept of logic that not only is the historically original and certainly predominant one, but that for purposes of leading upward to logic in a higher and specifically philosophical sense must also naturally be taken as the starting point. So, I am thereby already indicating the fact that different essentially intertwined disciplines can be discerned under the heading of logic, that from an initial, natural concept of logic one is driven upward to the demarcation of new disciplines growing out of one another in an essential manner. We shall ultimately also come to see that highest philosophical interests exist that demand the systematic clarification and development and of ever new clusters of problems and disciplines out of one another. That will be a main part of the task we set for ourselves in the following considerations.

The first concept of logic as a normative, practical discipline of thinking is so prevalent that it is already familiar to the general cultural consciousness. All of you, in any case, have from the start associated with the term logic the idea that it is a matter there of a discipline that our thinking, provided it is truth-seeking thinking, would set norms for and also help us practically to attain the goal
9 of truth. You are thereby also familiar with the idea that truth is precisely a goal, therefore something not given from the start, something that must first be worked for using what are called methods, using certain solidly forged procedures of thinking. Logic should then be helpful to us in that respect owing to its general investigations oriented toward truth and methods of searching for truth. It would set general norms, what are called logical laws, logical rules, to which one must be bound if one is not to fail to attain the goal of truth, the observance of which is in any case conducive to striving after truth.

<§ 3. The Difference Between Psychology of Knowing and Normative Logic. The Contrast Between Natural Law and Logical Norm>

However much such thoughts flow in vague generality, they suffice to make a somewhat clearer distinction between the normatively and practically oriented logic we are talking about here and psychology, more specifically psychology of knowing. Let us be clear about the following: Psychology is not a normative science. It does not aim to set norms for the flow of mental acts and the development of states of mind. It would explore how they are, in terms of their law-governed nature, but not how they should be conformable to any governing ideals.

Psychology is the natural science of what is psychic. Psychic experiences and the corresponding psychic capabilities, dispositions, in short, states of mind, taken as real facts within spatiotemporal nature are its field of investigation. Belonging in it are, therefore, also the particular kinds of pychic functions of interest to us here, the kinds we include under the multifaceted heading of "thinking" or "knowing". Exploring how they come and go within the context of human inner life, which empirical laws hold for that, how they spring from innate and acquired mental states by which they are determined and again react to the formation or transformation of what is peculiarly intellectual, and other such things, is a matter for psychology. Naturally, the consciousness of correctness, of validity, of truth and probability, of necessity, and so forth, also belongs in its domain, including also the consciousness that we have Evidenz, Evidenz of truth, insight into truth, and so on. However, for psychology, all that comes into consideration as a current psychic fact within the context of the living mind, within the consciousness of the living person, who is there, as he or she is, within the context of the whole of nature.

Just as physicists explain to us why some factual things behave, for example, optically in just this way and not differently, and explain that to us in terms of causality and laws using the laws of nature that govern the opt<ical> behavior of the things within the context of spatiotemporal nature, so psychologists explain to us–or at least would like to explain to us–why a person, for example, as a thinking person, on a given occasion, actually behaves in just this way and not differently, and for the purposes of such an explanation, they search for the laws of nature that causally govern the coming and going, the coming into being and the self-transformation of the psychic thought-acts. If, for example, the insight someone has is an event of his or her consciousness, speaking more broadly, of his or her inner life, even more broadly, of this psychophysical subject that we are calling this human being, then explaining in accordance with which psychophysical natural laws the occurrence of precisely this Evidenz-act in this human being can be grasped as causally necessary is a problem. Proper, insightful thinking therefore comes into consideration here as a natural fact among other facts called perceptions, presentations, inner experiences, passions, and so forth, as they occur in a variety of forms in humans and other beings endowed with inner lives.

The position of logicians is, as we readily observe, however, completely different. While their normative and practical interest certainly also holds for actual people, and while they aim to subject actual people and their actual judgments, modes of inferring, methods to criticism or improvement in conformity with reason, what they strive after is nevertheless something completely different from a natural scientific explanation of these actual thinking-events in actual people. So they seek laws, however, not natural laws presiding over the factual coming into being and passing away of psychic experiences with unconditional necessity, but normative laws by which one can measure whether or not knowledge
11 is actually knowledge, truths are actually truths, epistemological methods actually epistemological methods, proofs actually proofs. Here "actually" does not mean the same as fact of nature. The chief thing for logicians is an Idea, the Idea of truth, or the Idea of correctness, and all the particular Ideas that are included in it, or stand in an ideal relationship to the Idea of truth. With this Idea of truth and correctness as being adequate to truth, something fundamentally new comes into logic in comparison to psychology, therefore, something psychology lacks as long as it is psychology, the natural science of inner life. And, with this Idea, the new law-concept at the same time arises for logic, that of ideal norm, whose opposite in natural science, but not its equal, is natural law. I stress:

its opposite! For, provided they are expressed in the pure generality that distinguishes them from merely empirical utilizations, logical norms do not, by their very meaning, say anything about facts of nature, about any actual existence anywhere in the world.

It is initially good to keep in mind that the particular statements we make about validity, though they may also refer to factual judging–granting or denying justice–, nevertheless, insofar as they speak of validity, do not thereby refer to anything psychological-factual. If, for example, I state, "The judgment that a decahedron is a regular body is false", indeed, even if <I> thereby speak to another person and say, "Your judgment <a decahedron is a regular body> is false", then the little word "false" does not express any psychological predicate at all about the judgment-experience itself. The same is so if I, for instance, say to the other person, "Both judgments you are setting forth there as different are logically equivalent, those other two are mutually self-exclusive, are incompatible with one another, and so on". Incompatibility in the logical sense is something totally different from incompatibility in the psychological <sense>, i.e., as factually not-being-compatible in an animal consciousness. Contradictions are compatible factually, and are only too well so in our actual judging, but they are not compatible when it comes to validity, in the truth unit. Judgment-experiences have their psychological predicates. A presumption or firm conviction that I nurture has its specific temporal form, its objective temporal duration, its degree of liveliness, its changing wealth of associated intuitions, and so forth. Those are psychologi- 12
cal predicates. If, however, I call the conviction concerning what is judged in it the consequence of such and such arguments, or if I call it equivalent to that other conviction, or I call it downright false or true (or right), then such predicates do not express anything that could be shown as being something constituting the factual state-of-affairs for the factual judgment-experience concerned.

If we now look at very general logical norms, then they do not in general imply anything about facts of a psychical nature. They are not even weighed down by factual content in the sense that they deal with people's judging, though they may also serve to set norms for people's judging. At any rate, pure norms, unlike some practical logical rules, are free of factualness in the sense indicated.

It is of the greatest importance for beginners to make these things fully clear to themselves by means of an abundance of examples. In so doing, they naturally have to abide by norms that are directly intelligible in terms of their unconditionally valid truth and as a result belong to the sphere of those already familiar to them prior to scientific logic. Belonging there are the examples given earlier. <Belonging> there is, for instance, also the trivial rule that what universally holds for all triangles, for all people, for all numbers, in short–symbolized–for every A, also necessarily holds for every given triangle, for every specific person, and so forth, in short, for every particular A_0. And correlatively, if it is generally true that All A are b, then it is false that this A given here is not b, and thus for every A_0. If anything in the whole world is at all true it is this rule. And, if the meaning of a proposition is at all comprehensible, then the meaning of this proposition is. It harbors no mysteries. And now I ask whether this rule in the least talks about any factual living being, for instance, about a human being and his or her factual judging. Does it talk about human judging in empirical generality about what all people on earth judge and always will judge? Most certainly not. It is doubtlessly a law that what holds in general, holds in every particular case, and does not fail to hold in any particular case. But that is not a natural law. A natural law speaks about nature, about factual existence, be it in particular or in general. Here, though, we do not find anything at all about factual spatiotemporal existence in the meaning of the
13 statement.

It could enter into someone's mind to interpret the rule as an abbreviated expression for the fact that no one could judge that "All A are b" and simultaneously deny that the A_0 given is not b. Certainly, no one can do that. However, does this not-being-able amount to real inability? Then, the assertion would be false. It happens often enough that people contradict themselves and even do so in such a way that they state a general judgment and yet deny the subsequently corresponding particular judgment, for instance, by overlooking the fact that the particular A in general falls under the general concept of A, while at the same time some other motive for judging recommends to them the denial: "This A is not b".

In the logical sense, every contradiction is tantamount to saying that no one *can* judge in that way. And yet, it is factually judged that way. Human judging is swarming with contradictions. That no one "can" does not, therefore, signify psychological incompatibility or psychological inability, as if it was a matter of something that could or could not occur in a human mind. Rather the "can" has precisely a "logical" meaning. Anyone who judges otherwise, precisely judges incorrectly. Analogy with "must": Everyone *must* judge unanimously, not because nature only allows unanimous judging in accordance with natural laws to occur, but because judging unanimously alone can be correct.

<§ 4. The Absolute Validity of Logical Norms. The Ambiguous Nature of Talk of Understanding: Understanding as Psychic Faculty and Understanding as Idea of a Non-empirical Kind of Normality>

In view of the natural psychologistic movements nurtured by modern philosophical prejudices, it is most necessary to think through this situation in the most thorough way. For what I have set forth here is very hotly combated as being mysticism and scholasticism. What I must require here is nothing more than the intellectual integrity that allows the things that are prior to any theory because they are the most evident of all evident facts to count as being precisely what they proclaim themselves to be. If one gives a faithful account of what is given, it is not necessary to follow up on
all the often quite pompous twists and turns that psychology has 14
brought into logic in order forcibly to attribute a psychological meaning to everything logical. Of course, the words we have available to us to express what is logical are ambiguous, as already earlier became apparent in the case of the little words "can" and "must". It is also to be noted that we often use words in unclear ways, that we are for the most part content with a completely vague execution of meaning-conferring presentations, and that for various reasons, meaning-clarification and -analysis are not always easy matters. There is there one, and only one, way to settle contro-

versial questions that concern the meaning of words or of our statements: going back to clarifying Evidenz. Stated more fully and clearly: If the meaning of a statement is called into question, if we ourselves are perhaps in doubt about it, then a first, obvious methodological rule is to consider the statement within the context it objectively belongs. Words often change their meanings according to the context. The question is, however, what the meaning is within the given context, within a given sphere of application. Therefore, we have to take a logical proposition within its logical context. We, therefore, have to bear in mind the fact that we are taking an approach to the validity or non-validity of judging that is perhaps different from that we use to gather knowledge and want to make statements about existence or non-existence of judging as experiences of individuals.

The approach here is completely different and completely differently oriented, not oriented toward existence, but toward a certain ideal justification, toward truth, toward validity. It is obviously similar to that of ethics, and to a certain degree, even to that of jurisprudence. There too it is twofold: the ascertainment of the facts and the judgment of the law, of what is right in the sense of what is ethical and of what is right in the sense of the normative juridical laws. On the other hand, logical norms are once more after all not norms in the sense of laws of the land, whose particular facticity lies in their having been stipulated arbitrarily by kings and parliaments and which also could have been stipulated differently. On the other hand, however, no one has it within his or her power to create logical norms and to posit differently from obtaining norms, any more than someone can make natural laws and alter the ones obtaining. Norms of correctness and falseness are valid in themselves and absolute when they are in general valid. And they are valid in a sense that just does not involve facticity. Only when they are somewhat transformed for practical purposes of setting norms does something factual enter in an extra-essential way. In this extra-essential mode of application, we can draw actual or supposed facticity into every purely logical law. Here is an example! With a view to possible application, one can state a primitive logical norm in this manner: Whoever judges there, "If A is valid, B is valid", and whoever is then convinced of the fact that B is not valid can then, if he or she wants to judge correctly, only judge, "A is not valid".

Naturally, the talk of judging persons is not intended psychologically here, or the rule not intended in the sense of a rule restricted to factually judging persons. Then, the intent would have had to have been that that was only valid of the people and the couple of animals that we know through experience. Obviously, though, we may apply the rule to people, make-believe centaurs, angels, etc. The meaning is simply and plainly that if the judgment "If A takes place, B also takes place" is valid, and if the judgment that B takes place is not valid, then it is also not valid that A takes place. Which judgment is being talked about there? Certainly not of something human, in general, not of something factual. What is being talked about is a judgment of the content A quite simply in unconditional generality. Anyone who wishes to judge there may—in this or in a different world, in the real world or in fictional worlds. If he or she in general judges, namely, makes a judgment about the content given, then what is judged is true when it does not contradict this rule, and false when it does. It is clear there that, in general, talk of the person judging has to be left out of account and that, in general, it is not at all a question of the actual or imagined judging person here, but of the ideally grasped judgment in itself that as that which it means, which it asserts, is false or true.

Now, one may indeed find a problem in the fact that in our factual judging, we humans are supposed to be allowed to assert justifiably that there are absolute norms. We put forward the norms while we are judging. Of course, in judging we do not put them forward haphazardly. We do this while seeing their unconditional necessity and general validity. But this seeing is again itself something subjective. One may therefore find difficulty with this.

But, however matters may stand in this respect, that changes
nothing about the fact that the approach of setting of logical norms 16
is different from the approach of the natural scientific establishment of facts of existence. And it again remains certain that logical norms, logical laws of judgment differ in meaning from natural laws. We must hold fast to this insight from now on and not let ourselves be confused by appearances.

Consequently, I do not see the fact that, already thousands of years ago, logic established itself as a distinct scientific discipline–and not within psychology as a part, as a chapter of the psychology of knowledge–as a historical accident or a result of some convenient division of labor. Rather, the idea of the truth of logic–and the circumstance that, by their very nature, cognitive acts can be considered from the ideal standpoint of accuracy and inaccuracy–refers to a fundamentally different domain, <a> fundamentally different inquiry and, accordingly, fundamentally different findings than any psychology, than any natural science in general.

Thus, to be correct, one would also have to say that talk of the understanding or of reason is ambiguous. Understanding as a psychic faculty, i.e., as a term for a class of psychic experiences and empirical states of mind is a fact of nature and is to be distinguished from the understanding as an Idea, i.e., a term for certain kinds of normality, namely, of non-empirical normality. It <is> a term for certain connected clusters of ideal norms, norms that, so to speak, address themselves to judgments and other acts of understanding, but once more nevertheless do not address themselves to the factual acts of understanding of human beings, but to acts of understanding in general in unconditional generality, whether they are realities or imagined. That is, however, tantamount to saying that they are norms that are grounded in the ideal essence of acts of understanding and their contents. It is merely turning this thought around when we further say: It is basically false to consider the division of acts of thought into logical and unlogical to be an empirical-psychological division.[1]

As time goes by, we shall come to understand what has been
17 explained here ever more fully and learn to appreciate its significance. To begin with, it is enough for us to have correctly grasped the main point. If we have adopted it, then we shall be able to resist psychologistic temptations and naturalistic temptations in every guise, however much the prejudices of our times may have inclined us toward them. It will not also then be able to enter our minds, for example, to confuse the sense of logical normality with

---

[1] But the division is psychologically significant, as ideal particularities are for what is empirical of the corresponding essence.

that of psychological normality and so, in general, place it on a par with natural scientific normality. So it is, when one speaks of a normal organism, of a normally or abnormally formed plant and, from the mental point of view, of a psychically normal or abnormal person or animal. Naturally, what is now called psychic abnormality, especially what is called intellectual abnormality, would be transformed into normality if the vastly overwhelming portion of humanity were to take on the inferior characteristic concerned, if this were therefore to be changed into the defining distinctive mark of the human species. What is logically normal and abnormal does not, however, have the least thing to do with that. That $2 \times 2 = 4$ is, is a truth, even if all the people in the world, indeed every judging being in all times past and future were in general to judge *unisono* "$2 \times 2 \neq 4$". And like every truth, every norm of truth remains unaffected by empirical facticity of that kind. One also easily sees that in us humans, the capacity to see what is true as true, what is false as false co-determines, at least within certain bounds, the concept of intellectual normality whereby truth and falsehood in the logical sense are presupposed beforehand.

<§ 5. The Truth Motive in Psychologism>

<a) Excursus into the Historical and Factual Background of Psychologism>

In the reflections we have engaged in, the connection between logic and psychology seems to be broken in consideration of the fundamentally different approach that in the one aims at ideal norm-setting, in the other, at facts of existence. In certain ways, that is right, but in certain ways, as we shall quickly see, it is nevertheless again going too far. It also really needs to be said before- 18
hand that very powerful motives must speak in favor of the psychologizing interpretation of logic, since not just isolated eminent scholars, but a great, influential philosophical school of thought champions such an interpretation. Much in science is also a matter of fashion, but what is in fashion in science cannot be completely rational. A bit of rationality, a one-sided, even though inadequate, truth-motive must be involved there.

So, let us take up the controversy about so-called psychologism. Naturalistic philosophy, commonly called empirical philosophy, is characterized by the fact that only *one* area of possible knowledge exists for it, and that is precisely nature, the entirety of spatiotemporal existence. This is only another way of saying, as I do, that for naturalism, the concepts of empirical science and science in general coincide. For, experience is the consciousness in which all knowledge of nature ultimately terminates, from which every assertion about spatiotemporal existence ultimately draws its justification. In the broader sense, the whole entirety of scientific thought processes in which natural existence reveals itself in terms of existence and properties, namely, in a singular as well as a general, law-governed respect, has certainly been called experience since *Kant*. However, broadening the concept of experience only has its basis and justification in the fact that precisely everywhere–even where a natural law is to be grounded–empirical grounds in the narrower sense must necessarily form the point of departure and the groundwork, i.e., perceptions, memories, and direct experiences of that kind. For naturalism, there is therefore only one field of possible knowledge in general, nature. And since we can also say in accordance with what has just been explained that naturalism only recognizes one method of grounding knowledge, that of experience, or that its cognitive principle is that all knowledge rests on experience, we also now understand why naturalism and empiricism are synonymous concepts.

This naturalistic worldview gained decisive supremacy in the nineteenth century, particularly ever since the collapse of *Hegel*ian philosophy, which after of period of unparalleled triumph (from the mid-century on) fell into general contempt, contempt that car-
19 ried over to all idealistic philosophy in general. The magnificent progress made by the natural sciences, the immense practical significance that natural scientific technology acquired, the boom in international trade, and a series of other interwoven factors explain the fact that in Germany itself, this classic land of idealism, all interests and inclinations turned toward the real order. Culture-humanity not only lost lively interest in the realm of Ideas and of pure ideals, it became, so to speak, mentally blind to them. Pure mathematics, a genuine science of Ideas, admittedly continued to

flourish during that time, indeed with its magnificent discoveries achieved unprecedented development. But the intimate alliance it had entered into with the exact natural sciences because it served as an endlessly fruitful methodological tool in them, fostered the general propensity for naturalistic misinterpretations to which pure mathematics then actually fell prey. Generally, it regarded itself, and for the most part it still even today is regarded, as a natural science.

Historical idealism was not able to erect any dams to stem the floodtide of empiricism after the disintegration of *Hegel*ian philosophy. It was itself much too divided for that, also too inconsistent, too unscientific. What it offered by way of delineating something new was methodologically and factually not as significant as what it would have had to have been able to offer to compete with the exact factual sciences. And as competition, idealistic philosophies were generally respected. The exactness of the natural sciences was regarded without further ado as evidence in favor of the empiricist worldview generally propagated by the natural scientists. Those who combated empiricism were suspected of being enemies of exact science itself or shoved aside as not being in step with the times. *Comte* and both *Mill*s were regarded as great philosophers by that naturalistic age, and even *Kant*, who, however, was praised only because of certain comments apparently making concessions to empiricism that natural scientists preferred to misinterpret in an empiricist sense–he, the radical idealist.

In relation to this account, one must not point to the great blossoming of what are called the humanities and social sciences and not allow oneself be misled by the war between their proponents and natural scientists that has flared up surrounding the question 20
of educational reform, nor by the fact that the long prevailing penchant in circles of natural scientists for naïve materialism, which explained everything psychical as actually being physical, was vigorously contested and especially contested by the so-called humanists. The contrast between naturalism and idealism, as we understand the words here, must not be confused with the contrast between materialism and anti-materialism. The naturalism of our times also actually dominated (in spite of the just mentioned con-

trasts) what are called the humanities and social sciences: history, political science, sociology, in the broadest sense. It dominated everything one is also accustomed to referring to under the heading of cultural studies and to contrasting with natural science in the usual sense. Even the proponents of the humanities and social sciences stood, I say, on naturalistic ground, i.e., experience and induction were also regarded by them as being the only means of substantiation. The world of Ideas also remained closed to them. And so then on all sides, in the same way, anyone who advocated giving Ideas their due was berated for being a reactionary, mystic, and scholastic. Scholastic and mystic are in general the strongest scientific terms of abuse in this naturalistic age.

As, then, everything ideal in the mathematical sphere was reinterpreted empirically, so too was even the ideal in the sphere of the ideal, normative philosophical disciplines, of normative logic, ethics and esthetics. Logic was expressly viewed by *Mill* as a "part or branch of psychology", just as were the other two disciplines just now named. What I am calling psychologism in logic is therefore nothing more than a particular form of naturalism. By "naturalism" in the broadest sense, we understood that false basic point of view that only recognizes and sees the one sort of objectivities that I am calling nature and accordingly misconstrues all other objectivities falling within the purview of science (despite their existence-free meaning) as objectivities of nature. If one understands naturalism in this way, then, leaving composite notions out of account, two sorts of naturalistic misinterpretations are obviously possible. Either the existence-free objectivities are misinterpreted as being physical in nature or psychical in nature. The latter defines so-called psychologism. We have no name for the former.
21 The word physiologism, very suitable in terms of its original meaning, is problematic, because "physiology" is not tantamount to saying general science of physis, of physical nature, but has the well-known, very narrow meaning, and we generally lack a suitable word for physical nature as such. (Narrowing the word's meaning, we ordinarily say "nature plain and simple".) An example of physiologism would be *Mill*'s interpretation of arithmetic as a science of physical facts of nature. When this view is rather often contested in empiricist quarters, and for that reason the other

one–that, *au fond*, number is an expression not of physical, but of psychic fact–is preferred, then we immediately have in that an example of psychologism. And this is precisely where the generally popular theory that the philosophical normative sciences–logic, ethics, esthetics–are nothing more than offshoots of psychology then belongs.

<b) Logic as Technology of Human Knowing Must Be Grounded in Psychological Knowledge>

After this useful excursus that situates our reflections within a general historical, factual background, let us go back to our point of departure. I said that powerful motives speak in favor of logical psychologism. A certain plausibility must indeed attach to it for true, serious thinkers of the likes of *Mill* to have been able to embrace the error and reconcile themselves to it in a lasting way. I have even conceded that a bit of truth, in which the source of the false illusion may lie, lies on the side of those I am combating. I must talk about that now. This will at the same time take us a step further.

Logic in the sense I am now dealing with is thought of as a normative, practical discipline of knowing. I now ask whether any discussion of nature and Ideas, no matter how beautiful, can change anything about the fact that, logic inasmuch as it would set norms and practically govern human cognitive states of mind, has to do with psychic functions. If, however, what <it> would govern is something psychic, then logic of course rests on psychology. If it universally holds that whoever would establish rules for a kind of 22
activities must first of all study those activities in terms of their essence and their actual requisites, then that also holds here for the kind of psychic activities that we in general ways at times call thinking, at times call knowing.

That is actually the entire argument of psychologism and, as you see, it is plausible enough. Accordingly, one says that logic stands in the same relation to psychology as the physical technologies do to the corresponding parts of physics. We speak in vague ways of chemical technology, of mechanical technology, and so forth. They are obviously based on theoretical chemistry, on theoretical mechanics, etc. As there are physical and physiological technologies (with regard to the latter, the different branches of medicine come to mind), so there are also psychological and psychophysical ones, for example, psychiatry. So pedagogy belongs here. Regulating the inner lives of children in conformity with certain objectives naturally assumes knowledge of that mental life, and if the regulating is to be scientific, then scientific psychology is necessary. Obviously, with a suitable broadening of the concept of pedagogy (which admittedly would no longer be in tune with the name of the discipline, but however with the term science of education), one could subsume all of logic under the broadened concept. One could well characterize its task as educating our thinking in the sense of the Idea of rightness. All tasks that normative, practical logic has to set for itself can be positioned under this general educational task.

These obvious thoughts are so manifestly justified that nothing can actually be invoked in opposition to them. Have I not, however, on the other hand, made it fully evident that specifically logical predicates are not real predicates, that specifically and purely logical norms are not natural laws, indeed do not in general imply any factual content? Is not the requirement to recognize that logic and psychology are fundamentally different disciplines– the one normative and existence-free, the other not normative and connected with existence– implied as a consequence of this realization?

23 <§ 6. The Difference Between Logic as Pure, Existence-Free Theory of Norms for Knowledge and Its Application to Human Knowing. Exposing the Fundamental Error of Psychologism>

To find the right course granting legitimacy to the Evidenzen on both sides, one can take different paths, and they lead in the same way to the result that the concept of normative logic is to be delimited in a twofold manner. The one logic is narrower in scope. It is a pure, existence-free theory of norms of knowledge. The other <is> broader in scope. Besides this pure logic, it also embraces a wealth of empirical material in connection with the fact that it would regulate empirical human knowing in the empirically given sciences in keeping with pure norms.

To show this, one can, as I tried to do in the first volume of my *Logical Investigations*, pose the question radically in terms of the theoretical fundaments of the logical theory of the art of something. Let us recall the psychologistic argument, etc. One can namely ask whether the psychological argument implies that psychology is the sole theoretical foundation of logic. Each theory of the art of something presupposes theoretical knowledge of the tasks and outcomes to be subject to norms. It is therefore based on one or several theoretical sciences that deal with the nature and laws of the tasks concerned, of the materials, etc. to be formed in the manner of the natural sciences.

On the other hand, each theory of the art of something is, however, also governed by a normative Idea. If, however, a priori sources for theoretical knowledge then also lie in these sources, then aside from those natural sciences, we shall find an a priori science, as theoretical foundation, a pure science of Ideas.

So it is, for example, with the scientific surveying, or the land area surveying, nowadays called geodesy. It obviously rests upon the natural scientific disciplines. It works with optical apparatuses, with theodolites, and such things. In pursuance of its goals, it will take the tools it needs from those disciplines. On the other hand, its ideal of exact surface measurement of the earth's surfaces leads to the science of exact, ideal measurement of surfaces in general, to pure geometry. As long as ideal and empirical measurement of surfaces were still attached to one another (as in the early stages 24
of "geometry"; the word geometry indeed signified surveying, therefore, still points back to that era), as long as ideal geometrical theorems had not yet been detached and treated in the form of a systematic ideal science in its own right, one could also have taken it for granted that geometry or surveying was a natural science. The art of practical surveying is obviously based on theoretical knowledge of nature. It really deals with what is empirical. And for the naturalistic mentality, which after all is always that of *common sense*,[1] one would have concluded from that that it was the sole foundation of surveying–natural science. One would have overlooked the fact that lying behind it is pure ideal geometry.

[1] In English in the original. (Translator's note)

The situation with logic has been similar up until the present time. As normative and practical discipline of human knowing, it naturally points back to psychology. Precisely in this respect, it connects with a general theory of education. It is, then, just intellectual art of education. We could also say that it is then human art of science, for practically training and educating ourselves for logical knowing amounts to educating ourselves for successfully practicing the sciences. Logic as theory of the art of knowing is naturally directed, not merely in general toward knowing, but toward knowing in the highest, richest, most successful forms that *eo ipso* comprise all other forms. It is therefore theory of the art of science. It is theory of science understood practically and pedagogically and as such it naturally rests on empirical psychology.

Let us now take a look at the other side. Norms for cognitive acts ought to be set by logic. Their norms ought to be set in conformity with the Idea of correctness and of intrinsic scientificity. And now we see that norms in an ideal sense are connected with this Idea, laws which–regardless of any empirical particularities of human knowing and of human cognitive states of mind–state conditions of the possibility of correctness in existence-free ways. The systematic unfolding of the Idea of truth and of everything essentially connected with it leads to a pure theory of norms that, without any reference to a thought praxis, to human knowledge and human sciences to be regulated, explores the logical Ideas, norms for purely theoretical reasons, and does so precisely in their purely factual context. It therefore marks off its own "theoretical" discipline that is a prerequisite, indeed, the most essential prerequisite, for any theory of the art of knowledge.

The error of psychologism lies, therefore, in the fact that, blind to the intrinsic legitimacy of the ideal, it believes it has already proven the psychological nature of logic in general by the obvious proof that, as practical methodology of knowing, a logical theory of the art of something must be grounded in psychology. It does not take notice of the fact that related groups of logical laws–and precisely those that have made up the core content of logic since *Aristotle*'s time–have nothing in themselves of a psychological nature and that they stand in precisely the same relationship to practical logic as pure geometry does to practical geodesy.

What seems even more especially misleading in this regard is the fact that, as I already indicated earlier, for purposes of application, logical norms customarily acquire an extra-essential content, that one speaks of someone judging and of his or her being convinced, and so on. This is thereby precisely the situation in pure mathematics, which naturally does not turn into a psychological science due to the fact that one can normatively apply its theoretical propositions in connection with human mathematizing. Thus, for example, $(a + b)(a - b) = a^2 - b^2$ is obviously a theoretical proposition. One can, however, also state it in the applied form: Whoever would multiply the sum by the difference can also instead of that give the difference of the squares.

It is clear that instead of starting by inquiring into the theoretical sciences upon which a theory of the art of knowledge must be based (or instead of, as I am in the habit of saying, by inquiring into the "theoretical fundaments"), one can also directly point to the fact that talk of a normative theory of knowledge is ambiguous, inasmuch as a norm could be a pure and an empirical norm. Empirical norms aim at what is empirical-practical, at empirical judgment and practical conformation and so has what is empirical in view. Pure norms, however, are existence-free. Logic as theory of the art of human knowing is empirical. It would supply us with 26
useful norms in our scientific endeavors. But such empirical norms presuppose pure norms and so again demonstrate that dwelling in the logical theory of an art of something is a body of ideal norms, which taken in their own right and explored in their systematic context, yield an independent, ideal science: pure logic.

Through the just finished succession of ideas, we have taken an important step forward, because, to begin with, we exposed the great error of empirical logic. It consists in the belief that only aiming at practical norm-setting for logic, as opposed to psychology, lends it its own justification as a discipline and that consequently the standpoint of a methodology of knowledge is to be put first in every treatment of logic, indeed could be the sole authoritative one. This fundamental error dominates the most influential logical work of the nineteenth century, the logic of *Mill* and the logic of *Sigwart*. Second, it emerged that two sorts of scientific disciplines are intertwined under the heading of logic: the more comprehensive practically oriented logic–the art of knowledge, the methodology of knowledge–and pure logic. That has major consequences. It already has a consequence for the entire treatment of the former, of the theory of the art of knowledge, because it can only generate confusion if, in the presentation of the theoretical foundations of the theory of the art, in the compilation of the theoretical insights useful for the practical goals, one allows the psychological and the ideal to run together, as if having one single sort of nature and function. In fact, connected with that is the unsatisfying and confused impression everyone who has the sense of clarity and perspicuity in carrying out problems has of, for example, the premier logical work of the last decade, *Sigwart*'s logic.

But we are not interested here in finding fault with deserving thinkers and in introducing better treatments of the methodology of knowledge, because considered in its own right, such a methodology has merited so much interest that such interest is nothing specifically philosophical. This interest pertains exclusively to pure logic, which alone deserves to be called a philosophical discipline or deserves this in the primary sense. The methodology is philosophical only to the extent that it includes pure <logic> in itself.
27 And, consequently, it is actually an important and major step that we as philosophers have separated off what is specifically philosophically meaningful, therefore pure logic.

In his philosophical masterpiece *Critique of Pure Reason*, *Kant* made this much cited, yet little heeded, statement, "It is not an enlargement but a disfigurement of the sciences when one allows their boundaries to run into each other".[1] And, meaningfully enough, he made this statement in connection with logic. He knew why he did that. He knew what great philosophical concerns are attached to the clearness of the segregation of an ideal theory of norms. For *Kant*, however, philosophical concerns in the first place signified transcendental-philosophical, epistemological concerns. We shall yet hear enough about what kind of concerns those are in the course of our studies.

---

[1] In the second paragraph of the preface to the second edition of 1787. (Translator's note)

What have we then achieved? Have we already sufficiently defined and delimited "pure" logic as pure science of Ideas through the reflections up to this point? And, is the theory of science to which we were to devote these lectures thereby already realized with this? Looking closely, we still know too little. The examples of logical norms determined by our reflections sufficed to delimit an ideal sphere of knowledge as opposed to the psychology of knowledge and the practical methodology of knowledge, the theory of the art of knowledge. However, they gave us but a vague notion of a purely logical field in which such pure norms have their place. A scientific characterization of this field is however very much lacking. Pure logic should also have to do with knowledge. It should, therefore, be a theory of knowledge,[1] and one often enough hears those who separate pure logic and psychology call logic theory of knowledge. But, it is not excluded that knowledge as Idea, or understanding as Idea, makes possible and calls for ideal investigations along different lines, so that yet more, essentially interrelated–but, however, to be separated again–ideal disciplines of the understanding would be contrasted with one another. The course of our further reflections is thereby in general indicated.

[1] Compare Appendix III, *Distinction between Explanatory Theory and Theory of Knowledge.* (Editor's note)

<Chapter 2 28

Clarification of the Ideas Thought-Act and Thought-Meaning>

<§ 7. The Outline for the Following Reflections. The Noematic Perspective as Turning Toward What Is Known as Such. Apophantic Logic, Universal Formal Ontology, and the Idea of a Pure Theory of Science>

To start with, I shall briefly sketch the outline for our next series of reflections. I shall describe in advance the milestones along the path to be travelled, each one of which at the same time indicates an important subgoal. The orientation we have adopted up until now–and naturally have adopted because we are beginners–has been toward cognitive acts, <toward> what is somehow contained in "knowing" in the broadest sense of the word: judging asserting as considering-true, also considering-possible and considering-probable, questioning, and doubting. All that, I say, belongs here, as well as the varied forms of substantiations with the intuiting experiences that play an essential role here, thus experiencing as perceiving, remembering, and so on, grasping the sub-concept, inferring. For the orientation we have adopted, cognitive acts do not, however, come into consideration as psychological facts, but in an ideal connection. If we now understand the logical approach as an approach to ideal norm-setting for cognitive acts from <the> standpoint of correctness and incorrectness, then we are led to a normative logic in the sense of a noetic theory of justification, of a theory of justification of knowing. We shall not, however, consider this logic more closely now, but rather completely postpone that for the time being. We shall rather turn our gaze, as it were, in the opposite direction. Instead of contemplating the forms of knowing in pure generality from the standpoint of correctness, we shall rather turn to consider what is known as such. I call this orientation

E. Husserl, *Logic and General Theory of Science*,
Husserliana: Edmund Husserl – Collected Works 15,
https://doi.org/10.1007/978-3-030-14529-3_2

noematic. The word knowledge is ambiguous. On one occasion, it means knowing, on another, what is known, as it is known. In just
29 the same way, as I shall discuss in greater detail, the word statement, judgment, is ambiguous: on one occasion, stating, judging, predicating, on another, the statement, the judgment, the predication, the proposition.

Taking a look at what is stated in the stating, at what is known in the knowing, as it is known, meant, as it were, in knowledge, we shall ask to what extent it may, from the ideal perspective, be cause for investigation. We shall then encounter a pure logic in a new sense, first of all, apophantic logic. *Aristotle* called judgment, apophansis. Without making the distinctions I have indicated, and which are basic to philosophy, by apophansis, he did not (as one may suppose) however mean the judgment-experience, but the judgment as proposition, and so the convenient expression "apophantic logic" is tantamount to saying logic of the statement (not of stating), or logic of asserting propositions. If we look at what is essential, then it is this logic that *Aristotle* founded (under the heading Analytics), and the *Aristotelian* groups of laws of syllogistic logic are again those making up the core content of traditional logic.

If, apart from judging as asserting-with-certainty, stating, considering-true, we also consider the modalities essentially intertwined with it, thus presuming, considering-possible, then what is conscious there is not supposed truth or proposition, but supposed probability or possibility. That gives rise to the expansion of the idea of a pure logic by a pure logic of possibilities and probabilities.

For essential reasons, pure arithmetic–and, beyond it, the whole of formal mathematics or theory of manifolds–also proves to be intertwined with the logic of affirmations, with apophantic logic, though in a completely different sense. These disciplines form, so to speak, a higher storey of apophantics, and recognizing and characterizing them in this regard is of the greatest importance philosophically.

All the noematic and mathematical disciplines just now fleetingly indicated have intrinsic unity and constitute the Idea of a universal ideal science, namely, of a science of everything that can be stated with unconditional generality about knowable objectivi-

ties in general, or, what amounts to the same thing, about objectivities in general–regardless of the particularity of their content–what is in general to be valid for them, if it in general is to 30
be able to be said that they exist. I shall speak of formal ontology in this regard.[1]

It ultimately becomes apparent that all the disciplines designated fit in with the idea of a general, and primarily, formal theory of science. One word about this: I have upon occasion connected the theory of the art of knowing with the Idea of a theory of science. I have explained that the logical theory of the art of something can also be called the theory of the art of science, therefore, practical theory of science. The question now arises as to whether–apart from all practical aims–science cannot also be established as an Idea and a pure theory of science constituted as unfolding of this Idea. Science as such also actually has its "a priori". What is meant is not science as a cultural phenomenon, but science in the ideal sense which we contrast to pseudo-sciences like astrology and spiritism. Now this "genuine" science is an Idea, and the question as to what lies in this Idea, what determines it, is obviously tantamount to the inquiry into all the principles to which, as ideal norms, every science is bound and to which every statement, statement substantiation, every inference, and every theory to be fit into science in the genuine sense is bound. One already realizes that all normative logical laws of every narrower sense of logic and all the disciplines I have named up until now will fit in with the Idea of a "pure" theory of science.

<§ 8. The Pivotal Point of All Theory of Knowledge: The Recognition of Ideas as Atemporal, Supraempirical Objects>

Let us now turn to the implementation.

First of all, I shall give a general explanation of empirical and ideal objects, which I also more briefly call Ideas. I admittedly already talked about them when I contrasted empirical and ideal generalities, empirical and ideal sciences, with one another. But, it

[1] Ontology means theory of being, etc.

is at times necessary to bring Ideas for their own sake as objects in their own right to givenness. It is imperative to force people once and for all to concede that Ideas are genuine, actual objects by
31 contrasting them with the empirical objects that alone we are inclined to see in the natural attitude and alone to recognize as objects. And that has not yet manifested itself in a clear-cut way in our passing reflections.

Therefore, Ideas as objects, not as if it were merely a matter of some special way of looking at things that we accorded the objects of experience the status of being the sole objects, a matter of something like a change of the relative standpoint that did not change anything about the objects themselves and only viewed them from other standpoints. Therefore not, for example, that on one occasion we take the same objects as they are there and then on another occasion make claims about them, evaluate them, and so on. No, it is completely strictly meant in absolute earnest: new kinds of objects. This is a pivotal point of all theory of knowledge. A proper grasp of this point is decisive for all further considerations. We shall be dealing over and over again with Ideas.

Let us reflect! When we speak of sounds, then precisely sounds are the object of our statement, and we shall hardly have cause and desire to doubt the fact that such objects exist and true statements can be made about them. You will surely also grant me as a general principle that if objects exist, then true statements that state the properties of these objects must hold for them–and vice versa, wherever true statements hold that say affirmatively how objects are made up, the objects (but, taken precisely in the sense in which the statement posits them) must also exist. I am obviously taking the word object here in a broadest and really accurate sense. I call every subject of a true predication an object. I apply this to the talk about sounds. We then notice that two kinds of statements that are incompatible with one another are directed to sounds, and are in actual fact directed to them, so that talk of sounds has two kinds of meaning, must therefore concern two kinds of objectivities.

In one case, we speak of sounds that are ringing out now or have rung out earlier, in short, of sounds as individual facts, as events in spatiotemporal nature. These sounds have their delimited duration. They begin and stop. They change during their duration, be it quali-

tatively, be it with respect to their intensity or timbre, etc. Those are 32
the sounds spoken of in empirical statements.

In another case, however, we speak of sounds of octave $c' - c''$ within an ideal sound sequence. A sound occurs in it, sound *c*, as a sound, sound *d*, etc. And even the entire octave is a numerical unit. Of these sounds of the sound sequence, we declare that they have this or that definite order. One will say they are qualities *in abstracto*, are abstractions. That may be. One may then thereby understand by abstractions anything at all. However that may be, these pure qualities are not anything empirical. They have no place in time. They do not ring out. They do not start and do not stop. And it makes no sense to say that they change. In the sound sequence, we can go from c′ to d, etc. We can also think of the discrete sound sequence of the octave completed by continuous flowing of the qualities. But flowing as a real occurrence makes no sense here. The one identical sound c of the ideal sound sequence does not move from any place. It does not change. Only something real can change, only something existing in time. But sound c no more exists in time than a pure number of the number series exists in time, and no more lasts than the pure number lasts in the number series. The number series provides us in general with a second example of purely ideal objects. Each cardinal number occurs only once in it. There are infinitely many possible empirical sets that we can count: cardinal numbers of horses, of carrots, etc. But these empirical cardinal numbers come into being and pass away, start and stop, etc. That does not, however, affect the pure cardinal numbers. If no concrete cardinal number n were to exist in the real world from a certain point in time on, then the pure number series would not for that reason have a hole between $n - 1$ and $n + 1$. Similarly, if by chance, sound c did not ring out in the world and would never ring out again, then the sound sequence would not then perhaps have a hole–just for the simple reason that the Idea c is just not itself a c. Just as the Idea 2 and each of the Ideas that I grasp as members of the number series is nothing occurring in the world. Whether there is in general a world, an existence in time and space has nothing to do with it.

If someone asks me how I know that, then I answer that I after all know what is being talked about when I speak of the numbers

33 1, 2, 3,... in a number series or of the sound qualities in a sound sequence. I know that in so doing one is not talking about nothing, but always about an unreal-existing something. And this is given to me when I reflect and make evident to myself what I posit there. Indeed, perhaps absolutely given! Two, I make absolutely clear to myself, is one and one. And I thereby grasp it, have it itself, so that any doubt as to whether it exists would be totally absurd.

We need not wrack our brains over which kind of being this is and which problems it raises for human knowledge. The only thing important here is that it is something to be had, something that one can grasp and about which one can make evident statements. It is an indubitable, absolute truth that $2 < 3$, that in the cardinal number series, 2 has its place between 1 and 3 etc. Now, if absolute truths hold for two, then it is precisely something as subject of these truths themselves, or it is an object, because that implies absolutely the same thing. Consequently, there are objects of insightful givenness that are not things and not existential moments in the spatiotemporal world. I am merely giving these objects a label when I call them Ideas or ideal objects. In doing so, I am not thereby giving the word idea any sort of mystical meaning, any interpretation to be fetched from somewhere or another. For me, it is a mere label for what was indubitably given with those evident statements. If someone therefore asks “What are Ideas?”, then I do nothing but point to such evident givens as the cardinal number series, or to such absolutely evident statements about members of the number series as I have just now used them. If someone asked me, “What is ‘thing’, and is there anything like a ‘thing’?”, then I cannot really after all act otherwise than to point. I perhaps say, “Do you see that there, that house”. If the answer is, “yes”, then I go on to say: Well, <a> thing is just something of the kind, everything intuited in plain sensory perception of that kind. And something of the kind is then the subject of true empirical statements in which truth about things reaches us. Therefore, precisely the same for Ideas. I embrace Ideas and classes of idealities, not perhaps because I pride myself on my own intellectual intuition and want to penetrate into a mystical transcendental world by means of it, or because I boast about “idealism” and want to look down upon vile empiricism and psychologism. I do not embrace Ideas because with them it is easier for me to devise a "nobler" sort of philosophy that at the same time brings me the reputation of being a noble 34 soul. No, I embrace ideal objects for the same mundane reason that I embrace things, just because I see them, looking at them grasp them myself. I even maintain that ideal objects are in no way anything especially lofty that one could flaunt, but what is the very most ordinary, just like ordinary stones on the road. All people know them in a certain naïve way since they indeed talk of numbers and sounds and so on in ideal ways. Only philosophers do not wish to know them. They dismiss them, calling them *Platonic* Ideas. Granted, I say. In fact, if one recognizes givens like the series of natural numbers, or like the sequence of the sound species, as objectivities, one cannot at all describe them in any other way than with the words that *Plato* used to describe them in his theory of Ideas: as eternal, selfsame, as non-temporal and non-spatial, as unmoved, as unchangeable, and so on. But, instantly, *Platonic* Ideas –nothing but hypostatizations of abstractions–goes buzzing through the heads of those trained in traditional philosophy. *Platonic* realism would at the same time be mysticism. And people remember how Neo-Platonism teams up with a magical view of nature, the most extreme opposite of genuine natural science, how in the Middle Ages it merged into scholastic realism, etc. No, people say to themselves there, we must keep mysticism and scholasticism at bay. Therefore, people no longer get involved in disputes about it at all. The disputes are considered to have been fought out in the battles since the Renaissance. If someone ventures to recognize ideal objects, then people therefore have but one thing to say: That is scholasticism.

In opposition to this one may answer: What do theories–be they rational or non-rational–that were elaborated over millennia about the objectivities that I have presented as Ideas have to do with us? Have I theorized here? Have I not limited myself to recognizing objects that are given to us as subjects of evident statements as just that and to describing their general features just as they were to be grasped in absolute givenness? I have not spoken about theories, but about what is prior to all theories. That it is meaningful and right to speak about the number series and to make absolutely true statements about those objects that are called numbers there, no one can seriously wish to contest. And, what we are urging is nothing more than the recognition of the fact that one holds fast to what is stated there exactly in the sense in which these statements posit it. What I must urge is that that is not to be interpreted away and reinterpreted out of any historical reasons and obscure emotions, out of fear of long buried errors and such things, precisely thereby contravening the sense in which it figures as given. Every time we make an evident statement about numbers of the number series and thus grasp an objectively valid truth, precisely these numbers, and not something else, are the objects to which the truth is referring. They are therefore objects. And they are then exactly what they give themselves to be. They are not things and not moments of things. They are non-spatial, non-temporal, and so forth.

Proceeding in such a way suffices to incur the reproach of being scholastics. So I say, well fine. What is preferable, to have integrity and be so-called scholastics or to lack integrity and be modern empiricists? I am pleading in favor of integrity and do not fall flat on my face when I am called a scholastic. I also know a good saying that reads, “Integrity stands the test of time.”

Naturally, one may not come to us either with the objection that admitting Ideas is embracing innate Ideas, a priori Ideas, that mockery of all genuine psychology. That is no objection, because one is again laying a theory as a foundation for the givens. We however refuse to accept any theory where we stand prior to all theories. I have not said one little word about how human beings as natural beings have come to have the ability to grasp ideal objectivities, for example, that human nature was originally endowed by God with the ability to see Ideas, let alone that, since human beings are God’s handiwork, they were instantly endowed with ready-made a priori presentations, namely, presentations of pure Ideas, and so on. Those are theories, perhaps infantile, unclear, utterly mistaken theories. They are of no concern to us at all. I have no reason to criticize them. We only need one thing: that we actually speak about Ideas, that our statements about them occasionally have absolute Evidenz, and that we have given them in this Evidenz. Only thus, as they are given there, do we want to take them and may we take them.

One must, besides, also surely keep in mind the fact that the word Idea has taken on many meanings and that, especially in the 36 parlance of English philosophy, it is an expression for subjective experience, for presentations, that, however, Ideas in my Platonic sense are not presentations, but atemporal, supraempirical objectivities.

Finally, be again advised that it is a peculiar fact that for each Idea, for instance <number> 2, sound c, we can form the thoughts a two, a multiplicity of the form 2, an empirical sound of kind c, of species c, and so on, respectively, and that, accordingly, belonging to every such Idea is an "extension" of possible singular particulars, of which we say that, if they in truth exist, they fall under the Idea in question as its general species or genus. Every pure statement of Ideas is equivalent to an unconditionally general statement about singular particulars in general that fall under the Ideas, and such general statements are also called ideally valid statements, namely, in contrast to empirically valid statements about posited existence. The world of Ideas and the world of individual existence are not foreign to one another. Corresponding to every predicate is its specific content and with it an Idea, thus to the predicate "red", the Idea red, to the predicate "circular" the Idea circle, and so forth.

<§ 9. The Noetic Theory of Norms as Theory of Norms of Judgments in the Sense of Judgment-Idea. The Difference Between Judging and Judgment as Idea of Judging>[1]

Now that I have provided the foundations, I shall proceed to the execution of my plan. From the perspective of norm-setting, we look toward knowledge or thought-acts, but in a certain ideal way, because we already know that knowledge as actual experiences of living people belongs in the sphere of psychological research.

Therefore, we only hold fast to the pure norms themselves. They are norms for knowledge. <If we take> for example <the

[1] Compare Appendix IV, *Supplement to the General Characterization of the Noetic Theory of Justification. The Ideas of Judgment-Events Relevant from the Viewpoint of Theory of Knowledge.* (Editor's note)

norm>: Of two contradictory judgments, one is true and one false–or the norm: If two judgments of the form “All A are B and all B
37 are C” are valid, then the associated judgment of the form “All A are C” holds–or: Out of two judgments of the form “if A holds, B holds” and “B does not hold”, it follows validly that A does not hold, and so on. We are talking there normatively of judgments or, as we can also say, of assertions. We understand by that certain occurrences from the so-called thought-sphere very familiar to all of us. Every time, we make a predicating statement and in so doing say not merely the words, but also mean, assert what we are saying, we have judged. If we read out a sentence from a book without endorsing the author’s view, without therefore ourselves asserting what we are saying there, then we are just not judging. Judging, as I am now using the word, is tantamount to asserting really–or inwardly without saying the words out loud, expressing them to others facing us–, to really holding the conviction in question, being conscious that that is how it is! That would be sufficient here for understanding.

Now it is easy to see that it would be false to say that those judgment norms that I have drawn upon as examples make statements about judgments as psychological facts. Rather, it is clear that the case is the same as with those ideal statements about sounds in which talk is not of real sounds, of long and short, faintly and loudly ringing out sounds of the existential sphere, but of sounds as such in ideal purity or of the Ideas themselves. It is precisely the same here. Pure norms refer to judgments, but not to judgments that occur in reality, that are actually made by any persons–even if remaining undetermined–, but to judgment-Ideas or ideal particulars of such Ideas. The purely logical approach is everywhere an approach to Ideas or ideal particulars, here to Ideas of the psychic sphere, i.e., Ideas in whose ideal extension, possible empirical judgments, possible cognitive experiences belong. Already, when logicians give the judgment “All men are mortal” is true as an example, or when they say, “The judgment ‘A body is extensionless’ is contradictory and for that reason false”, they are not talking about judgments in the sense of actual experiences that take place in any living consciousness as an event of our empirical reality. That becomes instantly apparent in the fact that for this

approach the judgment "All men are mortal" is a single judgment, even if "All men are mortal" is judged a hundred times. One and the same judgment can, as we are in the habit of saying, be "repeat- 38
edly made", can thereby be made today or tomorrow, by myself or another person, on one occasion faster, on another occasion more slowly, insightfully, and not insightfully, and so forth. But in contrast to those actual judgment-experiences, the logical judgment is a single thing, and as this, it is one, right or wrong judgment, and has other logical properties. The property of being contradictory, and for that reason false, is not a particularity of the contingent judgment-experience that is just now taking place in me if I would in all earnest have said and asserted that a body is extensionless, but the judgment "A body is extensionless" is contradictory and false. Even if no one had judged thus, and no one in the whole world would judge thus, what is true would remain true and what is false would remain false. The judgment is obviously a supratemporal Idea, existing precisely in the mind, existing nowhere at all and at no time at all and yet existing like a number of the number series. In the realm of the ideal units that we call judgments or clear judgment-Ideas, this judgment-Idea has its place, and quite a few indubitable truths hold for it, for example, the judgment "$2 \times 2 = 4$" is a general judgment, has a subject and a predicate, is a judgment about numbers, a judgment about equality, and so forth.

Corresponding to every such Idea is, however, an extension of ideally possible particulars, and if we start from an actual judging that I (or someone else) may make, then it belongs in the extension of the judgment-Idea concerned. Now, if the judgment as Idea is a validity, then every actual judgment to be brought under the Idea is a correct, a correctly made, one. We can then state laws of validity with regard to pure Ideas, for example, laws that perhaps say that every ideal judgment of such and such a form is false or that, if they are validities, every pair of judgments of such and such a form entails the validity of a judgment of such and such correlative form. It is thereby to be kept in mind that all formal properties belong to the Idea, as, for example, when we speak of the judgment-form of the universal judgment "All A are b" and say the judgment "All humans are mortal" has this form of universal judgment. What the form has there is in the primary, genuine sense not somebody's experience, but the one identically-ideal judgment, "All humans are mortal".

I shall soon say more about forms. The only important thing here is to make it evident that a distinction is to be made between
39 judging and judgment as Idea of judging, in addition, that pure norms for judgments aim at judgment-Ideas and only thus acquire an unconditional, generally valid normative relationship to real or actual judging as possibly thought.

Thus, just as we bring judgment-experiences as experiences of "That is how it is!" under Idea and, in the attitude of positing Ideas, set norms of validity and non-validity, so <also> judgment-experiences in a broadened sense. I mean here experiences of taking-to-be-possible and -probable, of questioning and doubting. So, in this regard, there are forms and norms of validity that correlate with these forms, and one can interpret several norms of the purely mathematical theory of probability in the sense indicated. And finally, one can also bring under Idea and thoroughly investigate all cognitive experiences in general in accordance with all immanent qualities with reference to the jurisdictions of truth, possibility, probability, questionability.

Everything that is to be scientifically accomplished from the point of view of setting norms for rightness of cognitive acts –i.e. of acts that are either themselves judgments in the broadest sense or, as rightness-conferring perceptions, memories, and so forth are called to play an essential role in showing the rightness of judgments–, we allot to the noetic theory of norms, and in accordance with what has been said, strictly speaking, it does not have to do with acts as human or other experiences, but with the corresponding act-Ideas.

Were this pure discipline realized, we would therefore be in a position to trace every case of actual judging (in the narrow sense of actual presuming or considering-probable, and so on), as well as every case of actual substantiating, actual deductive inferring and theoretical explaining, inductive inference… and so forth, back to ideal principles and, in principle, to judge in terms of its basic normality.

From the outset, we want at the same time to validate the scientific-theoretical nature of our investigations. Therefore, we want to make clear to ourselves from the outset in each step the fact that the area of pure knowledge that we are presently scientifically delimiting must be a basic piece of a general, pure theory of science. By this we understand a science that systematically explores truths pertaining to the Idea of genuine science, consequently everything that must necessarily be valid if a science in general is to be able to be valid as science in the genuine sense of validity.

In keeping with my plan, I am setting this noetic theory of justification aside. Here, I have only had to characterize it generally, because it is the most obvious Idea of a theory of norms for knowledge. My first objective for the purposes of my plan is the pure logic of affirmative statements, essentially involving the area that *Aristotle* dealt with under the heading of Analytics. I myself call it apophantic logic.

<§ 10. The Judgment-Idea, Proposition, and State-of-Affairs>

<a) Judgment as Proposition as Opposed to Judgment as Judgment-Idea. Noetic and Noematic Reflection>

We now want to turn our attention to a certain striking objectivity that is proper to these statements and combinations of statements and confronts us in especially clear ways in scientific propositions and combinations of propositions. The theorems of both a science and its systematic theories pass themselves off as, as it were, creations of the mind, as the random result of research endeavors. We also characterize them as that. We speak, for example, of successful and failed theories. We say that *Pythagoras* postulated the theorem named after him, such and such false theorem originated with *Descartes,* and so on. This and similar talk seems to point to that fact that under the heading "proposition" or "proof" or "theory" it is a matter of facts of existence, especially of psychological creations. We distinguish, of course, between the proposition postulated, between the theory postulated and the one who <has> postulated them, and we refer to what has been postulated as something intersubjectively accessible, as a common possession of science and her researchers. That can, however, be accounted

for by the fact that the possibility of intersubjective understanding exists by virtue of language, as a result of which each person can take cognizance of the thought-findings that the other person has produced in him- or herself and can produce them later with the same content in him- or herself. It seems therefore that we are in
41 psychology when we speak of propositions, inferences, proofs, theories. If, however, we have already acquired the ability to adopt the ideal perspective, then it seems that we only reach a sphere of Ideas that separates empirically into psychic experiences and particularly into certain psychic products. In this respect, we would not go beyond a noetic theory of norms that we indeed precisely wanted to exclude.

Let us, though, take a closer look at the matter. If we state a scientific theorem, then in doing so we have an example of an affirmative statement. On the other hand, we do not make the statement in the normal way as our judgment-supposition when we hear and understand the proposition affirmed by others, but consider it false, or when we grow doubtful of it. In each case, we can make the following distinctions regarding the normally or not normally made statement:

1. The verbal garb: The same *Pythagorean* Theorem can be stated in the Greek, German, French languages, and so forth. Such differences are of interest to empirical linguists.
2. Bound to sensorily phenomenal visual or acoustic etc. verbal expressions–if it is really to be a matter of words, and more precisely of a coherent statement–are certain psychic experiences–judging affirming, if the statement is stated in the normal way. (For that, other experiences can, however, also enter in, mere understanding, without affirming, doubting behavior, and so on. All combined, judging).
3. It is now important to clear up a fundamental ambiguity in the concept of judgment. Belonging to each judging, for example, each judging coming to be stated verbally, is what is judged–what we are stating. If by judgment, we understand judging, by statement stating, then we call this What the judgment-content, the statement-content. This content is, however, what is simply called judgment (or statement) in innumerable cases of ordi-

nary and scientific discourse, and so the word judgment is subject to ambiguity—and this is of great importance to us. The fact that it is actually a matter of two sorts of things becomes clear in the fact that the judging, the subjective consciousness of the "That is how it is", involves an entirely different way of looking and consequently becomes a completely different object for us than judgment in the second sense, what is judged, the supposed What.

It must be kept in mind from the start that if we judge naïvely, 42
we are looking at the objectivities about which we are judging there. So, for example, when we judge; "This bench is yellow", then we look at the bench and the state it is in. That lies objectively before our eyes. The ray of our attentiveness and our grasping- and judging-as-actual-object aims at that. The judging itself is not thereby objective for us. It turns into an object in a so-called act of reflection. We must look away from the object and look back here to the ego and its act. And, on this reflective ground, so to speak, we perhaps then behold the corresponding idea of judging as an essentially general thing of reflection.

How is it then if we want to turn the "judgment-content", what we suppose in judging, into an object? This What is precisely what we are in the habit of calling a proposition within the context of stating, where it is obviously not a matter of the verbal expression itself, for example, the theorem of the sum of the angles of a triangle in the sense in which it is the same theorem whether it is stated in French or in German.

Now, keep in mind the way of looking when you speak of this theorem in this manner and not grammatically, but for logical purposes. When, for example, we state, "The sum of the angles of a triangle is equal to two right <angles>" and then continue on, "This theorem is fundamental to *Euclidean* geometry", what does the word "theorem" direct us toward? Obviously, not toward the judging—neither toward judging as experience, nor toward the Idea of judging. If we start from the way of looking at the judging, then the moment we look at what the judgment judges, what the assertion asserts, we immediately turn to look, so to speak, in an opposite direction. So it is, for example, when we say, "What you are judging there is a general proposition"—or, "What you are stating (your statement) has the form of a general statement". Here statement is obviously not stating, nor is it the Idea of the stating, but precisely something new, the stated as such, the proposition. So, I hope, it will be completely clear that proposition and judging must be separated. Judging as Idea is to the actual experience of judging as a general Idea is to particular cases, as the general Idea red is to the contingent empirical red of intuition. The proposition is, however, not the judgment-Idea.

43 Traditional logic did not take this distinction into consideration. (In *Bolzano*'s *Wissenschaftslehre*?) Traditionally, no distinction at all was made between judgment and proposition. Not only did traditional logic fall short in this respect (and fell short regarding a matter of fundamental significance) by mixing the empirical and the ideal together, by not distinguishing between judgment in the psychological sense and judgment in the logically ideal sense, but also by not distinguishing clearly between judging and proposition. It did not see that two Ideas correlatively related to one another <were to be distinguished> there.

I said "Ideas" because I surely do not need to demonstrate at length that what I just recently explained about logical judgments as Ideas of psychic acts also holds for propositions, therefore, that when we speak of the theorem of the sum of the angles of a triangle, we are speaking about it as an Idea. It is clear without further ado that this theorem, which is a basic mathematical truth, is not an occurrence in the world, that it does not refer back to contingent judging in which we—or anybody—have this theorem as content of a statement. However often we may speak of this theorem, it is still numerically one, and in so doing we do not thereby in the least think of the millions of people who have grasped it, taken it to be true, and used its truth in reasoning. In general, it is meaningless to ascribe existence in the real world, existence in space, size, duration, change or rest, and so forth to the theorem. In short, it appears to have all the essential features of an Idea. However, doubt could arise.

In point of fact, we are, however, not yet finished with the fundamental considerations. Proposition is "the content" of the judging. Is judgment-content the object-about-which or the state-of-affairs judged?

<b) The Distinction Between Proposition and State-of-Affairs. The Objectifying Way of Looking at States-of-Affairs, the Reflective Way of Looking at Propositions>

At first glance, it might seem that the state-of-affairs judged is the proposition. Let us just attempt to carry through with this perception. While I am stating something, I am turned toward certain things. Stating, I state something about something, about any object and state such and such about it. Therefore, while I am judg- 44
ing something, judging the triangle in general and judging that the sum of its angles is equal to two right <angles>, then the mathematical state-of-affairs is the What judged. Similarly, if I judge that the sun is at the center of our planetary system, then this natural state-of-affairs is the What stated, therefore, something real here, to which the real sun belongs as substrate, but nevertheless surely not an Idea. One could also continue, that is also compatible with our earlier talk of changed way of looking. "Judging" is having consciousness of the being of the state-of-affairs. In this consciousness, we are turned toward the state-of-affairs, and when that is the case, on the basis of this way of looking, we can judge what falls within this gazing, just as when we judge about a perceived thing on the basis of an ordinary perception. It figures in the perception, and about this we say, this thing, this table has such and such properties. In precisely the same way, a state-of-affairs figures in the judging here, and we make a statement about the <state-of-affairs> grasped in this judgment-gaze: This proposition has such and such properties, this subject, this form, the property of truth, and so forth. We must, however, position our gaze entirely differently–namely reflectively–when we want to judge about the judging itself, about the consciousness of the figuring of the state-of-affairs.

This view is, however, false and rests on defective analysis. We must rather distinguish–and I am noting that as new point 4–between proposition and the state-of-affairs in truth corresponding to it. And, in precisely the same way, we must distinguish between the nominal naming of the subject within a proposition, or the subject-meaning in which an object turns into the proposition's object-about-which, and the named object itself.

Accordingly, we must distinguish between different possible ways of looking:

1. the objectifying way of looking that we have in the stating, namely, the way of looking at the objectivity-about-which, or at the state-of-affairs; and
2. the way of looking we must engage in when we make the proposition into an object for ourselves. We have this way of looking in a new judging for us that states predicates, not of the original object, but of the proposition.

45 <c) The State-of-Affairs (as Supposed and as Actual) as the Objectivity Given in the Judging. The Proposition as What Is Judgingly Supposed as Such>

Let us therefore reflect upon the relationship between proposition and state-of-affairs. Let us start with an actual judging. "Today the weather is variable!" In this stating experiencing, something, as it were, stands before my mind's eye. It seems to me to be something. In my example simply, "Today the weather is variable!" The "That is how it is!" stands before my mind's eye. There is obviously an analogy present here between the judgment-consciousness and, for instance, the plain perception-consciousness. Perceiving is a certain experiencing, and something figures <in> it, in a way of existing, perhaps, the table.

If, therefore, I am simply perceiving, I am not reflecting on the perceiving, but I am completely naïvely living in the performance of the perceiving, then what my gaze rests on is the perception-object. Perceiving is nothing other than having an object in sight in this first-hand way. Similarly, remembering is nothing other than having an object in sight in a different familiar way in bringing-to-mind-again sight. Now, judging is also a certain consciousness of something, is a certain having an objectivity in sight, namely, having in sight a "That is how it is", a "That exists and is thus and so!" It is precisely this objectivity, as taken as being, as it is taken in the naïve attitude of judging that I call state-of-affairs.

But now, the remarkable thing is that judging–therefore having directed the judging gaze toward a state-of-affairs–in no way amounts to signifying that the state-of-affairs that the judgment is positing judgingly actually exists, any more than perceiving, having directed the sensorily seeing gaze toward a thing, signifies that what has been seen exists in reality. The same holds for every kind of so-called object-consciousness, for every experience, that, in the manner of a *doxa*, has the remarkable characteristic that, in engaging in it, we simply, as it were, have something before the mind's eye that figures there as being–while on the other hand this figuring-as-existing does not however signify actually existing, actually being in any sense.

Hence, we have different expressions to account for this evident situation. Any time we perceive, we say an object may appear, 46
a thing or process may appear as physically present. Perception "may suppose" grasping an object in its self-presence. For this reason, it need not "really" exist. Bringing to mind again by remembering may make something past appear to us in the manner of a re-appearance, it may be a supposing, the past thing may have really been, and so on. But "supposed past" does not amount to signifying that actual past must correspond to it. And similarly, in judging, we also say that in judging, this or that appears to us to exist, a state-of-affairs, an "It is thus and so" is "supposed." Supposing does not yet signify that it may really be so, that the state-of-affairs may obtain in "reality" or truth.[1]

Naturally, we especially speak about supposing when this contrast becomes particularly perceptible to us, therefore, in cases in which we become doubtful. Precisely for this reason, lying in the meaning of talk of supposing, of seeming, and so on is at the same time as a rule an indication that there may be doubt there, or what is meant there may be downright false, as we then stress "mere" semblance, mere "supposing."

Let us, however, rule out these added thoughts. Under the heading "supposing judgingly," "positing," we shall retain only the fact that with just the judgment-consciousness as such it has not yet been decided whether what is supposed, judgingly posited may really be.

---

[1] Obviously, the "really" does not signify natural reality. In mathematical judging, what is figuring as existing is a mathematical state-of-affairs. In false mathematical judging, it does not exist in "reality," i.e., not among truly existing mathematical facts.

Whether, then, what is judged, what is figuring as state-of-affairs in the judging really is or not, in other words, whether the judging is legitimate or wrong, one thing is absolutely certain, the fact that in the judging itself this and nothing else is posited, supposed. We can, then, precisely look at this supposed, posited thing as such. Instead of living naïvely in the performance of the judgment, instead, therefore, of simply judging that the sky is cloudy, the Earth is round, and so on, we can look at what is posited as such by this judgment (or "what is supposed as such") or, as I also say, at the judgment as judgment-noema. If someone else makes a judgment, we may not agree with him or her at all, but right when we want to refute him or her or to test his or her claim, we bring what he or she is positing as existing, has supposed by way of judg-
47 ment, to givenness for ourselves. We orient our grasping gaze toward what is posited as such. If we set aside the fact that it is something posited by this person in his or her judging, if we take it purely in itself, then we have a proposition. And, the proposition thereby figures for us as something ideal because with regard to what others are positing judgingly, when they all also judge, "The Earth is round", we at once say the same thing is judged universally, the proposition is the same.

A new way of looking and grasping therefore brings the proposition to objective givenness for us, makes it into our object and especially then into the object of our judging about the proposition. With Evidenz, we grasp in such a manner that a proposition belongs to every judgment as what it has supposed. But *nota bene*, the judging is not itself the directedness toward its proposition, as if in judging we had turned the proposition into the object. If I presently judge the sky is cloudy, then I am focused on the sky and the cloudiness, that is, on the unit "The sky is cloudy". Therefore, the state-of-affairs of nature is the objectivity upon which the judging is focused here. The proposition "The sky is cloudy" that I extract as object in a new orienting of my gaze is, however, something different from that natural state-of-affairs. The state-of-affairs is said to obtain only when the judgment is correct. No state-of-affairs in reality corresponds to incorrect judging. It refers to a state-of-affairs, but the latter precisely does not exist. Whether it is correct or not, it refers under all circumstances to a state-of-affairs, and that precisely comes to givenness in the form of our saying that each judging–and that is an irrevocable part of its essence–is the positing of a proposition. Even contradictory judging such as "The square is round", "There exists in space regular decahedrons" has as what it has referred to, has as what is supposed, a proposition, even a contradictory one. A proposition has its place in the world of propositions, is an existing thing. Even a contradictory, a ludicrous, false proposition is an existing thing. One can state many sorts of things about it. One can study in detail how generally to make the world of propositions into an object of study. And one can do this without making acts of judgment–or noetic judgment-Ideas–into an object of study. The situation is very different here for a noetic and for a noematic science. Belonging to the essence of each judging is a proposition as what has been supposed in it and, consequently, ideal involvement with judgments as act-Ideas also requires drawing in propositions. A noetics of judgment therefore also deals with propositions, differently, though, in the opposite direction. 48

<d) The Analogy of the Judgment with Acts Essentially Related to It: The Difference Between Objective Supposing and Actually Being>

First, it is good to pursue the analogy of the judgment with other act experiences essentially related to it. In judging, we have a consciousness that, based on simple cases, we can express with the words: That is thus and so. A "That is how it is" comes into view. Also, in perceiving, something also figures in mode of being. I see, for example, the table. I perceive it. The ray of the perceiving I's gaze is directed toward the table. Belonging to the very essence of the perception-consciousness itself is a directedness toward a perceived object, thereby, an object one is conscious of as existing. As, then, an actual thing or a tangible process is conscious to perceiving in the mode of being, so I say a "Such and such is thus and so" is conscious to us in the judgment in the mode of being. It figures as being. It is an objectivity. And I call this judgment-objectivity state-of-affairs and take it precisely as that, <as> what it is referred to in the judging at the moment.

Are the states-of-affairs the same as what I called propositions, or – in the sense defined earlier–are they the judgment-contents? Contents of the judging? The parallel with perceiving refers us back to the parallel question regarding this. Are perception-content and perception-object the same thing? Similarly, for all parallel acts. <Are>, for example, memory-content and memory-object, expectation-content and expectation-object <the same>? A separation is in fact to be made here and everywhere, although however such a close relationship obtains that it is surely understandable that both ordinary language and scientific thinking constantly waver here and are unwilling to make an exact distinction.

I tie my reflections in with cases in which it becomes evident to us that the perception, that the memory, that any judgment at all has no object at all. The acts involved do not for this reason stop supposing an objectivity, being conscious of an objectivity in the mode of being. But, in these cases, it is a "mere supposing". It merely "appears" to be something and we rightly say that it does not exist. And, just as we have found that in certain cases, we understandably say that judging and actual obtaining of the judged
49 state-of-affairs are two different things. Remembering, and actual having-beenness of the remembered process are two different things. Perceiving and real being of what was perceived are two different things.

We also understandably say that in each and every case judging is first a supposing that such and such may be. Whether, however, the state-of-affairs, what is supposed by way of judging, obtains, is a question in its own right. Actually, the critical, and in a certain sense, the logical question is the question of validity. What it presupposes in order to be meaningful–and it is surely always and necessarily meaningful–is a matter for our consideration. Actually, what constantly lies in the judgment while we aim to answer the question of legitimacy by substantiating or refuting is what we place under the heading of judgment-content, proposition, also meaning of the judgment, meaning-content of the statement, the supposed What. Whether a judgment is correct or not, correlatively speaking, whether the state-of-affairs obtains or does not obtain, the judgment is supposing that it may obtain, supposing that it may be such and such. The judging is, as it were, a positing, and the posited as such is precisely to be grasped as such, to be secured, to be dissected as such, in short, to be grasped as object about which one can in turn oneself make true statements. If we take the meaning lying in a judging (and analogously in a perceiving or remembering) or, what is the same, the proposition, as our subject in this way, then we obviously effect a distinctive change of our mental attitude in comparison with that effected in naïve judging. Judging naïvely, we simply engage in the confident supposing that it be such and such. In doing so, our consciousness is thereby that of the truthfulness of the state-of-affairs involved, just as our naïve perceiving is the plain consciousness of the existence, and even physical existence, of the perception-object. What is conscious in this way is quite simply called the object or state-of-affairs.

Essential change proceeds in our mode of consciousness when we go from naïve perceiving or judging to the attitude in which the supposed What, the meaning of the still simply naïvely performed act, is made into a subject, therefore, a distinctive kind of reflection is practiced. This change of attitude very commonly takes place when we are wavering in our conviction and have doubts, see ourselves pressed to ask our logical conscience whether what we perceive, judge, there then also actually obtains. The person we just now naïvely saw still stands there before our eyes, still the state-of- 50
affairs that was just now conscious to us as reality, but it is now plainly no longer object, a state-of-affairs, since it now no longer simply figures as existing reality. Out of the thing seen has come a "merely supposed", merely phenomenal thing, out of the existing state-of-affairs, something "merely supposed". And now we look and make the supposed What into its own subject and that is the meaning (<so to speak> in quotation marks).

To grasp the pure meaning, moreover, we do not exactly need to question. We need not be critically oriented, aiming to figure out whether we can accept the proposed judgment, whether we can go along with it, believe it. We do not need to doubt, deny, or ultimately be bent on refutation. Where that is the case, the contrast between proposition and state-of-affairs will be able to be especially satisfactorily made. But it is not necessary, for example, when we have focused on the meaning analyses beforehand, when, for example, we are dealing scientifically with different components and forms of propositions, since we are looking at the pure propositions without any focus on criticism. One grasps the difference most clearly when one repeatedly looks at perceptions, memories, and similar acts and pursues the precise analogies with the change of focus.

Perceiving is looking at an object and grasping it <as> being, for instance, this table, or also looking separately at the parts and moments of the object, at the form, the color etc. In this consciousness, objects are simply there for us, are posited objects. But, it is something different to position oneself and ask what is the perceived as such in this perceiving? Then, the perceiving is modified. We are no longer performing the perception-believing now. In certain ways it is still experience, but in changed <ways>. The believing-assumption is now out of action. We are merely considering the perceived as such, the "perception-sense". And were the perception still very much a fiction, were we still so very convinced that what was perceived was nothing, we could however still describe with the fullest Evidenz what shows itself in this illusion, what the perception supposes, or we can direct our gaze purely
51 toward the perceived as such, posit it as objectivity, analyze it, compare it with other such objectivities, and so forth. The mind showing itself in the illusion does not exist. The phenomenal mind as such, the mind as what is supposed is very well something. No mind, but mind-phenomenal. And as that, it necessarily belongs to the experience concerned that we cannot at all describe without describing what it has supposed.

Now, it is, however, thereby by no means necessary always to direct the reflecting gaze first toward the act and then toward its content, its sense, what it has supposed, therefore, in dealing with the things that have been supposed, continually to swing our gaze back and forth between it and the act. We can hold our gaze fixed purely upon what is supposed as such. This focus is simply noematic. In the judgment sphere, this focus is, therefore, that on propositions, on senses, meanings (the judgment). If someone makes a foolish statement, nothing however stands in the way of our directing our gaze toward the sense of his or her statement. The state-of-affairs that the Earth is a big flat surface does not obtain in all of nature. But we make the proposition–what the foolish person making the judgment supposed–into an object. It is not nothing, but an item in the world of propositions that includes all true and false propositions, and we can easily operate in this quasi-world. We can ask what kinds of forms the propositions have. We can, for instance, say that some of these propositions would have the form of hypothetical propositions, others that of categorical propositions. Or, we can position ourselves logical-critically and ask when propositions may be true and when they may be false, which conditions must be fulfilled with respect to their form, and so forth.

A purely ideal investigation can turn toward the propositions, the noematic correlates of the judgments in the sense of noeses, just as well as scientific consideration can turn toward the acts of judgment and the Ideas noetically belonging to the sphere of those acts. But, what is noetic has its correlate in noematic logic.

The study of acts is of course not possible without taking into account their noematic content, their What. Inversely, however, the study of propositions does not require entering into acts of judgment and act-Ideas. The fact that they are supposed in judgments has been established once and for all. But, whatever else by way of ideal characteristics belongs to the sphere of acts of judgment does not in the least need to be taken into consideration in the pure study of propositions. 52

<e) The Meaning of the Statement as Idea of Judging or as Proposition>

I have repeatedly used the term sense or meaning for what I have called proposition or judgment-noema. An explanation is required here. Judging and stating as a rule are conjoined for me. In particular, every scientific judgment is a statement.

Now, in every statement, a distinction is commonly made between word and meaning or sense. Now, we know the ambiguities of the word meaning. Let us first start with the following. In stating the same word, we can state it with a different meaning, as when on one occasion we understand the word "cock" as a bird called cock and on another as the cock of a gun (equivocation). Obviously, every statement with "cock" will then be correspondingly ambiguous. Now, ambiguity stands in contrast to univocality. The same word is used in different contexts, as we say, with the same meaning, in the same sense. Why is that? What makes this distinction comprehensible? One will immediately say that the word-consciousness is connected with a meaning-consciousness. Belonging to the word-phenomena, to the visual, acoustical etc. phenomena are in each case certain additional sense-giving, meaning-giving psychic experiences, therefore, those making actual words of speech out of the merely phenomenal visual, acoustical, etc. imagery. Let us explore this more closely.

"Sense-giving experiences" is, however, again ambiguous. As a rule, we address ourselves to others through speech, want to be understood by them, want to make known that we are persuaded of this or that, and so on. Those are the phenomena of making known. Let us set them aside. They are not genuinely sense-giving phenomena inasmuch as we really make statements in large stretches in understanding, rational deliberation and reflection, therefore, think in statements without communicating our thoughts. After taking away these communication, making-known, functions, the specifically sense-giving ones remain left over–in the case of stating judging, precisely the experiences of judging.

Can one then say that we used the statement unambiguously every time the same or fully identical acts of judgment are associated with the words, and ambiguously when not? That cannot be
53 right, because we very easily observe that when we repeatedly use a word or a statement in the same sense, the judgment-experiences do not by any means in any way remain the same <in terms> of their actual experience-content. On one occasion, we judge insightfully, on the other lacking insight, on one occasion with lively conviction, on the other lacking conviction, on one occasion with a clear logical conscience, on another occasion aware that everything may not be so completely in order, and so forth. However, we always state something in the same sense precisely when we always say "the same" thing, for example, state over and over again, the Germans will remain victorious. So, the identical meaning, not the full concrete judgment-experience, but only one component, is taken into consideration for constituting the statement's

identical sense. Only what we "suppose", what we posit as being, is important. Therefore, exclusively of importance is: first, the fact that we are in general judging, that we are affirming; second, that the affirming is affirming of such and such, therefore, the fact that it has precisely the proposition and no other proposition as content, i.e., as what was supposed.

If we now speak of the sense of the statement, if we in fact treat the statement itself as something identically ideal that is only singled out in the repeating, then the meaning is also obviously an ideal unit, namely, either the Idea of judging as judging the supposition-content concerned can be understood by this or, by meaning, simply the proposition itself can be understood, i.e., what is judgingly supposed, posited as such–but as Idea. And obviously, the latter is the normal and certainly indispensable concept of meaning wherever we speak of meanings of affirmative statements. In asking what this sequence of words, this grammatical sentence, means, we immediately look at what is supposed as such. In uttering or reading the sentence, we ourselves may judge in conformity with what it says, or we do not by any means judge so, consider that false. Or, finally, we can take no position, not judge at all, only engage in merely understanding. In so doing, we have not yet brought the proposition in the logical sense, i.e., the meaning of the statement, to givenness for ourselves. If we want that, then we have to direct our gaze judgingly–or thinking ourselves into the judging–toward what is posited or quasi-posited in the judging and to grasp that posited thing, the proposition as pure idea, precisely as the sense of such a judging supposing in general, sense that is identically the same, however it may stand in relationship to the 54
actual judging, and the same whether the judging may be right or wrong. The proposition as Idea comes to givenness in a consciousness that is built upon judging or quasi-judging, and for this reason, when the Idea proposition is grasped, the Idea of the corresponding judgment, and again of the essential connection of both ideas–proposition = "content" of a possible judgment in general–can be grasped through the mere turning of one's gaze. "Content" signifies the What of the supposing as affirmative supposing.

<§ 11. Proposition and Proposition-Thought. The Propositional Content as What Is Common in Affirmative Judging and in Mere Propositional Imagining>

Let us now continue extending the reflection further. The term "statement" or "declarative sentence" is commonly used in a limited sense, namely, the sense of an affirmative statement. One distinguishes, for example, between declarative sentences and interrogative sentences, optative sentences, imperative sentences, possibility sentences, probability sentences <sentences that express> imagining. Sentences such as "Is the weather beautiful today?", "Heaven help us!", and so on are, therefore, not declarative sentences in the narrow sense. They are not affirmative statements. How do things stand with meaning there?

So with those modifications, as well as with all consciousness in general, we can contrast the Idea of consciousness involved as noematic correlate with the Idea of what is conscious in it. Everywhere, therefore, we find something analogous to what I call proposition in contrast to the judging. We shall later hear more about that. As an exemplification of this, it is enough for you to direct your attention to the fact that we can place mere imagining opposite to thoughts, namely, in the sense of what has been thought as such. We imagine, for instance, that Sirius is a diamond. We think that. It does not at all occur to us to judge it affirmatively, really even just to consider it probable, to presume it in the least. We are only just thinking. About this thinking, we then say something may be thought in it, and we can then again express this What with the words of the declarative sentence, "What we are 55 thinking is Sirius is a diamond". Again, we can further correlatively set what is thought as such in the manner of an Idea opposite the imagining as Idea. The word "thought", or "mere thought", is occasionally used ambiguously for both. Nevertheless, the word distinctly better suits what is noematic, the What thought.

Therefore, just as corresponding to the judging is the proposition (what is judged as such), which can be grasped as Idea, so corresponding to the idea of the imagining is what is thought as such–again as Idea. That is not the proposition, because nothing is posited there. It is the mere proposition-thought (ontically speaking). Let us now compare the proposition and the mere propositional thought.

Corresponding *idealiter* to every proposition is obviously a proposition-thought and vice versa, just as–we shall be equally persuaded of this–corresponding *idealiter* to every judgment is an imagining, more precisely a propositional imagining. We are not judging, we are not really positing that Sirius is a diamond. But, we can obviously fantasize ourselves into such a judging. And consequently, it is evident that judging this same content is possible. The judgment-fantasy can also help us grasp the Idea of judging "Sirius is a diamond" with Evidenz. Obviously, belonging thus <to> every mere imagining of propositional content is a possible judging of this same content, and vice versa.

Empirically, it often moreover happens that we go from the actual judging "A is b" to the mere imagining "A is b," and perhaps also vice versa. We can, for instance, postpone our judgment and first form the mere thought in order to consider it without prejudgment. On the other hand, we do not, however, need to establish by means of experience that a transition of this kind happens and that *de facto* a judgment and a mere thought with the same propositional content can occasionally be contrasted, because in purely ideal reflection, we can really see that necessarily corresponding to every propositional imagining must be a possible judging–precisely as Idea in ideal possibility–and vice versa. Thus, in purely ideal reflection, we can also reflect upon what judging "A is b" and mere imagining "A is b" then have in common, and correlatively the judged as such, or proposition, and the merely thought as such, or proposition-thought. These same words, the same grammatical sentence, thereby always play a role. The person judging states, "A is b!"–if he or she has in mind the mere thought without affirming it inwardly, then this thought is again expressed with these words "A is b." In inner thinking, we have this proposition in mind. When 56
the readers or listeners do not join in the judging, but merely understand, then they engage in mere imagining. But the propositional expression is the bearer of these experiences, in a precisely analogous way, as if they had joined in the judging.

To the question then as to what is common everywhere, one can obviously only answer that the judging person affirms "A is b," the merely thinking person thinks "A is b". In both cases, something identical is evidently present there that can be singled out and grasped as something ideal. This same "A is b" is affirmed in the judgment and then has the character of what is posited, affirmed, alleged-to-be-true. In mere imagining, the same "A is b" does not have this character and, on that account, the character of the mere thought.

We naturally do not want to say that the "A is b" common to the two, this as it were characterless thing, is something in its own right that would in general be thinkable without characterization. Obviously, this new Idea, that of characterless propositional content is a dependent Idea, just as an actual judgment and an actually occurring imagining do not *in concreto* have a separable part in common either when they have the same character-free propositional content in common. Rather, *in concreto*, the "A is b" is necessarily given characterized, but with changing character, and in such a way that something common emerges abstractively. It may serve as a terminological distinction for us to speak of the mere propositional content (propositional matter of the *Logical Investigations*) as opposed to the logical proposition (or judgment). The logical proposition, the "judgment", consists of the mere propositional content and the character of positing qualifying it. The mere propositional content furthermore also makes up a concept of meaning.

Let us go further. It is clear that not only an affirmative judging and a mere propositional imagining can have a corresponding propositional content in common, but also a considering-possible, a presuming, a questioning, a doubting. The same propositional content can be content with different qualifications that, provided they are generically related, ground a comprehensive Idea of judg-
57 ment in general.

<§ 12. Proposition, Propositional Matter, and Positing Quality. The Qualifications as a Kind of Assessment. The Position-Taking Consciousness in the Broadest Sense>

At the close of the last lecture, I contrasted mere imagining to judging. And, just as standing in contrast to judging is the judged as such—and ideally speaking the proposition—, so standing in contrast to the imagining let A be b is the thought as such, the thought, more precisely, the propositional thought (proposition-thought). In this contrasting, it is <for> us a matter of detaching a fundamental concept in order to realize a certain disassociation within the Idea of proposition that only becomes visible to us when we realize the parallel disassociation within the sphere of mere imagining.

Obviously, corresponding to every possible judgment "A is b" is a possible thought "A is b"; and vice versa. In each case, the two Ideas are so to speak, like brother and sister. It is therefore not for this reason a matter of the fact that for every judgment "A is b" that someone judges in psychological reality, a thought "A is b" may actually occur, and for every actual thought "A is b"; a judgment "A is b". Rather, it is a matter of belonging together a priori. If we imagine "Sirius is a diamond"; then a possible judgment surely belongs to this that reads in exactly the same way, even if no one in the whole world actually conceives of that particular notion and could ever judge that. We grasp the possibility of such a judgment with Evidenz, just as we do the idea of this judgment, the moment we simulate such a judging in our imagination.[1] An imagination-intuition makes it possible and is the sufficient foundation for grasping the pure Idea of what was intuited in imagination.

Thus, for every judgment "A is b" (and whatever other content), we therefore find a parallel Idea: the mere thinking "A is b"; and vice versa.

What do the two then have in common? What does it matter that belonging to every possible judgment is a mere thought that is stated using the same words as the judgment, and likewise, vice versa? With the difference of the sense-conferring experiences, 58
something common must however correspond to the same wording. It is obvious to say that what is common is the state-of-affairs. In the one case, the same state-of-affairs is posited as a judgment, in the other, merely thought. However, let us rather avoid this tempting and ambiguously used word, because the state-of-affairs

[1] It scarcely even needs to be said that the possibility of imagining into the judgment belongs *a priori* to the Idea judgment in general.

"A is b" is only what corresponds to the judgment "A is b" when the judgment is a true, correct one. We cannot intelligently say of the state-of-affairs that Sirius is a diamond that such a judgment would be false. And yet here too, as everywhere, there is something common that affirms in the possible judgment "Sirius is a diamond" and is thought in our mere imagining. We call this something, this common content that can only be designated by the verbal proposition itself, the propositional matter, or <the> propositional content. The logical proposition is more than the propositional matter. The latter has a certain characterization in the proposition, by means of which it figures as posited, and the word proposition expresses that splendidly. In the thought, this very propositional matter has a different characterization, as what has been thought, to be precise.

With the fixing of the concepts proposition, propositional matter, and propositional quality of positing, we would then already have so much that we could approach the defining of apophantic logic as logic of pure propositions. However, it appears more advantageous first to make still further contrasts that in the same manner allow us to survey the entire field of formal logic and at the same time make clearer the relationships to noetic logic.

One and the same propositional matter–"A is b"; "Sirius is a diamond"–can still be characterized differently. As good as it is as the content of a proposition and content of a mere thought, it can also be the content of something questionable, of a possibility, of a presumption. Let us go back to the corresponding forms of consciousness.[1] The question as to whether the moon has an atmosphere obviously has the same propositional content as the corresponding judgment "The moon has an atmosphere"; and it is also evident that for every question a judgment is thinkable that agrees with it, and also vice versa. And the same thing holds *a priori* with respect to the possibility of every other mode of con-
59 sciousness that can in general be concerned with propositional matter, for example doubting, presuming, considering-possible. It is certainly not here a matter of empirical possibilities, but of purely ideal possibilities evidently realizable in imaginative activ-

[1] Compare Appendix V, *The Enigma of "Intentionality"*. (Editor's note)

ity that absolutely establish the general universality of the validity of what has just been stated.

It is now easy to see that considering-possible and considering-probable, and also questioning and doubting, are kinds of consciousness that are generically related to the positing, affirmative judgments, so that a series of acts having propositional content fall under a higher generic Idea, under the Idea of predicative judgment or, even more generally, of a predicative act in a greatly broadened sense. (Whether imagining belongs in this same series or whether it does not require separate treatment, I do not wish to determine here.)

Starting out from the acts of predicative affirming of a propositional matter "A is b", we were led to a genus of intellectual acts that as a whole also have a related structure <from> the noematic point of view. In all of them, a distinction is to be made between noematic "matter" and "noematic quality". The noematic matter is everywhere propositional, the qualifying, though, different, and nevertheless setting itself apart within a genus, for example: the certitude that the moon has an atmosphere; the presuming that the moon <has an atmosphere>; the question whether the moon...; the doubt.... One and the same pure propositional matter "A is b" can therefore occur in different acts on the side of the What, and it is called–depending upon the nature of these acts–affirmation-matter ="supposed to be true", presumption-matter = <supposed> to be probable, to be possible, question-matter = <supposed> to be questionable. And with this, the different "qualifyings" of the noematic side are expressed, and in all our examples the qualifyings, as it were, possess the character of assessments, the character of considering-to-be-something. Judging as affirming is considering-true, presuming is considering-probable, and so forth. Present everywhere is a "considering-as", a supposing *par excellence*. And all these sorts of supposing are obviously intrinsically related.

But the circle of intrinsic kinship can still be expanded further. To be talked about then is the broader class of position-taking acts in general, in which, then, acts of genuine valuating, of rejoicing and being sad, of wishing, desiring, willing also belong. But, com- 60
pared to these, the specifically intellectual acts form a tighter unity and form a sublayer in the structure of the position-taking con-

sciousness. For, as we can, for example, ascertain, rejoicing that A is b already presupposes being convinced or presuming that A is b, therefore, an act of the sublayer.

<§ 13. Questions of Reason and Rightness Belong to All Position-Taking Acts as Qualifyings of Propositional Matter>

In order to make the meaning of these analyses, of this definition of a broadest concept of assessment, of thesis, of position-taking somewhat clear, I point to the fact that the setting of norms for reason, and thereby the questions of validity or rightness, is connected to the particularity of the position-taking. It makes no sense to say of a table, of a thing, or material process that it is correct or not correct, it may or may not be valid. Similarly, a sensation of red is not rational or irrational, and that red sensed is not true or false, correct or incorrect. By virtue of their meaning, such predicates belong exclusively to the position-taking acts and correlatively to their noemata. A judging is correct or rational or non-rational. What has been judged, the proposition, is true or false. But also a presuming, a considering-possible, asking a question or doubting is rational or non-rational. And correlatively, corresponding predicates belong to the noemata of these position-takings. Supposed probability, supposed possibility, questionableness, doubtfulness are valid or invalid, similarly, however, also for wishing and willing, for all position-takings in general. They are rational or non-rational.

Everywhere, and especially in the basic intellectual sphere, with respect to the overall noema, we can speak of a proposition, thereby broadening the concept of proposition. Within the broadened Idea of proposition we can therefore very well distinguish truth-proposition, probability-proposition, questionableness-proposition, and so forth. It is everywhere common for the concept of the proposition to have a matter nearby, but with different qualifyings—as true, as probable, and so forth. And, the proposi-
61 tions can everywhere be valid or not. That is, starting from the standpoint of "reason", the qualifying can belong or not belong to the propositional content. And according to the circumstances, the corresponding state-of-affairs or state-of-truth or, state-of-

probability, state-of-question and so forth actually obtains or does not obtain.

With all these acts, we also speak of meanings, that is, of propositional meanings. We therefore carry over the talk of meanings from the verbal experiences and experience-Ideas to the meaning conferring acts themselves bound to the word-presentations. These acts confer meanings on the words because they having meaning in their very selves. We therefore say of judging, even where it occurs without words, that it is an act giving meaning or having meaning in itself. Namely, every judgment posits something. Out of each one we can extract a proposition. That is the meaning. The same thing holds of presuming. In presuming something figures as presumable, as probable. Presuming is considering-probable in a sense similar to the way judging is considering-true. The meaning is the propositional matter with the quality of supposed-to-be-"probable". Likewise, for questioning or doubting, provided we but exclude the emotional moments intertwined with it, for example, the moment of the wish, and limit ourselves in general to the, so to speak, theoretical question, which is really something different from questioning in practical life. Furthermore, on countless occasions, inauthentic questioning is also mixed with genuine questioning there, as when, for example, in the course of questioning, examiners pretend to be aware that something is questionable to them and would like to know how that might be. That is naturally not a question, any more than playful or deceitful pretending as if one affirmed is affirming.

Regarding this sphere of propositional thought-acts, it is important for you to make it completely clear to yourselves that meaning really must be spoken of everywhere in a strictly analogous sense and, correlatively, all the associated acts really have essential generic unity. In all these acts, considering- is effected. As a result, we say in parallel fashion: considering-true, -probable, -questionable, -doubtful.

But, you will ask, to what extent should the generalization of the Idea of proposition be required very first here? Does, you would like to ask, something genuinely new really come to be placed on the same footing with affirmative propositions here? 62
Probability-statements are themselves affirmative propositions,

just ones with a special content. "It is probable that A is b; it is possible etc."–there, nevertheless, something is asserted, therefore, judged.

By way of contrast, I call attention to the fact that the essential difference between affirmative judging and presuming, questioning, and so forth is not eliminated by our ability to make affirmative statements on the basis of presumptions and questions, and more directly so by <our> predicatively separating and bringing to expression the positing of probability lying in the presuming itself in the form of an affirmative statement about probability This is rather circuitousness.

<§ 14. The Essential Parallelisms in the Different Intellectual Kinds of Consciousness and Broadening the Concept of Proposition. The Universal Dominance of Affirmative Statements and the Primacy of Truth-Logic>

With judging (as predicating affirming), we talk of states-of-affairs and, on the other hand, of proposition, more precisely of predicating proposition. Instead of state-of-affairs, we could also very indicatively say truth-state and call the corresponding proposition the truth-proposition. In that case, in presuming, we would have to speak of a state-of-probability and would call the associated proposition a probability-proposition.

Affirmative propositions find expression as statements, as apophansis. By the same token, probability-propositions no less find expression as statements. A statement, however, primarily owes its meaning to an affirmative judging. Therefore, what is proposition-like in the considering-probable indirectly finds expression, namely, on the basis of an affirmative positing of probability. As soon as we talk about probability, and similarly <about> possibility, we simply predicate affirmatively. If we state, "It is rain", then in doing so what has been posited affirmatively as such is expressed in completely straightforward manner. What stands before our eyes as being true in judging supposing is expressed, however, is not predicated about truth. If we wanted to do that, we would have to state a proposition more complex in terms of meaning: "The fact that rain is, is true, is actually so". If one calls judging considering-true, then the expression is questionable. It must not be understood as truth predicating. 63

If we then turn to the presuming, to the considering-probable of the propositional content "It is rain"; then the sequence of words "It is rain" cannot serve as an expression for us, because we use it where we affirm, where therefore raininess putatively stands as a reality before our eyes. Hence, we state, "It is probable that rain exists". And this is circuitous on our part. In an appropriate reflection, we stress and grasp the propositional content and state of it that it is the content of a probability. Then, instead of the plain probability-proposition, as it is in the original presuming as correlate, as presumption-content, we have an affirmative statement, a truth statement about a probability.

Deciding whether every expressing as such presupposes affirmative acts, acts of certainty of belief, though perhaps not predicating acts (therefore, for example, similar to the way the affirming in the nominal positing <is> a setting-forth-as-existing, though not positing of a complete predicative state-of-affairs) is one of the most difficult and most controversial issues in the philosophical theory of expression. The controversy concerns, for example, questions. In asking, something figures as questionable. The noema, what is questionable as such, would be the interrogative proposition. The questioning consciousness does not in itself presuppose any verbal expression and any conceptual formulation going hand in hand with it. When we expressly ask, "Is the weather rainy today?" it is hard to decide whether an affirmative positing of the question as such or an affirmation about questionableness is present there. We often clearly ask in predicative form, for example, "I am asking, whether–" or, "The question is whether–" or, "Whether there will be beautiful weather is open to question". What though then lies in the particular form of expression "Is there beautiful weather?"

We fortunately need not enter into this extremely difficult problem here. It is enough for us to see that behind every affirmative statement about probability, possibility, questionability (1) lies a consciousness of the questioning, presuming, considering-probable, and so on, very closely related to the judgment-consciousness, and that (2) every such consciousness corresponding

64 to that kinship also carries within itself its noematic What, its meaning-content or sense–a sense generically related to the sense of the judging that we are ordinarily in the habit of designating with the words judgment or proposition. As a result, extending the concept of proposition and ascribing a proposition as meaning-content to each of the different kinds of acts figuring under the genus of intellectual acts is definitely justifiable. We then have different kinds of propositions: truth-proposition (or proposition in the narrower sense), probability-proposition, interrogative-proposition, and so forth. But, <it is> to be kept in mind that this concept of proposition is entirely free of anything grammatical. Here, it really expresses a moment associated with intellectual acts, their supposed as such, whether or not the intellectual act occurs within the context of the discourse, is therefore bound to the words.

The exceptional position of predicative acts of judging, and in connection with that the universal dominance of affirmative statements, consists, however, in the fact that every proposition in the extended sense, and of predicative material—for example every question, every doubt, every probability-proposition—has its equivalent in an affirmative proposition, while the opposite is not the case. We can always–and necessarily–carry a mere presuming over into a judgment that such and such matter is presumable, probable. However, we can reflexively only transform a judgment again into a judgment, for example, the judgment "It is cold" into the new one "It is true that it is cold".

Only if one has at some point looked at these relationships scientifically, if one has made clear to oneself the essential parallelisms of the different intellectual kinds of consciousness, does one understand why not merely formal logic in the narrower sense, that is the a priori discipline of declarative sentences (of "truth propositions")–as I also readily say–can in general stand opposite the noetic theory of justification of position-taking intellectual acts. There must be as many noem<atic> disciplines of propositions as meanings as there are basic kinds of propositional meanings, and there are as many basic kinds of meanings as there are basic kinds of intellectual position-takings. It is to be expected beforehand that every domain of meaning must have its a priori

laws–be they many or few. Therefore, the old formal logic that upon closer reflection will prove to be an a priori discipline of affirmative propositions, of supposed truths, must certainly have 65
its parallels in an a priori discipline of possibilities and probabilities. One can, in opposition, parallelize the logic of truth, logic of possibility and probability. When one has gone that far, one soon notices that under the heading "mathematical theory of probabilities" (even if in a form contaminated with empirical constraints and in general very much in need of clarification), this postulated logic of presumption-meanings is developed to a significant extent, that is, as formal logic of probabilities in a sense similar to the old formal logic. It consequently becomes apparent that what seemingly merely runs side by side in the separate sciences is in truth intrinsically one, belonging to an essentially unitary logical science. People realize that logicians' interest in probabilities must not be merely methodological, as if it were merely their business to ponder the practical-methodological meaning of this and other mathematical disciplines within the context of human knowledge.

On the other hand, if one understands the merits of truth-logic, one comprehends the temptation to treat it as universal logic. In a certain sense, it is universal, inasmuch as likewise, where we scientifically presume and ask, scientific securing ultimately always ensues in the form of a statement that predicates on probability, possibility, questionability.

<§ 15. Extensions of the Meaning Concept and Definition of the Idea of a Thought-Act>

<a) The Partial Acts of the Propositional Thought-Act. The Concept of the Nominal>

I first wish to pursue a new extension of the Idea of meaning. Not all thought-acts (thought-theses)–to start from the side of consciousness–are propositional acts. But, the essentially generic community of consciousness-experiences included under the concise heading "thought-act" reaches just as far as the idea of logical meaning reaches, and just as far as the concept of logical proposition <reaches>. You notice that our investigation proceeds

66 analytically and that starting out from a relatively determinate and immediately available Idea of thought-act and meaning, we are working our way through to laying hold of the broadest range of the kinds of acts and kinds of meaning united in essential kinship and, thereby, at the same time, to laying hold of the corresponding determinate generic Idea, namely, to a pure, clear, well-defined Idea of thought-act in general and thought-meaning in general.

The first thing given is the statement. And in actual stating we laid hold of the Idea "judgment" in the concise sense and as correlate the Idea of the judgment-proposition, of what is being affirmed. Then we realized that an identical propositional content can always be secured as "matter"; and essentially related position-takings of another kind can occur for the judging. That yields, therefore, the Idea of a genus of acts in general that are essentially related to the propositional positing, to the affirming, and are also of propositional matter. And now, one further step, we realize that limitation to propositional matter can cease to apply, or in other words, we realize that there are acts that are qualitatively certainly at one with affirming, seeming, presuming, but do not have any propositional matter and yet have matter.

No great amount of skill is needed to make that obvious. We only need to take a look at the relatively independent partial acts of the propositional thought-acts, that is, at those partial acts that run parallel in the verbal structure of expressive discourse as a whole. Let us reflect upon the following: If we compose a grammatical declarative sentence within a developed language–"The king who came back from his journey took up affairs of state again"–then a certain structure and formation of the meaning-conferring experiences is at least to a certain degree mirrored in the verbal structure and formation. I say "to a certain degree" and at the same time emphasize that the talk of mirroring is only figurative. I need not explain here how what is verbal and what concerns meaning particularly stand in relation to one another. I would only like to direct your attention to the fact that, generally speaking, in these experiences something corresponds to the word structures and to the syntaxes, as you can assure yourselves of at any time by observing the meaning experiences. We thereby soon also notice that, for example, belonging to the nouns

and to the nominally consistent complexes are relatively delimited meaning experiences that are not in turn themselves of the nature of propositional acts. It is as a result beyond doubt that non-prop- 67
ositional experiences occur in the unit of propositional experiences as constructive constituent parts, only formed in the comprehensive wholes by unifying forms. Let us make that clearer.

Within the context of statements, I said, components are marked-off under the heading "nouns." For example, occurring in a sentence is the subject component "the Kaiser" or the more complex component "the Kaiser who has returned from a hunting trip." And the predicate perhaps reads "took up affairs of state again." There too is a certain, though dependent, unit–and obviously so with each complete predicate. You readily clearly see that all the distinctions I have made are also to be made for such statement parts *mutatis mutandis*. Here too, the meaning experiences—the ones being supposed—and the meaning itself as the supposed as such face one another. Here too the meaning experience can be called positing, affirmative in the good sense, but with some broadening of the ordinary concept. What is decisive is that what is posited as such is not a complete propositional sentence here, a "That is how it is!", an "A is b!". If we say, "the Kaiser", then an *object* is posited as existing, not as a *state-of-affairs* in a complete affirmative declarative sentence. Likewise also, when we say, "the Kaiser who has come back from a hunting trip."

One will perhaps find that a state-of-affairs is affirmatively posited here in the relative clause, therefore, a predicative judgment made, though in connection with a nominal positing. But, the change of consciousness, or the change of meaning, is surely to be heeded when we go from making a free predication to making the corresponding determination. The relative clause is a dependent component of the nominal whole. It determines the Kaiser as the one who has come back from the hunting trip. The consciousness "The Kaiser has come back from the hunting trip" is, however, something different. The predication "A is b" is rigorously differentiated in terms of consciousness and meaning from the determination of the attribution "A, which is b." In one case, we have an independent predicating proposition "A is b!." In the other, we have a nominal positing or, as we can really correlatively say, a

nominal clause that an independent relative clause called determi-
68 nation helps to construct. It also holds for such nominal clauses that a matter and a positing character can be abstractively distinguished in them, so that we must now also speak of nominal <matter> as opposed the propositional matter.

We thereby lay hold of the concept of nominal in a very broadened sense, similar to the way *Mill* did in his psychological interpretations. We not only call every simple grammatical noun a noun, but every expression however complex that can act as the subject of a predicative statement. The whole "the King who has come back from the hunting trip" is accordingly a noun, its meaning, a nominal meaning.

Now we also notice that all qualitative modifications that occur in complete propositions can occur in the partial acts of propositions, for example, seeming, doubt, etc. on nominal matter. It obviously makes a difference whether in engaging in naming–"the A", "the Kaiser", one accepts and posits the object named as existing without further ado, or whether one does not do that and instead has the object mentally in sight in a way that we cannot <then> call anything other than seeming-to-be or presuming-to-be, depending upon whether the object figures for the consciousness as possible or presumed.

I do not want to pursue what has been said in detail for all partial acts or partial meanings that can occur within the context of a propositional act, of an affirmation, of a propositional seeming, presumption, question. We really also lack here a systematic overview of the different acts of building blocks of thought-acts or thought-meanings. It is enough, if you can but convince yourselves of the fact that this pertains to all instances in which you find structures in the overall meaning.

<b) The Relationship of the Overall Quality to the Qualities of the Partial Acts>

The following important observation is related to what has just been discussed: Every component to be defined within the context of an independently complete full judgment and of a full propositional act in general has its quality and its matter–and that in both a noetic and a noematic respect. By the same token, the propositional whole has its overall quality and its overall matter. 69

Now, it is generally to be said that both factors, quality and matter, are not externally juxtaposed, but that the quality is something that in a way breathes life into, colors, all parts of the matter. The affirmative positing, for example, runs throughout the statement. Everything distinguishable in a statement participates in a certain way in the affirmative positing. The same is so for all other positing characters.

On the other hand, one cannot without further ado say that the qualitative characters that specifically belong to the partial meanings and are present in them themselves are mere rays of an overall quality of the overall meaning. Thus, for example, the grammatical sentence "It is presumable that Mars is alive with organic beings" is an expression of a presumption. But: Mars. Sirius is constantly considered by us to be existing. Of course, inasmuch as it is thought of as a diamond, it leaves room for thought. But in itself, it remains as good <as> being posited, as if we had stated affirmatively, "Sirius is a fixed star".

Nominal affirmative positings can naturally occur within the context of presumptions (and that, I say, is the rule). Presumptions pertain to objects held to be actual and named in the presumption-propositions, as when, for example, we presume that Sirius is bigger than the sun. That at the same time shows that we in fact have serious reason to treat any kind of distinguishable partial meanings, notably nominal ones, as characteristic meanings. Although they are mere components of more comprehensive wholes, they have their own qualitative characters and their own matter in their own right, only just not propositional ones.

New light is likewise shed on the difference between qualitative character and matter. We saw that modifying the overall quality does not affect or must not affect the qualities of the partial acts. Through abstraction, let us focus on the mere matter of a suitable complex example. I judge, for example, "The Kaiser is staying in Berlin". But the same thing can be conscious to me as possible, as probable, as actual, as questionable. In the changing of these qualitative characters, the identical matter remains. Let us abstract just the matter, "The Kaiser is staying in Berlin", which retains nothing

70 of qualitative characters in this abstraction. There we have something remarkable. In this abstract pure matter, we encounter two affirmative positings, “the Kaiser” and “Berlin.” Therefore, we see that positings can occur within the matter. This can be explained thus: Thought-acts can, as it were, build themselves up syntactically provided that lower acts intertwine themselves with lower positings in certain forms, and then, as complex ones, undergo higher positing by which means a higher act then arises that can itself again act as the substrate for connections and new positings. In relation to such syntaxes, it is conceivable that what acts as matter for the higher positing already itself implies positings and full acts.

<c) Combining Simple Predications to New Judgments. Conclusive Determination of the Idea of a Thought-Act>

The question then arises as to whether the extension that the Ideas “thought-act” and “thought-meaning” <have undergone> through the reflections up to this point is the greatest possible one prescribed by the requirement for generic unity with the predicative affirmations. We took as a basis for predicative acts the form of the apophansis in a primary and original sense as formal type symbolized by “A is b.” Something is stated about something.

After we have appealed to the possible partial acts of such propositional thought-acts, the question arises as to whether there might not be new thought-acts yet to be discovered along different lines. One would like to ask whether propositional thought-acts do not themselves combine again and in so doing can perhaps undergo modifications that represent new kinds of thought-acts with new overall qualities and new overall matter. It is now to be noted in this regard that something new actually proceeds out of the combining of the plain predications. Let us predicate “A is b,” something is said about something. In doing so, the A and <the> b may be as complex as ever. If, however, we state, “Because A is b, then C is d” or “A is b and C is d,” then the whole is no longer a mere predication in which something is stated about something, or as *Aristotle* says <an> apophansis. Nevertheless, everyone will say, “That is a judgment” or “That is a proposition.” Every inference,

every proof belongs here. An inference is an assemblage of predicative judgments, each of which can occur independently in its 71
own right. The assembled whole does not, however, have the form “A is b” or “A is not b,” therefore not the form *Aristotle* defined as apophansis. Inferring, however, is manifestly not a mere sum of judgments, but a unit of judging, and as a unit itself a *single* judging. Correlatively speaking, a coherent proposition is there that weaves the independent proposition into a propositional whole. When it comes to the combinations, something actually new therefore occurs when we begin with elementary predicative forms, just as the other way around, when we go from the predications to the parts that are no longer predications. If, however, we take a closer look, we observe that through these ways of increasing complexity and of partition only new matter arises, not however new qualitative characters. Whichever propositional meanings we look at, whichever meanings we may single out through partition or produce through combination, we always find the qualities–the affirmative positing, presumption, question, and so forth, therefore the same qualities–that had occurred in full propositional matter. Whichever act with whichever matter and quality we may encounter, we can always think of its quality exchanged with any other arbitrary quality that we find actualized in any other arbitrary matter.

We now need only to go one step further to make evident to ourselves what gives unity to the Idea of “logical thought-act” and to the correlative Ideas of “logical proposition,” “logical truth.” Above all, it is clear that here a consistently complete class actually stands out from the totality of intentional experiences as <those> experiences that are in themselves characterized as consciousness of something.

Let us go over the course of our reflection. We started from the Idea of predicative judging. Corresponding to it correlatively was the Idea of the affirmative predicative proposition, of the apophansis. It proved to be a union of a quality and a predicative matter. We changed the quality and found that a series of other qualities had a very close generic essential kinship with the affirmative quality. So, we extended the Idea of the thought-act beyond affirmations to presumptions, questions, and so forth, as far as the

qualitative kinship reached. We then changed the matter also, once again holding on to the generic essence. In this respect, we pro-
72 ceeded to the consideration of constituent parts of a predicative matter or to the formation of complex matter. Here too a generic unit emerged, and that means that here a class of matter can be isolated that does not include all possible matter of intellectual position-takings. If, however, this is the case for them (I shall substantiate this finding in a more detailed way in a moment!), then we can obviously rigorously determine the generic Idea of a thought-act or logical act using the genus of qualities I called doxic qualities in my *Ideas*[1] and, at the same time, using the genus of matter that we can call logical matter. Every act doxically qualifying logical matter there would therefore be a thought-act.

<§ 16. The Doxic Qualities Are Also Found in the Sphere of Intuition. Extension of the Concept of Meaning to Acts of Intuition: The Intuited as Such>

Now for the required supplement. In accordance with their genera, both components, quality and matter, served to define the Idea of the thought-act. We now first observe that we do not find doxic qualities in the specifically logical acts only, but also within the manifold acts of the sphere of intuition, of perceptions, memories, and so on, which we are not even in the habit of calling logical acts, although we know that intuitions play a major role as substratum in specific thinking. If we compare a perception–with its plain consciousness of actual existence–with a predicative affirmation, then common ground obviously exists when it comes to the quality–and likewise if we draw upon a nominal proposition. If I say, "the king," "the cathedral," then I am conscious of something truly actual, as if I saw the cathedral. Each one is consciousness of something existing, a considering-as-existing, considering-true. There the word "*Wahr-Nehmung*"[2] is quite nicely appropriate.

[1] Regarding the bibliographical information compare p. XXII of the "Einleitung der Herausgeberin" to Hua XXX. Further compare p. 426 f. of the *Textkritischen Anhang* to that edition. (Editor's note)

[2] The German word for perception is "*Wahrnehmung*," which literally means true-taking (Translator's note).

Since *Hume* (*Mill*, *Brentano*), the common ground has also been designated as belief.[1] In predicative judgments, we have the 73
believing or positing supposing of predicative truth. What is truly existing here is a state-of-affairs. In the nominal positing, the true thing that is supposed there is an object. Objects and states-of-affairs can, however, also be sensorily-intuitively perceived, remembered, presented. What does the difference then consist of on either side? On one side, predicative judgment and nominal positing belong. There a state-of-affairs or the subject-object of a state-of-affairs is thought. On the other side, we have mere "intuition" and mere presentation in general in the meaning that is contrasted with the thinking, as for example in *Kant*. An object and a state-of-affairs can be thought, but they can also be intuited in a mere intuition (although in a synthetic intuition) and in general be presented. In <the> one case, we have thought-matter, in the other intuition-matter, and when belief-positing takes place (true opinion), then with respect to the meaning-like essence, we in one case have a logical proposition, a nominal or complete proposition, in the other, as it were, an intuitive proposition, something intuitively supposed as such, perception. At one time, for example, we think and believe that the table is red. At another time, we see the table, but we do not think of it as a table in conformity with the general-conceptual meaning of the word "table". And on the basis of the plain intuition of the object, we proceed to the grasping of the red-moment and grasp the red on the object or the table as the red-having thing, but also without our thinking the red as red in conformity with the general word meaning, therefore, knowing it intellectually at the same time.

Or, we contrast nominal thinking of the thing "this table" with the mere perceiving of the same table. In the one case, the object is thought. With regard to the verbal expression we can also say that it is named. For all naming is thinking; all word-meaning is thought-meaning. In the other case, the same object is merely intuited and intuitively thought. And again the matter–hence the qualitative meanings–is different in both cases. If we extend, which is absolutely proper, the talk of "meaning" itself to the intuitive

[1] In English in the original (Translator's note).

acts, then it pertains to "the intuited as such". The intuited table as
74 such–i.e. not simply the table, which is supposed in perception and which, if the perception is not deceptive, exists in reality and, besides the few properties directly grasped in perception, has innumerably many physical and chemical properties to be established empirically. Rather, when we speak of what is perceptually supposed as such, we have in view the definite meaning in which the thing appearing in sensorial appearing, and appearing as table, stands before our eyes–a meaning that can be adequately grasped in concepts and hence described. Obviously, the meaning of the perception, the appearing thing as such, is essentially distinguished from the meaning of the nominal thinking, from the nominal meaning.

<§ 17. The Difference Between Meaning- and Cognitive-Essence with Respect to Determining the Idea of the Thought-Act. The Generic Makeup of Matter as Thought-Matter>

Are, however, the full concrete acts merely determined by quality and matter?

Let us reflect upon the following. In judgment in the ordinary sense of the word, i.e., as a predicative affirmation, it is clear that ordinary talk about the same judgment—for example, $2 \times 2 = 4$—does not amount to identity of the concrete judgment-experience in accordance with all its component parts, but just "only" identity of judgment-quality and judgment-content. What does the only signify? Now, while we continually make "the same" judgment, even if in repeated acts, various differences can be found in its concrete psychic makeup. And if we pay attention to this, we observe that not all moments of the experience are incorporated into the Idea "judgment", for example, not the length, which can at times be longer, at times shorter, likewise, not the difference between the actually-judging-now and the subscribing-to-the-judgment made, the judging-again. We exclude such differences, which are not proper to the field of judgment and which, especially, are irrelevant in a purely logical and noetic respect.

Added to that are, however, certain, differences highly important for every complete logic that are also not included in the

ordinary Idea of judgment, namely, those differences of intuitiveness, and perhaps of Evidenz, repeatedly mentioned by me, that are absolutely determinate for the cognitive value of the acts con- 75
cerned and, consequently, for deciding the question of rightness. In spite of such important differences, we speak of one and the same judgment; we say it is the same judgment only repeatedly made—whether in one case there is Evidenz or in another case not, in one case confusion, in the other clarity, and so forth. The same judgment just means that the same thing is posited in it, that the same affirmative proposition is extracted from it.

Precisely the same thing holds of the parallel acts: the same presumption, the same considering-probable, i.e., one that continually supposes the same probability. The probability proposition is the same. The acts are of the same meaning-like essence, but they differ, as I am in the habit of saying, perhaps very greatly, due to their cognitive essence.

The following is to be borne in mind here. The characteristics of clarity and Evidenz so relevant for knowledge accrue to the judgment through certain underlying acts, for instance, through fantasy-intuition, memories, perceptions, and so forth. However, if the judgment becomes clear through them, if they confer intuitive fullness upon what judgments suppose and bring the state-of-affairs before the perceiving eyes of the mind, achieving this does not, for instance, signify that predicative judgment and intuiting acts lie side by side but somehow glued together psychically. Rather, the judgment itself undergoes something. It is then something different, shot through by a ray of light, which does not remain something alien to it, but is something inwardly modifying it. And yet we say "the same judgment"—namely in the sense that what is supposed is the same; it is continuously supposed "A is b". And, likewise in the other cases, where the same thing would be to be said with respect to the shining through. As long as in judging, presuming, questioning, we mean the same thing, then what is meant—i.e. the judgment, the presumption, the question—is the same. We hold neatly on to the meaning-like essence.

In all thought-acts—and likewise for the genus "logical proposition" in general—, this essence is generically one. And at the same time, it is clear that by means of the same essence, we can rigorously,

and for essential reasons, define the class of concrete acts that are to be called thought-acts there. Every intentional experience whose intentional essence is of the genus "logical proposition" is hence defined as thought-act or logical act. It is truly an abstrac-
76 tion inasmuch as each actual judgment-experience, <each> presumption, <each> question-experience must somehow be individuated with regard to the kind of intuitiveness or lack of intuitiveness (of fullness and emptiness), as well as with regard to those previously excluded moments of immanent duration, and so forth. The possibility of adding such individuations pertains *a priori* to the meaning-like essence, but it does not contain them in itself.

We therefore see—and this is a finding of fundamental importance—that the strict unity of the Idea of the thought-act is, on the one hand, indeed determined by the meaning-like essence, by the union of matter and quality, but that, on the other hand, the decisive uniqueness of the thought-act as such lies in the generic makeup of the matter. The same qualities already also occur in intuiting acts free of everything specific to thinking. And, the same is so–as can be easily shown–in empty-presenting acts that, as it were, lack the fullness of intuitiveness and are of the same generic essence as the intuiting acts.

The qualitative moments are not specifically characteristic of the thought-sphere there. On the contrary, it became apparent that the matter in the sphere of predication—and thus in the sphere of thought-acts in general—is of a basically essentially different kind than the matter of plain intuitions and of equivalent acts. If one sees what is proper to thinking in "what is conceptual", then the word concept would have to be understood so broadly as specifically to designate what is specific to the thought-matter as such. If we pursue the qualities, then we have the highest genus "intellectual act", and then the matter is differentiated into thought-matter and presentation-matter. With the strict parallelism between acts and meanings is hence also said that the thought-meanings, the logical meanings in the concise sense, form a complete, rigorously differentiated genus, namely, in contrast to other similar qualitative characterizations of accessible meanings.

I would, however, like to emphasize yet another important point. There are deep underlying reasons why we form the Idea of thought-act in the specific way that has repeatedly been discussed, namely, the fact that in so doing we give priority to the meaning-like essence and speak of the same judgment as long as this is
77 retained, however greatly the cognitive state may change. Namely, it becomes apparent that with respect to their ideal possibility, all other intentional characteristics of acts, all characteristics that concern them as concrete acts, as concrete consciousness of something are dependent on the meaning-like essence in conformity with a priori laws. And precisely that is reflected in the fact of a logic itself as <a> logic in the narrow sense of science of the logos. The cognitive value of the judgments that *idealiter* we only think determined by matter and quality essentially depends on the possibility of demonstrating those judgments and, perhaps, of making them evident. The remarkable thing, however, is that law-governed conditions of the possibility of such demonstration—and, in further consequence, conditions of the possibility of validity—lie in the pure meaning-like essence. So, for example, what we call contradiction is a certain form of judgment-matter, and belonging to it is the ideal law that a judgment of such matter does not admit of a concordant demonstration, i.e. overall, consistent demonstration, but only to a, so to speak, discordant demonstration in the form of an conflicting combination of partial intuitions.

Something similar moreover holds of all acts, of all kinds of consciousness of something. The meaning-like essence occupies its analogous singular position everywhere, which is why I also called it intentional essence in the *Logical Investigations*. And, from here on out, it is understandable why a radical classification of acts is absolutely conducted and must be conducted by the classification in terms of meanings.

<Chapter 3

Formal Logic and Theory of Forms of Meanings>

<§ 18. Logic as a Priori Discipline of Thinking in General Formal logic with its two levels and as basic piece of an a priori theory of science>

I must, however, resist the temptation of entering into problems of consciousness greatly overlapping the logical sphere. With the clarification of the Idea of "thought-act" and "thought- 78
meaning" or "logical meaning" I have reached my goal. And now clarity is spreading on all sides.

Thought-acts as such, namely, as concerns their specific essence, constitute the domain proper to logic in general and in the broadest sense–the whole domain of thinking in pure ideality and unconditional generality. Logic does not investigate thought-experiences as empirical facts of nature–that is a matter for psychology–, but rather thinking as Idea in accordance with all ideal relationships pertinent there and the unconditionally valid truths grounded in the Ideas. The words a priori discipline of thinking signify nothing other than that.

Scientific investigation can thereby focus purely on the meanings of thought-acts, that is, <it can> remain on the level of "formal generality", i.e., it leaves in ideal indeterminacy what confers upon the meanings determinacy of the relationship to particular spheres of knowledge. For example, from two propositions of the form "All A are B", "All B are C", a proposition of the form "All A are C" validly follows. The logical discipline that deals with meanings in formal generality is formal logic.

First, the qualities. There are as many basic kinds of logical meanings–determined by the qualitative differences into which

E. Husserl, *Logic and General Theory of Science*,
Husserliana: Edmund Husserl – Collected Works 15,
https://doi.org/10.1007/978-3-030-14529-3_3

the general genus thinking breaks down qualitatively– as we have lines of investigation. The first of these qualities is that of the truth-positing called "affirmative-positing" in the thought-sphere. The associated meanings are the apophantic ones, therefore, the predicating affirmative propositions (meanings of declarative propositions), their partial meanings and, on the other hand, the possible combinations of such propositions. With that apophantic logic is defined, in which, if it is purely apophantic, no other qualities are admitted.[1]

Standing in contrast to it when we move on to the other qualities
79 ties is the formal logic <of> qualitative modalities, of possibilities and probabilities, and likewise the formal logic of uncertainties, which however says nothing about the fact that logical disciplines are purely oriented towards qualities and do not interweave the investigations.

All these disciplines remain within the purely ideal spheres. The true propositions that they formulate would be ideal-propositions, truths that do not refer to empirical facts–about which there is no talk at all there–, but to Ideas, or in unconditional generality to the range of conceivable particulars that fall under these Ideas, provided all truths belonging there have the nature of laws for everything actually or possibly thought as such that falls under the Ideas in question.

Let us go further. In pure logic we, however, have two tiers. When we speak of logical laws, we in the first place understand by this laws to which supposed truths, probabilities, possibilities must conform if they actually are to be able to be truths or probabilities. Now if such laws are actually determined by pure meanings, then for them it can only be a question of the qualities or the matter. If we hold fast to the quality–for example as affirmative quality–, then all differences lie in the matter. Therefore, laws that prescribe conditions of possible truth for supposed truths hinge on merely formal characteristics of the proposition-matter. The fact that affir-

[1] Subsequent note by *Husserl*, "It is a law that every thought-meaning–every logical one–can only have affirmative quality. We have affirmation as soon as the expression and its thought-meaning are there. As a result, there is but one logic, apophantic logic. All remaining noemata are expressed logically. Cf. about negation 104 ff." Compare §29 (Editor's note).

mative propositions can only be true, or must be false if their matter takes such and such general forms, as for example, "Affirmative propositions whose matter has the form of a contradiction are false", or "When affirmative propositions of such and such a form are true of matter, then propositions of such and such correlative forms are true", and so on, will, therefore, be a matter of laws of content.

Along these lines, there are actually many kinds of laws that speak of the dependency of the possibility of truth on <the> form of the proposition-matter. And something similar <holds> for possibility and probability.

I shall, however, have to demonstrate that by their very nature such laws of validity (that speak of conditions of possibility of truth and probability) must be prior to other laws that relate to the meanings or to their matter without inquiring into validity or non-validity of meanings. 80

Within formal logic, I shall distinguish between two levels of investigation. The lower level is the mere theory of forms of meanings or, what essentially amounts to the same thing, the <level> of the mere theory of forms of thought-matter. It forms the first and basic piece of an a priori grammar understood as a discipline that systematically explores the a priori pertaining to expressions. Moreover–detached from any interest in grammatical issues–it forms a discipline in its own right. Apophantic logic as the formal logic of possible truth and the logic of probability as formal logic of possibilities and probabilities are then built as higher levels upon this theory of forms.

The general science-theoretical nature of this discipline is clear. Since every science aims at validity, would make statements whose truth—or reasonable probability and initially potential possibility—occurs in valid ways, it is clear that any such *a priori* is bound to the formal logical laws and that this constitutes a basic piece of an a priori theory of science.

While formal logic as logic of meanings does not concern itself with questions of Evidenz and intuitiveness, and thus disregards all differences of ways of being given, of cognitive fulfillment and cognitive emptiness–while it is indeed continuously immersed in thinking, but only in order to extract meanings from it–, the situation is the other way around with noetic logic. Its attention is directed toward full experience–not *in individuo*, but in the Idea–and <toward> the essence-relations that connect all ideal components with one another.

Finally, yet another remark. Since from the outset I have conducted the investigation in such a way that the thought of a most general essence-kinship of every kind of consciousness remained alive, I need not say that an analogue of logic must also be envisaged in the realm of inner life. However, I only say that by way of indication.

81 <§ 19. Characterization of Formal Logic and Its Essential Problems>

<a) The Theory of Forms of Meanings as a Theory of Forms of Possible Thought-Matter in General>

I shall now proceed to a more precise characterization of the disciplines of formal logic designated in general ways, of whose essential problems, boundaries, and reciprocal relationships, the dominant logic—both that of the majority of professional philosophers and also that of mathematizing logicians—scarcely has a hint. As far as the latter are concerned, in certain ways, they have brought the disciplines of logical validity to a higher level of technical perfection. They have realized the essential likeness between these disciplines and the formal mathematical disciplines, as a result carried over to them the same algebraic methods thoroughly suitable to them, and thus enlarged the sphere of the exact mathematical disciplines to include these new ones. But the greatest confusion reigns among them about the nature and meaning of these disciplines, about the content of the leading basic concepts–the same thing indeed also still holds as well for the remaining and old mathematical disciplines. The Idea of a theory of forms of meanings as a discipline that by its very nature is a comprehensive, difficult discipline prior to the disciplines of logical validity is also completely beyond their horizon. On the other hand, as far as the philosophical logicians are concerned, the psychologism so widespread among them has blurred their view of the essential concept-formations of formal logic and, consequently, of the set of problems proper to it. And that has gone so far that people have loved to vilify formal logic as being hollow scholasticism. Moreover, espousing idealism for a pure logic has gone undefended, surely because enough psychologism also remained operative in those quarters to prevent piercing through to the genuine basic concepts and basic problems. Philosophical training in formal logic has most especially suffered from this. Lashing out against psychologism, as has become common with the sudden change of fashion, is of little help when people remain bogged down in psychological confusion with regard to the decisive points, and out of anguish in face of psychologism reject the single source of insight, intuition.

I shall begin with the theory of forms of meanings. The fourth 82
of the essays making up the content of volume II of my *Logical Investigations* makes the first attempt to do justice to this logical discipline as a highly important underlying level of formal logic. The reading of this essay could be quite beneficial to you. I cannot indicate any other literature, except for *Marty*'s 1908 *<Untersuchungen zur> Grundlegung der allgemeinen Grammatik und Sprachphilosophie*, whose excellent author has indeed discussed my arguments, but unfortunately in a way that shows that, owing to his biases, he has not in general grasped the essential meaning of my arguments.

I wish to deal with the theory of forms of meanings as underlying level of apophantics and thus as a mere theory of the forms of apophantic meanings. The qualities are therefore absolutely qualities of affirmations.

Now we know that each matter occurring here can also be qualified differently, and so it seems that the meaning of such a theory of forms is by itself already much more far-reaching, inasmuch as we would have had to realize a theory of forms of all meanings in general along with it. We really only need to replace the quality "affirmation" by other qualities (considering-possible, -probable, -questionable). It is therefore not necessary in, for example, the sphere of the formal logic of probabilities, to work out a theory of forms all over again. The thought also readily comes to mind that we would do best not to speak of qualities in the theory of forms in general and rather develop it as a theory of forms of possible thought-matter.

Every logical meaning is a proposition, is a qualified matter. Nothing special is to be said about the qualities. The only investigation they require is that of determining them completely. That may present difficulties, but it is certain beforehand that it is at the same time a matter of few factors undifferentiated in themselves. All richness of thinking or all richness of varied thought-meanings originates from the matter–a principle that holds, not merely for thinking, but for all consciousness.

So one may raise the general question: How is a systematic
83 overview of possible kinds or forms of thought-matter obtainable? Can primitive material emerge upon which all thought-matter is based and a system of primitive forms that give the material coherent form? Can laws be revealed in accordance with which ever newer matter arises from the simplest, self-contained matter through increasing complexity?

Those are absolutely justified and very important questions. Only, one must just bear in mind–something I have upon occasion noted–that, insofar as the matter has segments, it must also already imply qualities precisely with respect these components themselves. If I say, 'The Kaiser has returned from the North Pole', if I am really acting affirmatively in so doing, then the whole would have the qualitative character of affirmation. I am not, however, then making any affirmative judgment, but expressing disbelief, and that is a different qualitative character. What is identical in both cases is the matter, which however itself already contains affirmation-qualities in the nominal components, because we all use the meanings of the words "the Kaiser" and "North Pole" with the being-consciousness that is a belief-consciousness.

<b) The Problem of the Compatibility of the Quality of the Overall Matter with the Qualities of the Part-Matter>

One could then proceed in such a way as to leave all qualities undetermined in the general theory of forms of thought-matter, but arbitrarily determinable within the quality-genus. In this way, one would actually arrive at a level of generality that would include all possible determined matter.

But a difficulty then arises. Without an investigation of our own, we must not take it for granted that the highest quality, namely, that belonging to the matter as a whole, is completely independent of the qualities attaching to the components of the matter. Thought-acts are really as a rule very complex with respect to their matter. If then, for example, the overall quality is to be that of considering-true, can one say there that any arbitrary quality can be in the components? That should in no way be the case. The qualities occurring within the matter will affect the overall qualities, and the manner in which they do this would have to be closely 84
investigated. Naturally, I cannot refer you to the literature on this problem, since the basic concepts making sense of it are as yet foreign to our contemporaries. Unfortunately, I cannot myself offer you anything conclusive either. In general, in formal logic, even though it belongs to one of the oldest sciences of mankind, everything is evolving–and right in our time, since the ossified concepts and formulas of the tradition are coming alive again, moving, evolving.

Were I to state what, without systematic, exhaustive investigation, I hold to be right, then it would perhaps be the following: The quality "considering-true", which makes independent matter into a complete affirmation, presupposes in all components of this matter either the quality "considering-true" (belief), or a conceivable modification of it, therefore, that imagining that corresponds to the belief. It is, however, to be noted that, as soon as it occurs in the unit of an affirmation, this imagining assumes a form, that of assumption, in predicative matter possibly that of hypothetical presupposing and positing a conclusion. It is thereby then actually said that mere imagining however undergoes something like a qualitative modification as soon as it occurs in the matter of an affirmation. What is characteristic of this modification is, however, that it only concerns the quality in the role of composition of a matter and that thereby a quality therefore arises that likewise cannot occur as main quality, as the highest quality. A hypothetical presupposition, such as, for example, "Let a triangle have two right angles..." is not mere imagining, but simply presupposing, and presupposing, I say, points to the positing of a conclusion and is what it is only within the whole of a hypothetical judgment.

Of course, one could object to that: Cannot we make a presupposition just in order to see afterward what would result there? As long as we do not have the conclusion, the presupposition is after all a free act. Cannot we hypothetically pretend that it is raining bread and then reflect about what would then be the case? However, we do not then really have a presupposition in the sense the hypothetical antecedent implies. Rather, we have the mere thought and in addition the aiming at hypothetical knowledge in
85 which that thought would take on the role of antecedent.

I therefore also think that we are dealing with characteristic qualitative modifications of the main qualities here–modifications specifically pertaining to components of thought-matter. The matter with these qualities can then take on the different possible main qualities. Like the highest quality affirmation, they can take on any other quality that can in general be a quality of independently complete matter. Thus, for example, the matter of a hypothetical statement can also be qualified as presumption or as question.

I would further venture to state the general principle that when an independent act has the main quality of presumption, partly affirmative qualities, partly seeming and presumptive ones, finally, though, the conceivable modifications pertaining specifically to one or the other (modifications of non-actuality) can also occur within its matter, in its different components.

Here, therefore, the situation is different than for affirmations, where presumptions really cannot occur in the components and conceivable modifications of presumptions.

Naturally, qualities of any kind and whole acts of any kind can in certain ways occur within the context of a comprehensive thought-act. Namely, it is to be kept in mind that here I am speaking exclusively of qualitative occurrences that can be found within the matter of a thought-act. If we switch from the act to its components, then we exclusively have in mind compositions that break the overall matter of the act up into its parts. That is however not the only manner of separating acts and, correspondingly, of joining acts. In their full concreteness, acts are really more than meaning-like essences, and separating concrete acts can run along lines other than in those that the meaning and its compositions prescribe. That is the way it is after what was done earlier, if the thinking is based upon intuiting and merges cognitively with it into a concrete unit.

Similarly, a predicative statement can then be made on the basis of a presuming, namely, in such a manner that the statement is not made about the presuming, but about being presumable, being probable. Presuming that A is b, we can state, "That A is b is probable." Inversely, if we really engage in believing such a state- 86
ment, then we must actually consider that A is b is probable. Presuming thus underlies predicative affirming. On the other hand, this, along with everything that is to be directly taken from it, does not however pertain to the judgment-matter, i.e., to the statement's thought-matter. Pertaining to the stated What is not the presuming–and what lies in the presuming–, but the thought-concept "presumably," "probably." It may be that this, like every thought-consciousness, refers back to a plain presenting. It may, therefore, be that, in order to realize the thought-concept "probable," we must have the presumptiveness present on the matter "A is b" as, so to speak, coloration. But this still unthought "presumably" is to the thought-concept "presumably" as the merely presented red is to the thought red.

The fact that the matter of the thought-acts–insofar as it is composite, can itself already contain qualities, and perhaps qualities of all kinds–introduces a problem into the general theory of forms of the matter of theoretical acts in general. It might seem that one would have to take all conceivable qualifyings into account here and that a purely apophantic theory of forms that also only admits affirmative <qualities> into the matter would be only a completely special part of the theory of forms in general. It is, however, to be seen that a theory of forms of apophantic meanings, provided we but understand the term suitably, nevertheless permits one to study all forms of meanings and must eventually yield all laws pertaining to them.

<c) Restriction of the Theory of Forms of Meanings to Statements. Non-affirmative Qualities as Form-Concepts>

Statement-meanings are *de facto* given priority in scientific thinking. In fact, all knowledge strives after statements. So possibilities and probabilities not only remain conscious in presuming, they are brought into predicative statements–indirectly through apophansis. Based on acts of seeming and presuming, and likewise on those of questioning and doubting, are new acts that permit the formation of statements about the being possible and being probable of this or that matter.

When, then, the formal logic of a general theory of forms
87 requires meanings, then only statement-meanings, however, come into consideration for it for the purposes of laws of validity for all kinds of meanings, since all other meanings–even if indirect–are also expressed through the medium of statements. If, for example, we reflect upon the possible composition of probability-meanings and we seek law-governed conditions for it, then an apophantic statement about probability suffices, since, if it is true, the presuming is *eo ipso* correct. Therefore, the presumption-proposition is valid and the other way around.

If one then formulates a theory of forms of statement-meanings, then to meet all requirements, the following is necessary: first, a completely general theory of forms is necessary that only accepts affirmation-qualities into the matter and, moreover, as pure theory of forms of matter, takes no account of the particularity of the objects, properties, relations of which it is a question in the concrete statement at any given time–completely analogously to the way in which in a pure theory of numbers, the particularity of what is counted, which is naturally something determinate in every actual counting, is left out of consideration, and it is a question of counting-units in completely undetermined generality.

If this purely formal theory of forms of statement meanings has been formulated, then something more is nevertheless needed. Theoretically speaking, presumption- and question-propositions surely also need to receive such purely formal consideration. They too can be considered in ideal generality–regardless of the determinate, particular matter that is different in the cases of actual seeming, presuming, questioning. When, then, such propositions become predicative statements, these statements have a content that from the standpoint of apophansis–the statement as such–is particular. "Probable", possible", "questionable" are predicates like "pleasant", "presented", "thought by someone", and the like–or also like "round", "red" etc.

Consequently, the concepts "possible", "probable", "questionable" no more occur in a pure apophantic theory of forms than the concepts "red" and "blue" do.

If, however, we think about the fact that, even though taking a detour via predication, the pure forms of probabilities and possibilities find expression in statements, then nothing after all stands in the way of studying them in the realm of statements. We therefore now explore all possible forms of probability-statements, that is, not the forms pertaining to them as statements, but those pertaining 88
to them as statements about probabilities. We do not now set aside the concept of probability. We do not take it as particular matter, but as an expression of a form, and the same for the concepts "possibility", "questionability", etc. We perhaps state the principle that corresponding to every possible apophantic form, such as "A is", "A is b", "If A is b, then C is d", and so forth, is a probability-form, namely, "It is probable that A is", "It is probable that A is b", etc. All of these are individuations from the standpoint of the apophantic theory of forms, but in them are expressed the possible forms of probability meanings that cannot be discerned from the merely apophantic forms. And that is the point.

Just as the formal laws of the apophantic theory of validity–laws that state the ways in which the proposition-forms affect the possible validity of propositions of those forms–then connect up with the purely apophantic theory of validity, so corresponding laws of validity connect up with the theory of forms of apophantic statements about the probabilities and possibilities–laws, therefore, that state what can be stated about the validity or non-validity of supposed probabilities and possibilities on the basis of mere form.

<d) The Forms of the Intentionality of Thinking Are Mirrored in the Forms of Thought-Meanings. The Shortcomings of Logic Up Until Now. The Task of a Logical Theory of Thinking>

Apart from that, one must not perhaps believe that the theory of forms may have the mere function of clearly spreading out before our eyes the possible forms for which the theory of validity, as the genuinely interesting logic, may establish laws of possible truth or probability. The theory of forms in no way has a merely subservient function. It is worthy of respect and has inherent value. If meanings are subject to set a priori laws in accordance with their intrinsic construction out of components and in accordance with the possibilities of their complexification or modifying transformation into new meanings, then learning about these basic constituents of the world of meaning is inherently a respectable field of scientific investigation.

89 On the other hand, the Idea of thought-meaning is, however, inseparably interwoven with the Idea of thinking, and with the knowledge of those basic constituents of the field of meaning, insight into the a priori essence of thinking is *eo ipso* also attained. Thinking has the obvious, and therefore wonderful, characteristic of being thinking about something, and that is tantamount to having inherent sense, meaning. Subjecting the remarkable characteristic of thinking that we call its intentionality to scientific study is in one sense tantamount to studying possible meanings in terms of the basic types, in terms of the forms, laws that are proper to it as its possible thought-meanings. Thinking in general, as thinking about something, has its immanent a priori constitution and from the noematic perspective has its necessary types and forms in which it alone can acquire a relationship to objectivity or, what amounts to the same thing, in which it can intend, can mean, something–and do so prior to any question as to whether that intending is valid or not.

The theory of forms of logical meanings is therefore at the same time correlatively a theory of forms of thinking with respect to its meaning-like essence, with respect to the possible forms of its intentionality. Studying the wonders of the intentionality of thinking is certainly a matter of utmost interest, and one studies it *eo ipso* by investigating the basic constitution of the realm of meaning, still prior to all questions of validity.

A major part of recent logic since *Brentano*, *Sigwart*, and *Bergmann* is so-called theory of judgment. It passes itself off as a psychology of judgment, as a scientific analysis of psychologistic characteristics of judgment. However, if one takes a closer look, it is thereby essentially, and as it were unconsciously, a matter of an elucidation of the intentional characteristics of affirmative judgments and then, in further consequence, of thought-acts in general–which certainly involves a psychological misunderstanding of the genuine goals and accomplishments of such an investigation, as well as a mixing of the empirically psychological and the eidetic, which belongs in an entirely different level of the investigation.

When nineteenth century logic opposed the technical virtuosity of traditional scholastic logic and instead called for a "psycho- 90
logical" understanding of thinking and a "psychological grounding" of the norms of correct thinking, there was then obviously in all that a great deal of error and confusion, on the other hand, though, an obscure longing for what is correct. It was keenly sensed that something was lacking in the old logic, only no one was able to pierce through to clarity about where it actually lay, what may actually be lacking there. Precisely the same deficiency was sensed with respect to the technique of the logical theory of laws reformed by mathematics, the so-called mathematizing logic. The same deficiency was sensed, and it was rejected as empty artifice. However, one could have been just as justified in also rejecting all of formal mathematics, for example, all of arithmetic, as technical artifice. The rigorous carrying out of purely deductive theory requires one and the same method everywhere, namely, what is called the algebraic method. Mathematicians are everywhere the technical experts qualified in deductive theory, and the development of the theory to such technical perfection is everywhere the requirement of exact science.

What apart from that is lacking in both the old mathematics and in formal logic is a scientific understanding of thinking in terms of its intentional constitution and in terms of the possibility of its cognitive attainments upon which an intrinsic understanding of the meaning of the laws that make valid intentionality possible

then further depends. I shall still have much to say about all the problems above and beyond the mathematical construction of deductive theories that are settled in what has just been indicated. But my basic reflections have already shed bright light on the main points: the distinction between eidetic and empirical investigation, the further distinction between noesis and noema, and in the latter the singling out of the intentional essence involving quality and matter. And from here on out, it is finally completely clear to us that a first and fundamental desideratum is directed toward the theory of forms of meanings and must be satisfied in developing it, because after the entire process we have engaged in, it has become obvious that even though the meanings in themselves are studied in pure ideality and objectivity in the theory of forms, the basic constitution of thinking with regard to its intentional essence is nevertheless at the same time studied with it. That is obvious, because the nature and form of the intentionality of thinking is
91 reflected in the nature and form of thought-meanings. If thinking is exclusively under consideration insofar as it is thinking about something, then it is exclusively characterized by its intentional essence, hence by the fact that it has meaning.

What can a theory of thinking, and consequently also more especially a theory of judgment rationally aspire to? There has been so much talk of theory of judgment in modern times, and according to many, elaboration of this theory should be the particular glory of the modern era. If, however, we take a closer look, then in all the so-called theories of judgment we fail to see clear objectives, and a theory without clear objectives cannot well be an actual accomplishment. If we therefore ask what a theory of judgment–and more generally a logical theory of thinking–can have in view, then according to my findings, my answer plainly reads:

1) a theory of the essence of thinking, a science of what indissolubly belongs to the idea of thinking;
2) an empirical theory of thinking, i.e., an investigation that explores the thought-experiences of experiencing humans in a natural scientific manner in terms of their empirical interweaving and empirical laws.

If we leave the latter out of account, then for an a priori theory of judgment we can again separate: In one case, the question aims at judging and thinking in general, if and only if, it is thinking of something, completely leaving out of account questions of Evidenz and non-Evidenz, of validity or non-validity. Thinking then differs exclusively through the different meanings that it has. The theory of forms of meanings then *eo ipso* gives us a theory of forms of thinking as that of logical meaning. But, of course, only in this respect. A second case, is then the formal logical theory of laws, which deals with the laws that correlate the meaning-form and possible validity. Certain formal characteristics of the thought-matter are in law-governed ways conditions of the possible validity of thinking of such matter.

Finally, remaining still is the study of the cognitive characteristics of thinking as opposed to mere meaning-like ones and the elucidation of the cognitive role of formal logical laws, and then in general the comprehensive a priori study of the essence-components of thinking in terms of its immanent essence. A scien- 92
tifically exact characterization of the connection between to mean and meaning, as well as that between to mean and to know would also belong in there. However, that is a dimension of investigations that presently lies outside of our line of inquiry. We are definitely standing in formal logic.

<SECTION II

THE SYSTEMATIC THEORY OF FORMS OF MEANINGS AND OF JUDGMENT. ITS PROBLEMS AND THE PARTICULAR NATURE OF THE THEORIES TO BE FORMULATED IN IT>[1]

[1] Compare Appendix VI, *On the Content of the Theory of Forms*, to Section II of the lecture course. (Editor's note)

<Chapter 4 93

The Most General Characteristics of the Structure of Meanings>

<§ 20. Independent and Dependent Meanings>

After having defined and characterized the theory of forms of logical meanings from the outside in, so to speak, we want to enter into the discipline itself. I cannot think of providing a systematic, complete survey of its content here. Such a task would take up the entire semester. But it is absolutely necessary for you to acquire a substantive picture of the particular nature of the specific theories to be formulated in it.

A grammatical distinction passed down from Scholasticism, and otherwise going back to the Stoics, can serve as our point of departure. This is the distinction between independent and dependent expressions. Every expression, every word, and every coherent word complex means something, and when expressions are pieced together out of expressions and vice versa, when new expressions are formed out of expressions by composition, then piecing them together also affects the meaning. For, it is always the 94
unit of meaning that characterizes the expression formed at the time as *one* expression. As a general rule, when expressions are pieced together out of several words, and the words themselves have their inflexions or other grammatical forms, then as a general rule, corresponding to this composition is also one in the associated unitary meaning. Generally speaking, something of the overall meaning fastens on to every word—and only because is it a word. However, it is clear that not every arbitrary syllable has its own meaning-function attached to it. No particular meaning-moment belongs to the syllable "ei" in, say, the <German> words *eilen*, *Eisen*, *eigen*, and so forth. In terms of its sensorial side,

E. Husserl, *Logic and General Theory of Science*,
Husserliana: Edmund Husserl – Collected Works 15,
https://doi.org/10.1007/978-3-030-14529-3_4

not every part of an expression is itself an expression and expression-part in the true sense. However, it is clear that grammatical forms, such as genitive or verbal forms, are meaningful, that something of meaning attaches to them within the context of the discourse.

If we now survey the field of composite and simple expressions, and if for that purpose, we also take the genuine expression-parts in the sense of the reflection just carried out, then the distinction between independent and dependent expressions is easily understood. Let us resort to some examples. An independent expression is any complete declarative proposition. For example, "The Kaiser betook himself to the soldiers in the field." It stands there in its own right. It does not point beyond itself. It is not in need of completion. If, however, we take the intrinsically interdependent part, "the troops in the field", or even individual words like "betook", "in", "field", out of this overall expression, then those are examples of dependent expressions. Obviously, this distinction between independent and dependent expressions traces back to an underlying distinction between independent and dependent meanings. Independent expressions are independently meaningful, and that is once more to say that their meanings are independent. And exactly the same thing <holds> for dependent expressions. Their meanings are dependent. Dependent meanings are meanings that are what they are only as parts of comprehensive meanings or as form-moments with respect to them. Independent meanings, however, exist in their own right. There is nothing in their essence that may point to a more comprehensive meaning into which they would have to fit as a part or moment.

95 Speaking more generally, objects in general–whether ideal or real–can be independent or dependent. Corresponding to the being-in-its-own-right of the former is their originally-being-able-to-be-given-in-their-own-right. Their givenness does not necessarily bring the givenness of other objects in a sense conjoined with them, as is the case with dependent objects. The Idea of sound-quality, for example, is dependent. It carries the Idea of sound-intensity with it into givenness in certain ways. Quality cannot exist without intensity. What this carrying-with-it signifies more specifically for Ideas–for singular or genus-Ideas–, what it involves,

would require a more precise, otherwise easy to carry out investigation.[1] It is enough here for it to be evident to you that–the moment we bring the dependent Idea to original givenness to ourselves–for the same intuitive reason, we can bring what is conjoined with it to givenness to ourselves and that a certain necessity holds sway here, whereby the one Idea refers to Ideas interwoven with it, is ideally bound to them–and in that sense it is dependent. Moreover, there are different sorts of dependencies determined by the nature of the objectivities concerned or their Ideas. And each dependency ultimately points to a certain kind of whole, whose components incorporate into an independent unit, into a concrete one, as one can also say.

This is also the case then in the meaning-sphere, and that is mirrored–at least to some degree–in the sphere of verbal expressions. A dependent expression is merely "connotative," as people are also in the habit of saying. Its meaning is only actually given in a comprehensive meaning-context. That means that a dependent expression can only be understood in a normal sense when it occurs in contexts of the kind and form that satisfy its dependent meaning's need for completion in their complete meaning. One must not allow oneself to be deceived by the fact that dependent expressions such as "the troops in the field," or even isolated particles such as "and," "over," are in a certain way understood even outside a complete context. The understanding is then, however, not genuine, i.e., in the detaching, the meaning of the act of understanding is different from that within the context. We quasi understand the 96
isolated little word "and" while we are in the process thinking: Aha, that is the familiar conjunction. But the meaning of this thinking, is naturally independent meaning, is that which attaches to the little word "and" when it functions normally, for instance: Marx and Lassalle are the founders of German Social Democracy.

The difference between independent and dependent meanings is ideal, therefore a priori, and pertaining to it is, as one soon sees, a wealth of a priori laws which, for easily understandable reasons, I called "purely" logical-grammatical in the *Logical Investigations*. The fact that laws obtain here–and laws that are very relevant,

[1] Compare Appendix VII, *Independent and Dependent Ideas*. (Editor's note)

indeed fundamental, for grammar–is easy to see. Dependent meanings are what they are in concrete, complete, independent meanings. We can especially consider them on concrete ground, single them out for our consideration. We can also generalize them provided that we think of what it fully determines in the meaning-concretum at hand as indeterminately variable. Let us, for example, consider the meaning of And in general. For this, we familiarize ourselves with a concrete meaning, for instance, "Marx and Lassalle etc."; but we now no longer tie the meaning <of the> And to "Marx and Lassalle", but we could just as well take "Socrates and Plato" or anything else. We thereby constitute the thought of an indefinite generality that encompasses the manifold possibilities of concrete meaning-wholes in which the And occurs with the same ideal content as form. We thereby also however grasp with Evidenz the fact that this generality that prescribes a rule restricting the possible ways of completing the And into a concrete full meaning is law-governed. Dependent meanings are of course never bound to the complementary pieces that complete them in the given concrete context. That already becomes evident through the ever present possibility of such generalization by holding fast to the identical essence proper to the dependent meaning "and" in our example. On the other hand, though, if we rise to the level of generality and, for instance, say, "A and B...", it nevertheless becomes evident that not just any arbitrary meaning can however be assigned to the "A and B". We can assign an arbitrary nominal
97 substrate-meaning such as "the dog and the cat" or "plume and ink", and so forth, but not, for instance, "but and or", "similar and when" etc. One realizes from the start that the kinds of meaning can in general, and in law-governed ways, be indicated that have to function here as complements.

<§ 21. The Idea of Law-Governed Compound-Forms>

Hence, we see that dependent meanings are not only in general in need of completion, but are in need of completion in specific, law-governed ways. Upon further reflection, it soon becomes apparent that all meanings fall under fixed types, that they are generally to be characterized in terms of substance and form, and that fixed a priori rules govern all possibilities of combinations of elements into completed wholes (then, however, even further, the combinations of such wholes into more comprehensive wholes), while the elements fall into fixed meaning categories. This is similar to a game of chess. Each piece has a meaning in the game, and a rule of the game pertains to this meaning. However, for the purpose of the game, the rules of the game determine every possible, meaningful arrangement of pieces on the board prior to any questions about good or bad ways of playing. Placing a chessman on a line between the squares is not bad, but goes against the meaning. The difference, though, lies in the fact that here in the realm of meaning, it is not a matter of combinations, but of an a priori typology of wholes and parts that is prescribed in a law-governed way by the ideal essence of the meanings as such.

This situation is then to a certain degree mirrored in the grammatical realm. Why cannot we arbitrarily gather random expressions together under random isolated words for the purpose of forming a coherent expression out of that? Why cannot we completely arbitrarily exchange the words in a sentence construction, but only do so in certain ways when the new construction is again to yield an expression, a meaning unit, even though with a changed meaning? Is that some chance grammatical fact? Does this lie in the fact of the historical development of our language? Or, is it a mere psychological fact that we humans are incapable of putting *disiecta membra* from different expression constructions together
arbitrarily in arbitrary order? 98

Of course, empirical considerations also come into question here. Of course, the grammar of the language developed historically in just that way also places its requirements on the form in which coherent discourse has to proceed. These requirements change from language to language and from period of development to period of development within one and the same language. Linguistic development is thereby essentially determined by psychological motives, and various peculiarities in the manner of using particular expressions, in the word order that is actually permissible, in the ways in which speakers and hearers in a linguistic community understand them, in the ways in which they impart various emotional nuances to the meaning-giving acts, in which they sound along with auxiliary presentations that can acquire great meaning for the esthetics of the language, and so forth, are explained psychologically. The whole great realm of intrinsic linguistic form that *Marty*, notably, has thoroughly studied belongs here. It is, however, a huge shortcoming of contemporary linguistics that, definitely owing to the psychologism reigning in modern psychology and logic, it has not seen the whole world of ideal meanings (therefore, as it were, the forest for the trees). What gives unity to the expression as such is naturally first and foremost its meaning, and it has magnificent characteristics that go beyond any empirical linguistic development and go beyond any empirical psychological features of human presentational activity. If unity of expression amounts to much the same thing as unity of meaning and if it prescribes the essence of the meanings as such in general generality, which types of wholes are possible there and which possibilities for the kinds and forms of meaning elements that the Idea of such wholes requires, then it is evident that the possibilities of grammatical combination and formation are restricted, not only on empirical, but also–indeed above all–on a priori grounds. The a priori of meaning must somehow be reflected in the grammar, in the language. What is possible or is not possible for unity of meaning to be able to prevail in discourse, and for it in general to be able to be *one* discourse, is determined purely *a priori*. And this a priori is just as important for the grammarian as it is for the logician.

Therefore, the first and most important thing that grammarians
99 must appropriate is this knowledge that unity of expression and unity of meaning are correlates and that unity of meaning is subject to fixed, very significant a priori laws that consequently thoroughly govern everything grammatical. The a priori theory of forms of meanings must first show grammarians the law-governed multiplicity of formal-possible meanings that should be able to be expressed in any language in general. (First, it must reveal to them the intrinsic structure of the meaning formations, display the categories of elements, the intricate pattern of connecting forms that have supraempirical being for it.) And only then can they ask how German, French, how any given language, *de facto* expresses such and such categories and forms, and which historical and psychological contingencies explain why the manner of expression

is precisely as it is, why it changes ambiguously depending on the circumstances, and so forth. It makes no sense to ask how the categories of meaning we call substantival meaning or property-meaning or relation-meaning, or how the form-meanings–say "and", "or" etc. have originated. One can only ask how people have come to express that and which sensory images were of use to them in the process, how one or another linguistic community acted in the process, and so forth.

<§ 22. *Unsinn-Widersinn*. In the Theory of Forms of Meaning, There Is No Talk of Truth or Falsehood>

The pure logical-grammatical a priori I am speaking about here, or what amounts to the same thing, the pure a priori that belongs to the logical realm of meaning in keeping with its distinctive essence-content, concerns all meaning categories as well as all types and forms in keeping with which new meanings can, as it were, be operationally formed out of given meanings. Our starting from dependent meanings first led us to the idea of law-governed compound-forms that permitted forming new meanings out of meanings of the given category. I wish to explain that still more precisely. You will, however, later hear that meanings can also be formed out of given meanings in a law-governed manner in other ways, not through compounds, but through certain modifications. 100

Let us therefore remain with compounds. Every statement-proposition that is freestanding and independent in its own right has an independent logical meaning, an independent logical proposition. For example, "Gold is yellow". Meaning parts are to be differentiated there. The proposition is in a certain sense a whole that has its structure. In whatever realm of knowledge in which we find a whole, however, we can always rise from the given whole to the general Idea of a whole of that kind, and the general Idea everywhere prescribes formal laws of compatibility to parts combinable into wholes of that kind. Elements that should be able to fit into a whole of that kind must conform precisely to the ideal requirements that the general Idea of such a whole spontaneously sets up. This is also the situation with meanings. The proposition "Gold is yellow" is a whole. We can easily see that it falls under a

general Idea of meaning-wholes that we can designate by the symbolic expression “S is p.” In it, a rule of combination is stated at the same time because the “S” and the “p” cannot create an empty place here that can be filled by any arbitrary meaning in general. We see that from our example. For “Gold is yellow,” I can say, “This paper is yellow,” I can say, “A triangle is yellow,” “A human being is yellow,” “The sky is yellow,” and so forth. What does I “can say” signify? Well, what is said is that through such substitutions the proposition is over and over again turned into a proposition. If, however, I wanted to substitute “bigger,” “similar,” “when” in the subject position, then no proposition would result. “Bigger is yellow” (“Similar is yellow,” “When is yellow”) is not a proposition. It is in general nothing. It is a sequence of words without a coherent meaning. If someone speaks in that way, then we have the indirect thought: He or she probably wants to say something coherent, a proposition, with these words. We perhaps think that he or she was mistaken about the words. But, these indirect thoughts are not the meaning of those words themselves and do not by themselves give unity of meaning. Originally, a nominal substantival meaning, “gold,” was in the subject position. We soon see that whatever nominal meaning we place in the subject position yields a coherent meaning, and we also see it generally. We can say that each expression of the form “S is yellow” yields then–but also only then–a meaning, a complete proposition, when we take “S” as a sign for an
101 arbitrary nominal meaning. Likewise, we can substitute any arbitrary adjectival meaning for “yellow.” A proposition always results.

And it is thereby to be noted that whether a true or false, a clever or ridiculous, a contradictory or logically possible proposition arises from such substitutions is not the point here. I can say, “Gold is reckless,” “Gold is a round square.” Those are still propositions. A natural scientific *Widersinn* is inherent in the first inasmuch as the psychic predicate necessarily calls for a person as subject and is *a priori* in conflict with a physical subject. The other proposition is mathematically *widersinnig* and may possibly be logically *widersinnig*, were “round” perhaps to be defined as “not having any corners.” It is, however, to be noted that *Widersinn* is also meaning. One does well to differentiate between *Unsinn* in the purely grammatical sense and every kind of *Widersinn*, including formal-logical *Widersinn*. Anyone who strings together words, even in protention, to say something coherent where the meanings of the words are not combinable into any coherent meaning, speaks without meaning, utters *Unsinn*, not-sense, in the strictest sense, like, for example, a lunatic's flight of ideas. However, someone who states a proposition, really a proposition, a unit of meaning, speaks with meaning. But for that reason the logical proposition stated can be false, wrong, *widersinnig* in one way or another. What is *widersinnig* cannot be valid, namely, because in the manner in which it is put together as a unit there, meaning is not compatible with meaning when it comes to validity. When it comes to validity! Therefore, the question there is what requirements are to be placed on the meaning if it is to be valid. Or, what must a proposition look like? Which requirements must it satisfy for it to be able to be a true statement that corresponds to a state-of-affairs in reality?

Obviously, however, when such questions are asked, the proposition must be a proposition, i.e., a unit of meaning. I already have coherent meaning and am only asking how matters stand regarding its truth. Can truth in general correspond to it in terms of its content or its form? Therefore, even false, *widersinnige*, contradictory propositions are propositions. In the theory of forms of meanings there is, however, no talk of truth and falsehood at all. So it is that in a game of chess, each play is a play provided it complies with the rules, while it can otherwise be a cleverly or foolishly played play, because the playing rules do not regulate cleverness or foolishness, but only the form of the moves that each chess piece of its category may execute to be in compliance. So, it is, 102
therefore, so to speak, a playing rule of the realm of meaning that where there is a nominal meaning, any arbitrary nominal meaning can be, where an adjectival meaning <is>, any arbitrary adjectival meaning <can be>, and so forth. And, one thereby becomes aware: There is something such as a grammatical category, form-types as well, to which all particular meanings are subject. Every kind of meaning-whole again yields a coherent meaning-whole when we allow any meaning within its category to vary arbitrarily. Accordingly, a would-be proposition-form like "S is p"—where S is the sign for a meaning of the category "nominal meaning", p a sign

for an adjectival meaning–not only tells us something tantamount to: Propositions of such a form actually (since we have really abstracted this form from the examples) occur in the realm of meaning. Rather, merely on <the> basis of the thought-form, it is to be seen as a law that by purely keeping to the prescribed categories, every individuation of this form yields a coherent proposition and not, say, an *Unsinn*. Though very primitive, that is, nevertheless, an important and enlightening insight. It naturally concerns not only this one form. Rather, everything in the realm of meaning is subject to form and form-law in the same way. The duty to study these laws consequently arises and, thus, to learn about the general structure of all meanings and meaning-complexes, the categories of all structure-levels, the laws for the operations of combination and transformation.

<§ 23. General Distinctions>

<a) Primitive–Compound Independence>

To find our way amid the multiplicity of possible meaning-forms, let us first make some general distinctions. We start with independent meanings in relationship to which all dependent meanings must indeed be found. We can then first distinguish between independent meanings, from which independent meanings can again be detached and independent meanings for which that is not the case. We therefore distinguish between what is primitive and independent, primitive meaning-concreta, and what is compound and
103 independent, compound meaning-concreta, that are compounded as concreta. The theoretical possibility of such differentiation is already apparent before its confirmation as a fact. For, it also otherwise happens that concrete wholes can be changed into parts or independent objects into comprehensive wholes from which they are then again detachable. However, the following is thereby to be borne in mind. If any concretum–in whatever sphere–enters into a combination, be it partially, then it *eo ipso* assumes a form that is prescribed by the kind <of> whole to be created. Even a physical part of a physical whole is, as what it is in the whole, never ever outside the whole precisely by virtue of form. If, however, we speak

of the same part that in one case exists in its own right and in another exists in the whole as a part, then we are saying that an identical essence of the independent content survives and assumes mere combination-form through inclusion in the whole. <This is the case> in the realm of meaning as well. An independent proposition is, for instance, "All human beings are mortal," and again "Socrates is a human being." Already when, taking both propositions together, we produce the conjunctive link as it is present, for instance, in the inference, "All human beings are mortal, and Socrates is a human being, therefore, Socrates is mortal," each proposition has, besides its own content with which it exists in its own right, taken on a combination-form. An overall meaning–"both take place"–indeed twines around both propositions. And likewise, when we consequently conclude, "therefore Socrates is mortal," then this conclusion as such has its indicated form through the little word "therefore." On the other hand, we can also do away with this form again, and do so without the proposition's own essence being changed, then "Socrates is mortal" figures there as something independent in its own right. Taken in the form that the function gives it as a part of a compound, every such proposition is obviously dependent. Yet, we speak of a compounded concretum, of a proposition compounded out of independent propositions because with the full retention of its essence, nothing else befalls propositions independent by themselves than to take on that form that makes them into parts. So, when in the realm of meaning, we speak of decomposition and distinguish between non-decomposable and decomposable independent meanings (concrete meanings), we definitely remain in the general, valid meaning for any kind of 104
decomposable wholes.–Every independent meaning is a proposition, that is, in the apophantic sphere of concern to us, an affirmative proposition, a judgment in the noematic sense.[1] When I afterward speak quite simply of a simple judgment or a simple proposition, I am always thinking of a simple independent proposition from which therefore no further proposition may be taken.

[1] One may state this as a truth. Cf. the argument below in §28c.

<b) Composition–Decomposition>

We only recently procured an initial distinction for purposes of a thorough investigation of the structure of meanings and more specifically of apophantic, affirmative meanings. As in other spheres of objective wholes, there is something like decomposition in the realm of meaning as well, and among meanings, one can distinguish those from which meanings can be detached and those for which this is not the case. (A distinction that, as I additionally noted, also occurs in dependent meanings. One may detach something from a dependent meaning, so to speak, cut an independent meaning out of it. For example, the antecedent of a causal judgment "because such and such holds" is truly taken as dependent in and with its because-form, but one can free the proposition itself that is functioning as an antecedent precisely from this form. It is, however, important for us to classify the independent meanings, namely, into ones out of which complete independent meanings can still be withdrawn and those for which that is not the case.) That is still a completely general distinction that does not specially address judgments, affirmative independent meanings. At the same time, a classification of judgments into compound and simple ones arises. It is thereby also equally to be said that every simple independent meaning has matter having a specific composition ("S is p" or "S is not p"). I further note <that> for this reason we cannot avoid the word judgment in theory of meaning, because this form of matter (*propositio*) is also equivocally called proposi-
105 tion, so that we do not manage well with one word. A simple independent judgment or a simple judgment is therefore quite simply an affirmation of such matter.

It then appears that in exploring the structure of affirmative meanings, we can allow ourselves to be guided by this distinction we have procured. Every affirmation must indeed be simple and compound in the sense indicated (*nota bene*, every independent affirmation; dependent ones must find themselves *eo ipso* in association with independent ones). Therefore, it appears that we would first have to explore simple, independent affirmations, simple judgments, in terms of their structure and then compound ones, for which only the question about the forms would be to be asked, in conformity with which compounds are formed into new judgments out of simple judgments.

Meanwhile, as natural as this initial thought is (already, because we shall first of all think about possible decomposing and piecing together of wholes of any kind in other respects as well), it nevertheless becomes apparent that in the realm of meaning decomposition is only a special case of a more general kind of compounding that I call composition.

How then is the difference between composition and decomposition to be characterized? Components are, for example, the nominal meanings that cannot be missing from any independent meaning, adjectival or relational meanings as well. Components are furthermore antecedents or consequents in hypothetical contexts, the propositions that occur in disjunctive propositions, in propositions of the form "Either A exists, or B exists, or C exists". –All parts are components, but not all components are parts. Component is therefore the more general concept. In a hypothetical judgment, "If A is B, then C is D", the hypothetical antecedent is a component, but not a part, since detachability is lacking. We encounter such a thing in causal judgments, "Because A B, then –". Here detachability yields a judgment of its own, "A is B!" This judgment is really hiding in the causal whole. If this judgment is simple, it is apart from that still composite. Every propositional matter of the form "A is b" points to composition. Before I pursue the matter further, I must still, though, discuss the following.

Discussion of the concepts composition and component. A meaning is composed. In the natural sense that means that it is a whole in which we can discern parts, "components" fitting together, combining with one another. What we discern there in the whole, 106
similar to our body and the parts of our body, we take just as it is in the whole, in its place, in its context and the form thus imprinted on it. For example, if the hypothetical judgment "If A is b, then C is d" breaks down into two distinct parts, I have, so to speak, a break. "If A is b–then C is d". Every such component again breaks down into parts. The antecedent has the propositional matter "A is b". It says, "Given that A is B". There we again have a break. "A–is b". The same naturally holds of the conclusion. The A can then

have a composition, perhaps more specifically read, "the re-opened Reichstag" or "the Reichstag that is again open". There we again have a break, that is, as it were, a main component and an auxiliary component. Reflecting generally, one easily notices that a unitary meaning can be broken down into more rich and less rich (be it, in other respects, independent or not), and at the same time we find that not all components have to figure on the same level. For example, hypothetical statements directly break down into antecedent and consequent. As components of the first level, the direct components themselves again have direct components that, in relationship to the whole, are components of the second level. And this can go on further in this manner to components of the third, fourth level, etc. With every meaning, we come to ultimate compositions and components.

Components in this sense are dependent under all circumstances. They are what they are in wholes, and different wholes can have equal components, but not the same component.

<§ 24. Syntactical Stuff and Syntactical Form>

<a) Propositional, Nominal, and Adjectival Syntagmas>

An important pair of concepts is, however, connected with this concept of component. We do well to adopt a special word for it. I talk of syntactical stuff or syntagmas, and in addition <I speak> of syntactical form. Syntactical stuff is always given in syntactical form that lends it the specific thought-function in the proposition unit and overall meaning in general. If we take up the concept of component, and if by that we understand every part within an
107 independent meaning that is, as it were, separated by a break vis-à-vis its co-component, then the fundamental fact is then precisely that every component is formed stuff, namely, syntagma in syntax. That becomes evident when we compare components with components (no matter whether they belong to the same or to different wholes) and, thereby, unconcerned about their intrinsic structure, compare them with one another in their complete unit. We then immediately notice that different components can have the same form and different stuff, and again different form and the

same stuff. We then learn to look at components in two respects and to bring them under concepts. In one case, with respect to syntactical form, we realize that certain forms of components run throughout the realm of meaning and are to be reduced to few concepts, for example, subject-form, object-form, predicate-form, attribute-form, form of antecedents of hypotheses, form of consequents of hypotheses, of the disjunctive component in disjunctions, and so forth. To these component-forms, which we call syntactical forms, then correspond the changing syntagmas as the syntactically formed stuff. Belonging to the concept of syntactical stuff is the fact that it can take on different forms while, on the other hand, it ought not be said that every such stuff is authorized to take on every arbitrary syntactical form. (And also the general concept of syntactical substance diversifies into definite kinds, as we shall immediately realize.) Examples are first needed with regard to such stuffs.

We are already acquainted with a group of examples. For, wherever we have a decomposable meaning–for example, a compound judgment, "Because foggy weather occurred, military operations were suspended"–, there the part is given in the whole as a syntactically formed part, as a component. And when the part is cut out, then the component does not, as it is, become independent, but an independent meaning having the same quality and, in terms of the nucleus, the same matter can be produced that precisely as a result has the same "content" in common with the component, exactly what I am now calling the syntactical stuff. So what is detachable can indeed occur within different wholes with the same content, but in a different functional form.

Let us now, however, consider other spheres of examples. In terms of syntactical form, an antecedent of a hypothesis is obviously separated from every consequent of a hypothesis (in a hypothetical judgment). The syntactical form thereby likewise 108
stands out for us from the syntactical stuff, the form of the antecedent itself from the proposition itself that is in this syntax. And the same stuff can change its syntax. As the same proposition-matter can have the form of the antecedent, it can also have that of the consequent of the hypothesis. For example, on one occasion, it can read, "If humans are ethical beings, then..."; and on another occa-

sion, “... then humans are ethical beings.” But, yet other forms are also possible. Thus, within the context of a disjunctive proposition, it can read, “Either humans are ethical beings....” What is common is therefore a stuff as against this changing form. If the view expressed earlier is correct, then what is common here is the proposition-thought, “Humans are ethical beings,” which is also detachable in its own right. Obviously, for antecedent and consequent, it is truly not a matter of judgments. No affirmation that they are ethical beings lies in the antecedent “if humans are ethical beings.” It is different with the putative causal judgment, “Because A is b, then C is d.” Here the antecedent is an affirmative proposition. It is affirmed that A is b, if <it> is also the special function of the causal proposition not to state this affirmation independently, but rather that from this affirmation the other one, “C is d,” follows. Underlying a modification is, however, this same affirmation, which as an independent affirmation states, “A is b!” I mean then that, just as the causal antecedent refers to a propositional affirmation that in a certain modification, in a certain functional form enters as well into the causal judgment, so the antecedent of a hypothesis points us to a propositional simple thought that is contained as a term in the hypothetical judgment, but precisely as a term with a functional form designated by the word antecedent.

Accordingly, a hypothetical judgment is a simple judgment, but an independent propositional thought can be detached from both of its components. Although a simple judgment, a hypothetical judgment is a compound independent meaning (compound meaning-concretum). Otherwise, a causal judgment is indeed structured completely analogously and is likewise a decomposable meaning, but where thoughts figure there, here judgments figure as
109 parts so that the causal judgment is a compound judgment.

In this way, we have acquainted ourselves with some examples of propositional compositions and propositional syntagmas. Non-propositional ones are indirectly in every propositional component, for example, as the syntactical stuffs belonging to the nominal compositions. A nominal meaning S figures in the subject position of simple propositions of the form “S is p.” We all say that the same noun must not merely act as the subject. For example, “The Kaiser visited Prince Heinrich,” reads thus on one occasion. On another occasion, it can read, "Prince Heinrich visited the Kaiser". The same noun "Kaiser", which at first acted as the subject, now has an entirely different role as the object in the predicate. If we then distinguish the nouns from the component and from the marked syntactical form in the latter, then under the heading nouns, we once again have a syntactical stuff, but now one of an entirely different type. It is not a propositional syntagma, but simply a nominal syntagma. That does not however rule out the fact that also propositions, even if dependent clauses, also pertain to a noun as an incorporated component. For example, it is a noun, or more correctly speaking, a nominal syntagma, when we say, "the Kaiser, who visited Prince Heinrich". But even this whole can, as it were, be declined. It can, with its identical content, function in different ways in judgments and has a different formation, as the case may be.

Predicates, for example, an adjectival predicate green, again offer us a new sort of syntagma. On one occasion, it really acts as a predicate, as <in the proposition>, "This tree is green". On another occasion, it figures attributively, as <in> "this green tree", in a markedly different role, therefore, in a different form. It can also occur within a comprehensive predicate–for example, "This is a green tree"–, which again modifies the meaning somewhat. But, "the adjective" is the same, i.e., what, in terms of meaning, we call an adjective there is a syntagma, namely, a syntagma of a new type as against nominal ones. But the general type extends far beyond the simple predicate. Every predicate (the whole component that figures after the "is" in the simple categorical proposition), however complex, implies a single syntagma as content. The whole can always change functional form and remain identical as regards what it signifies. It can at times function predicatively, at times attributively.

<b) The Syntactical Form of the Whole-Component and That of the Subordinated Components> 110

What I wish to call syntactical form here is–as must be obvious from my presentation–thought of as form of the component in question, consequently, as forming its syntactical stuff into the component. Since, then, in line with its breaks, every independent

meaning then shows itself to be a whole made up of components (where everything that the components bring to the unit in relationship to one another is already included in the component-forms), one sees that every form to be made to stand out within the independent meaning is shared among the components. A coherent judgment is a coherent syntax, a syntactical unit out of syntactically formed stuff. On the other hand, one however again sees that corresponding to the different composition proceeding in stages, the forms separate in accordance with the stages, and that forms that differ from the syntactical forms of the components subordinated to it can occur within a whole-component. For example, the conjunction or <the> disjunction–"the philosopher Socrates and the philosopher Plato", "the philosopher Socrates or the philosopher Plato"–can occur in a proposition as a component and then perhaps have the syntactical form of the subject component of a conjunctive or distributive predication. Components again occur in the component, "the philosopher Socrates", etc. And, each one has its syntactical form. Where then does the "and" belong, the "or" likewise? Obviously, not to the direct syntactical forms of both names. On the other hand, not to the syntactical form of the conjunctive connective either. In all changes of syntactical form of this connective, the And remains the same. It is clear that it belongs to the syntactical stuff. Ought one then say that no syntactical form corresponds to the collective whole? However, if we take the statement "The philosopher Socrates and the philosopher Plato were native Athenians" as an example, we however notice that a syntactical function is attached to the And there as well. We can express that in the following way: The nominative and the position within the context indicate that the meaning "the philosopher Socrates" refers to an object, is the object-about-which–an object of which something is stated there. The meaning is a relatively complete
111 component and has a syntactical form in its own right, namely, the same one it would have if a statement had merely been made about the philosopher Socrates. Here, however, it is, indirectly speaking, stated of him and the philosopher Plato in the plural that both were especially native Athenians. Obviously, in this way, each subject-meaning obtains not only connection, but along with it a syntactical function embracing both, so that the names in addition

have a higher syntactical form apart from the initially designated ones proper to them. While each individually functioning name has its individual syntactical form, each name functioning in the conjunction has–not in its own right, but together with the conjoined names–a form pertaining to them in their togetherness, which for its part already presupposes primary forms in them.

<§ 25. Nucleus-Stuff and Nucleus-Form. The Formation *in* the Syntagma.–Contrast with the Traditional Understanding of Term, Concept, and Presentation>

If we pursue the compositions of meanings in terms of the natural sequence of levels, therefore, go from the direct components to ones of the second, third levels, and so forth, then we finally come to the ultimate, simple components no longer to be decomposed. By reflecting comparatively upon those components and abstracting from the syntaxes forming them, we come to the primitive syntagmas. These are primitive, nominal syntagmas like "Napoleon", primitive adjectival syntagmas, for example, "green", and primitive relational syntagmas like "similar". In connection with the fact that syntagmas separate into fundamental types, for instance, substantival, adjectival, relational, we can speak of syntagmatic categories. These categories are, however, not anything proper to the primitive syntagmas. They occur just as well in higher, more complex ones. There are indeed compound nouns, compound adjectival and relational formations.

Apart from that, new kinds of syntagmatic categories also occur in the realm of complexes, thus, above all, those of propositional syntagmas, the sum syntagmas ("A and B") etc. If we now restrict ourselves to non-propositional ones, namely, to those syntagmas that are also found among primitive components, then a new fun- 112
damental distinction between stuff and form is to be indicated that forces itself with regard to them. If we compare the syntagmas "similarity" and "similar", "redness" and "red", then we see that in every such pair, syntagmas of a different category having an essence-nucleus in common with one another stand opposite one another. This stands out *in abstracto* when we compare "redness" and "red" and is, as I shall have to say, differently formed. This

concept of nucleus is obviously to be distinguished from the syntagma in a clear-cut way. We had formed the concept of syntagma in such a way that it concerns what is identical, which stands out as the same noun in a different predicative function, or as the same predicate–but in a different function–sometimes as actual predicate, sometimes as determining attribute. The change of the function does not alter the same noun, but rather makes it stand out as the same noun. Therefore, the syntagmatic category belongs firmly to the syntagma. In contrast, it is now apparent that syntagmas of different category can still have something identical and something different. What is identical is the nucleus. The syntagmas "similarity" and "similar" have the same nucleus. They differ in nucleus-form. The nucleus-form is what forms the pure nucleus into the syntagma of the specific category.

The difference between syntactical stuffs and forms is no more restricted to primitive components than the difference between nucleus-stuffs and nucleus-forms relates to primitive syntagmas. Rather, it pertains to all syntagmas in general. Of importance here is, however, the fact that every non-nominal syntagma can be nominalized, while obviously not every other syntagma can as a rule transfer to such a different category, therefore, not every syntagma can, for example, be made into an adjective. In contrast, every adjective can be nominalized, similar-similarity, algebraic-algebraicality, red-redness. But the propositional syntagma "S is p" can also be nominalized into "the fact that S is p," "this–the fact that S is p (entails etc.)." Similarly, the And-syntagma <can be nominalized> into the nominal sum-syntagma. Likewise, we can speak here of the nominalizing of the whole meaning itself. What is nominalized and what is not-nominalized then have the same nucleus-stuff in different nucleus-form, or – what amounts to the same thing–they have syntagmas having the same nucleus, but differing in syntactical category.

113 So the analysis of meanings in terms of their intrinsic structure leads to recognition of a remarkable conformity to laws. Every independent meaning displays the difference between matter and quality. Each one breaks down into parts and, with respect to its composition, displays a twofold difference between stuff and form. Every component at first has syntactical stuff and syntactical form, and then the syntactical stuff again has the difference between

nucleus-stuff and nucleus-form (i.e., syntactical category). That holds no matter how composite the matter may be–and accordingly the syntagmas and the nucleus-stuff. If, though, we go back to the primitive meaning components, then the intricate pattern of forms leads to the ultimate stuff, the ultimate nuclei, the ultimate bearers of all formation. Such an ultimate nucleus is, for example, "green" – taken, so to speak, as logical root, because what is adjectival is already nucleus-form.

With respect to the nucleus, the following is yet to be done. The distinction between nucleus and syntagma would not force itself upon us without certain cases of variability of the nuclei in identity of the syntagmatic category, namely, the particularity of the substantivization that occurs in adjectival and relational syntagmas. "Green" can be a predicate, as I said earlier. It can also be nominalized, as in "Green is a color". The same is so for similar-similarity. The situation for nuclei and nucleus-forms is, however, not analogous in the manner of the relationship between syntactical stuff and syntactical forms. Syntactical stuff can change its form to the greatest extent. In contrast, for nucleus-stuff, we only have the one transition to the nominal form and sometimes the reverse transition. It is already not possible for an adjectival nucleus like "green" to transition into the relational form and the other way around. I must furthermore say that for a main kind of syntagma in general varying between syntagma and nucleus is not to be done. That is so for all meanings of proper names, <such as> "Socrates". These are individual meanings. Their individual content cannot in general be brought into any syntactical category other than that of nominality. But only through changing this category could the nucleus and character of the syntagma vary. Nevertheless, also there, one retains the right to distinguish between the two, provided, however, that the syntagmatic category is a character that occurs in the most different syntagmas, and among them those having an entirely different nucleus-content. As a result, we also say for individual meanings like "Socrates" that they have a nucleus. And in the cat- 114
egory of nominality, this nucleus gives the syntagma in a unified manner. The syntagmatic category genuinely belongs to the nucleus everywhere. As opposed to the individual nuclei, the other nuclei, such as the adjectival and relation-nuclei, are also called

general-nuclei because, by means of theirs, the meanings refer in general ways to individuals that for their part could be meant by individual meanings. A general-nucleus is also each one of the kind "house", "tree" etc. that is general, but not adjectival. Such primitive nuclei cannot be made into adjectives in the genuine sense either.

Traditional logic has worked out next to nothing regarding these fundamental distinctions, although they also surface in it from time to time. I can in particular say that the concept of nucleus I have established essentially coincides with what, in a completely vague way, and without the slightest attempt at clarification or definition, traditional logic calls a term and moreover has only used in a narrowly limited sphere. Syllogistic logic talks of terms. The judgment-forms of universal, particular, singular judgments, and so on are expressed symbolically there by "All A are b", "Some A are b", and so on. Similarly, the hypothetical proposition-form is occasionally expressed by "If M is, then N is". If one asks oneself what the letters stand for there, it at first seems as if it is a question of syntactical stuff. Upon closer inspection, one is easily persuaded that it can only be the nucleus-stuff. Especially, if one is mindful of the fact that from the standpoint of syllogistic logic–for instance, the inference, "All humans are mortal, all mortals are ephemeral", and so forth–, "mortal" and "mortals" are considered to be the same term and are designated symbolically by the same letter. People very often also say concept instead of term. However, the word concept is so ambiguous and in particular <is> also used so ambiguously in the field of formal logic itself that we cannot use it without thinking twice. In any event, it may be said that by means of my analysis, an extraordinarily important meaning of the word concept as "nucleus-content" has been scientifically defined. And at the same time, the general-nucleus narrows down the meaning of concept in a clear-cut way. It is indeed often said that generality is part of the essence of concepts.

115 I have already mentioned that within that logical chapter in which meaning-forms are *de facto* dealt with, even though amid psychological confusion, the talk of presentation, namely, in contrast to judgment, is frequently common. Presentation thereby at times means something psychological–for example, tantamount to the intuition or thought-presentation underlying the thought-act, which therefore has nothing to look for here in the theory of

pure meanings–, at times tantamount to mere neutral thought, as neutrality-modification of affirmation, at times also, and mixed in with this, something different. For example, people say that nouns are expressions of mere presentations, not, however, complete judgments. *Bolzano*, the true discoverer of the world of pure thought-meanings, used presentation as an expression for every component, indeed moment, distinguished in a proposition, also <for> a moment of simple form. In my earlier lectures on logic, I used to define another, very important concept of presentation as opposed to this not very useful one. Namely, I called nominal presentation what I am now calling nominal syntagma. Likewise, I discerned as many kinds of presentations as syntagmas. In short, presentation = syntagma. In any case, I consider it indispensable to introduce a separate artificial word that can always be appealed to when danger exists of becoming entangled in the countless ambiguities of the words presentation and concept. (If the Idea of presentation is admitted into the realm of meaning, then the nucleus naturally suggests itself as being the specific meaning of "presentational content"–as distinct from the objectivity presented).

<§ 26. The Fundamental Distinction Between Primitive Nuclei in Full-Nuclei and Empty Nuclei>[1]

<a) The Difference Between Full-Nuclei and Empty Nuclei Is the Original Source of the Distinction Between Definite and Indefinite Object-Reference of Judgments>

Staying the same course set by our analyses up until now, we must now go one step further.

Let us first reflect upon the following. Independent meanings– 116
less broadly speaking, judgments–refer, people say, to a meant objectivity that in truth exists, *is*, if the judgments are true, and is not, if they are not true. This referring to objects by judgments is distributed in certain ways over the judgment-components, or is based on that of those components. Every component has its relationship to an object. For example, every individual noun names a

[1] Compare Appendix XIII, *On the Theory of Primitive Full-Nuclei.* (Editor's note)

thing to which the judgment refers, for instance, as the subject judged about there when the noun figures in the subject position, or as the object that is placed in relation to the subject-object, and so on. More specifically, the nominal syntagma, the nominal presentation, has this naming relationship to objectivity, and the same object can be presented by different nominal presentations, such as “our Kaiser”, “Wilhelm II”, “the present king of Prussia”, etc. Every such presentation has a different nucleus-content and different ultimate nuclei formed in such and such a way, and it is due, someone will say here, precisely to these nuclei and the way in which they are connected that the different syntagmas refer, or can refer to the same objectivity. Something similar holds, it seems, of all other syntagmas and their nuclei, whatever their syntactical category may be.

However, there is a fundamental difference in the reference of meanings, or of thought-presentations, to objects, inasmuch as the former refer to objects in definite ways, the latter in indefinite ways, and corresponding to this difference is a fundamental difference in the primitive nuclei, in which the original source of the distinction <between> meanings definitely and indefinitely referring to things actually lies.

A striking difference actually exists between primitive nuclei like “red”, “house”, “round”, “triangle”, “bigger”, etc. and the primitive nucleus “something”. In the one case, we have “*full-nuclei*”, in the other, a nucleus that in a certain sense is not a nucleus, but an empty shell of nucleus as nucleus. I am speaking of the *empty nucleus*. In the one case, we have, as it were, full-nucleus presentations, in the other, empty-nucleus or logically empty presentations.

To the extent that definiteness of reference to objects–namely, logical definiteness–prevails in presentation and judgment as what
117 imprints <the> meaning-content itself, the full-nuclei are to be held responsible for that. To the extent, however, that logical indefiniteness <prevails>, the different empty nuclei can occur scattered here and there in the meaning-wholes. If it is stated that “a lion has escaped”, then the indefiniteness lies in the little word “a” – a lion is a definite something due to the full-nucleus presentation lion. But indefiniteness is still found there. If it is stated “our Kaiser is in Breslau”, then there is no empty place, and everything

is determined, provided one is not of the opinion that the empty place resides hidden in the "in" and in the "is." If merely "something" is stated, then we have full indefiniteness. In the judgment "Something is red," something fully indefinite is determined as red. In this judgment, the "something" is a subject component in its own right–the component is obviously a nominal component–and like every other one has its syntactical stuff and its syntactical form. If we say "This house is similar to something or other," then the nominal something has a different function and is the same nominal something, the same syntagma. One can also speak of an adjectival and relational something. To the question "What is this house?" one can answer "something or other." Not a very brilliant answer, and actually an empty one. But, it nevertheless has a meaning that is to be paraphrased with the words: It is constituted somehow or another. And likewise for relations. Considering the fact that something identical is present in all these transformations of the syntactical category that we can again only designate by "something," here too one can speak of a nucleus, precisely the empty nucleus, the empty presentation "something or other." That is, namely, a primitive nucleus. Presentations having many components, or nuclei, then divide into those exclusively built out of full-nuclei, those built out of empty nuclei, and mixed ones.

<b) The Empty Presentation "Something"–What Is Specifically Formal in Formal Logic>[1]

The presentation "something" is of the very greatest importance for the theory of meaning, for all of formal logic. It is, as it 118
were, what is specifically mathematical in the mathesis, what is specifically formal in formal logic. It occupies a remarkable middle position between form and stuff. On the one hand, we see that it is the nucleus of separate syntagmas, separate thought-presentations. An entire judgment-component can be thrown into question by it alone. On the other hand, it is to be termed form, and I would like to consider everything that has been designated form up until now

[1] Subsequent note by *Husserl*, "The theory of propositional 'functions' in the mathematical sense and their arguments that is later advanced, 133ff. <§40b below > is to be borne in mind from the start and everything to be reworked accordingly." (Editor's note)

as being related. I would almost like to say that it is pure thought-function posing as stuff.

However, as always, one must indeed separate the distinction between form and stuff presenting itself here from the previous distinctions. If one says: The presentation "something" is "mere form without stuff," then that does not mean it is presentation without nucleus, but that it is presentation without any logical fullness, without any full-nucleus. Or: The form-stuff dichotomy is now that between empty nucleus and full-nucleus.

In general, definite nuclei do not belong in formal logic. A theory of forms of meanings need not be concerned with definite nuclei–"green," "hard," "triangle," and so forth–, not with their factual differences and the differences they require for meanings to be built out of them, any more than a theory of validity of meanings <has to be concerned with> the truths pertaining to "green things," "triangles," and so forth. In contrast, the empty nucleus is something purely logical. Indeed, it is, as I already suggested, the index of everything logical. That already follows from the fact that the presentation "something" can formatively co-determine the structure of meanings in every possible field of thought in general. It is obviously at home in all knowledge and every realm of thought as well. In its full indeterminacy, the "something or other" is simply determinable as everything and anything. In every field of knowledge, thinking as determining continuously proceeds from what is tentative and incompletely determined to further determining, and so the something must play its role everywhere, "Something is A," "Something that is A–or an A–is then further determined as B," and so forth.

This absolute emptiness of content, however, characterizes the formal logical. Obviously, the compound-forms in accordance with which what is indefinite keeps on being determined belong within the theory of forms, in the way then that, for this reason, the "something" already plays an overriding role in the theory of forms,
119 indeed is nowhere absent, because everything we state in it remains indeterminate in terms of stuff–namely with respect to the material individuation of the nuclei that would form the link with special fields of knowledge–and <because> everywhere the sole

determinate thing is the form. We think of a form like "A is b", however, as being something logically universal in such a way that the A and <the> b are presented as "something or other" through one syntactical category or another.

The empty nucleus "something" occurs in different formations that are very remarkable. Something essentially new occurs here in comparison with those syntactical forms that determine the difference between the nominal something, on the one hand, and the adjectival and relational something on the other hand. On the other hand, they themselves belong to the presentation-forms (within the syntagma). –It is a matter there of the difference between universality and particularity. Though belonging in specific ways to the "something", they acquire a general, form-giving meaning referring to entire presentations through the fact that precisely the something can function in indirect ways in a general form-giving manner by combining with any arbitrary presentation.

We therefore have, as is now to be explained, not only the distinction between the nominal and non-nominal something and, in what is identical, thereby, a nucleus, but–intersecting with that–the difference between the something in the universal and particular function. For example, if on one occasion we state, "A triangle has as the sum of its angles two right <angles>", there the "a" is the start of a universal meaning-function: of something determined as a triangle holds generally–, of triangles generally holds–. In other cases, it is also a matter of a triangle in general, but in a different sense, namely, that what we are stating is at least equivalent to the statement, "There is a triangle of such and such a kind." In every proposition of the form "There is an A", this "an" stands out. It is the bearer of a particular meaning-function. Closely related to that is the plural "some"–some triangles are right-angled, some acute-angled–, but only in certain connections. It is somewhat different with the suggestive, in the good or bad sense, "a certain something." For example, if someone caustically or scathingly says, "Certain people would do well to learn good manners", involved besides the "certain people" is also, "namely those there" or "among those there!" We therefore have a confusion with the curious "those", about which yet more will be said. The universally and

120 particularly functioning something or a, the universal function, is as a rule expressed by "every", by "all", even by the ambiguous "generally". This universal function can only come into play if a something is there as bearer of universality.

Many empty places can occur in a judgment and, accordingly, one and the same judgment can have several places of universality–or several "terms of universality"–and several of particularity. These places can be pure empty places, such as, for example, "something or other is red". There, a nominal something is in the subject position and bearer of a particular function. Likewise, universally, "Everything is red". But, as a rule, the "something" is determined as an A or an ABC.... In arithmetic, for example, $a + b = b + a$. There we have two terms functioning universally and determined as numbers: a, b. Everything formal is exclusively composed of terms of this kind. However, that already belongs in the theory of the judgment-forms for which these distinctions in general are fundamental.

<c) The Empty Presentation "This" as an Empty Thought-Form>

I have, however, to discuss yet another important presentation having an empty nucleus. Standing opposite the something in certain ways is the "this" which, when it stands alone, is an empty deixis. Yet, here, that is a completely different emptiness. It is an emptiness conceivably linked in a certain sense to a greatest fullness. I say, perhaps, "This is a mighty tree", and the tree stands perceptually before my eyes, mine and those of my interlocutors. For all of us, what is meant by "this", therefore, something definite, is something completely intuitively given and something meant in this definiteness, therefore not empty. Yet, is this fullness in our thought-matter thought fullness? Nothing at all, however, lies in the thought-meaning "this" that would give its reference precisely to this tree. We indeed also say on one occasion, "This is a tree", and on another "This is a triangle", "This is a number", and so forth. No one will, though, want to affirm that it is here a matter of superficial equivocation, as when the word "cock" is on one occasion understood as a rooster and on another occasion as the cock of a gun. We say "this" everywhere in a common sense. On the other hand, the matter cannot be left to rest there, for the little word nevertheless on one occasion means the tree that we see, and on 121 another occasion a house that we see in our memory, on a third occasion, an objectivity that is only present to us in thought, such as a proposition, a number, and so forth. The meant-This in general can be objective in any way and be present in any way. Depending on the underlying intuition or other presentation, the This may have a different objective direction, therefore, a different meaning. That is beyond doubt as well, and both must be taken into account. One will have to say that the This-presentation is the same presentation everywhere inasmuch as it has an identical thought-meaning everywhere. This thought-meaning, what is common in the deixis everywhere, is however a thought-form that has a different content from case to case. The content is, nevertheless, empty insofar as it does not have any thought imprint and determination. With the This, it is not stated, is not thought whether it is a matter of a tree or a triangle or anything else. The actual meaning that the This has in the given case differs due to the underlying presentation, which does not itself, however, occur in the unity of our stating thinking as an act constituting thought-meaning. If, for example, we remove the definite objectivity that we mean with the This from the tree perception, then the removing exists just simply in the indicating. The This has its orientation therewith, but what has been perceived is and remains perceived. It does not imprint itself in thought. It is not represented in our statement by a corresponding thought-meaning. And even if it is a matter of underlying thinking, as when we judgingly state the theorem of the sum of the angles of a triangle and then add, "This is an important theorem", the deixis refers to what was thought in this thinking, but what was thought itself does not thereby occur as meaning in the unit of the statement. The This is therefore what was displayed in the form of the deixis, but in terms of thought is indeterminate determinacy. The This-presentation is an empty presentation inasmuch as it does not have any full-nucleus when it comes to thought. Its nucleus is protean. It is something different in every new objective direction. It is, however, not a conceptual nucleus, not a nucleus that determines when it comes to thought, that grasps. Everywhere "this" occurs, something is lacking from the standpoint of thinking

and thought-meaning. Insofar as the This merely refers to something definitely grasped in thought, the lack is not a genuine one. It
122 only concerns the individual place, not the context though. One can indeed, when for example it was a matter of the number π, again later say “π”, or “this π” instead of “this”, therefore, link reference and thought-determination, unlike when intuition is what lends the This its meaning-concretion. The meaning of This is then a mixture of thought-meaning and intuition-meaning, therefore a blank when it comes to thinking.

<Chapter 5 122

Preliminary Remarks About the Systematic Theory of Forms of Meanings>

<§ 27. The Formal Theory of Meaning Stays Within the Sphere of Mathematical Generality>

<a) The Concept of Meaning-Form as Configuration-Idea>

Through our discussions up to this point we have acquired knowledge of the most general characteristics of the building of meanings, enough to be able now to approach the higher tasks of the systematic theory of forms. The goal is a systematic searching for and describing of the primitive forms of independent meanings closer to the apophantic thought-meanings and then, in addition, the systematic construction of all possible complexifications and modifications to be formed out of them. That presupposes that everything there is subject to laws, laws that govern the building of all meaning-forms, of all modifications and complexifications they can undergo, in the way laws for operations do. And this is actually the situation, as we shall soon become persuaded.

Needed here to begin with is a short explanation of the now standard concept of meaning-form or, in our domain, of judgment-form and presentation-form. The expression is ambiguous because form very often amounts to meaning a moment that formatively transforms stuff of some sort into formed substance–and in this sense it has constantly been a question of form up to this point, for example, syntactical form or And-form, and so on. If, however, it is
a question of judgment-forms, inference-forms, and so on, then as 123
a rule something else is meant, namely, the Idea of a general meaning-configuration. Meanings are wholes. In every sphere of wholes, we speak of form or configuration, and we can lay hold of the Idea of configuration everywhere, can also consider the question of possible configurations and configuration classes that are

E. Husserl, *Logic and General Theory of Science*,
Husserliana: Edmund Husserl – Collected Works 15,
https://doi.org/10.1007/978-3-030-14529-3_5

possible *a priori* in this object-sphere, such as, for example, sound configurations in the sound sphere, consequently then also for meanings.

If we proceed from a definite independent meaning–more precisely from a definite judgment–, then we acquire the general Idea of the configuration associated with it when <we> replace the stuff that <we> configured there with something indefinite and designate it with letters of the alphabet in such a way that identical stuff is designated by the same letters, different stuff by different letters. (What is indefinite thereby functions in particular meaning.) We call the generalization realized through this mathematical generalization, in contrast to that which is, for instance, present in the transition from species of stuff to genus of stuff–as from red to color in general. We therefore mathematize or formalize the judgment "All humans are mortal" when we form the idea "All A are b". That is no longer a judgment, but the Idea of a judgment-form or judgment-configuration, as we say more clearly. Through this mathematization, specifically formal-logical ideas emerge, namely, ones that have nothing left over of the particular distinctiveness of the full-nucleus and so precisely <nothing> of what distinguishes extra-logical ideas from logical Ideas. That is no chance difference. As long as we remain within the realm of pure form and follow up on the modifications that take place there, we dwell in a sphere of generic essence-kinship. All individuations of the nuclei, that is, of the ultimate full-nuclei, form a realm of their own. The individuations take place in a totally different dimension when we make the transition, for instance, from the concept-content "blue" to "green", or to the concept-content "human being" or "triangle" and so on. The genera and species pertaining to these full-nuclei are something totally different from the kinds of form pertaining to the configuration of such nuclei (or to nuclei in general), and again from the genera and species of configurations themselves that fill
124 any nuclei and hence can concretize.

The formal theory of meaning remains within the sphere of mathematical generality. Indeed, as fully pure theory of forms, it has to leave out of account even the most general differences in the full-nuclei that I have made–such as those between general-nucleus and individual nucleus. One can hold the view that this difference already extends beyond pure mathematical logic, formal logic in the concise sense (and that a universal ontology begins with it that may still be distinguished from the pure mathesis).

Why I am attached to speaking of mathematical, instead of, formal logic will become clear later on. Words are not void of interest. The vagueness and ambiguity <of> the concept "formal" have given rise to many pointless attacks on formal logic, to very much unclear chatter that could have been avoided if one had delved a little into the concept of purely mathematical generalization.

The Idea of the judgment-configuration–for instance, "All a are b", "something is a", and so forth–is mathematically precisely analogous to the Idea cardinal number, for instance "one and one". Here we have the Idea of a sum-type, of a sum-configuration, that can at the same time be used in the general presentation of any sum of that type. There we have the Idea, namely, <the> configuration-Idea of a judgment-type that can also be used in the indefinitely general presentation of any judgment of this form. And mathematics in general, especially the mathematics of the judgment-configurations, operates with such Ideas, or these general presentations.

<b) The Method of Iterable Operation>

In the introduction to the theory of forms, I already discussed the fact that every configuration-Idea acquired through mathematization brings with it a general law of meaning. We can now formulate the situation much more specifically once we have worked out for ourselves the concepts necessary to concreteness. Every independent meaning proves to be a complex of syntagmatic units that in the meaning-whole are brought to a greater comprehensive unit by means of the syntactical forms. And the unification may thereby be achieved in several levels built one upon the other. If this structure is expressed in the transition to the corresponding 125
judgment-form (as configuration-Idea), then every concrete individuation of this judgment-form (which is realized in such a way that a particular syntagma of the same form is substituted for a

syntagmatic form) yields a meaningful coherent meaning and not, say, nonsense. I can also say that every presentation is replaceable by any other one of the same category of presentation. <A> nominal syntagma can thus be replaced by any arbitrary nominal syntagma, any propositional syntagma by an arbitrary propositional one, etc., and so within every syntactical category. (We can naturally ideate the individuation in such a way that we hold on to all forms and replace the ultimate full-nuclei with something indefinite. Then these nuclei are not absolutely freely variable, but bound to the nucleus-categories, i.e., individual nucleus must remain individual nucleus, property-nucleus, property-nucleus, relation-nucleus, relation-nucleus.)

The prior securing of this important conformity to laws governing the entire field of thought-meanings makes it possible, starting from concrete meanings–I say concrete judgments–, to form through mathematization the corresponding meaning-forms, the general validity of which we are certain beforehand. In other words, we need not ask the "existential" question in each new case, namely, whether such a mathematical configuration really has general meaning.

It is time to inquire into the implementation of the mathematical theory of forms. Are we then directed to start randomly from concrete meanings, from judgments as examples, and through their mathematization pile up forms just to see afterward how a system might be introduced into these forms? Such an inductive procedure would not be very satisfying and not very far-reaching. It is, though, also unnecessary, or only necessary to a very limited degree, just as in arithmetic, where the basic operations are first abstracted, say, from examples, and yet the whole procedure (particularly that of producing new formations through iteration) proceeds *a priori* throughout. Precisely with respect to mathematical configurations, the realm of meaning is a realm of pure, continuous conformity to laws. It is part of the essence of meanings that, out of
126 the four basic kinds of full-nuclei, and in accordance with a fixed, circumscribed system of form-producing moments, they form fixed configurations into concrete meanings such that all meaning is bound, so to speak, to fixed crystal-configurations and only so crystallized can have concrete being. These crystal configurations

are the presentation- and judgment-configurations. Every judgment has a crystal structure–as it is itself a crystal. The crystal structure is the formal structure of the components which we have already looked at. Crystals as such have their crystal form and conform to a crystal system. The metaphor is actually appropriate with the one difference that in our sphere there is no amorphousness that the same matter can assume. In our sphere, every crystal must be from the fixed, coherent system.

Now, to find the system of judgment-forms, the following leading thought, which I already used as a basis for my courses many years ago, suggests itself. Proceeding lucidly, we first look for the syntactical primitive judgment-forms, the forms having the least possible syntactical complexity, in which therefore the syntactically simplest presentation-forms attain the unity of a judgment-form in the simplest ways conceivable. While holding on to the form-type of the whole as much as possible, we then subject every such form to an operational or constructive procedure that applies to the components as these are being modified or made more complex in conformity with laws. Therefore, to this end, one must look for the primitive operations and their laws in accordance with which just such transformations of the form of presentations can take place. Since every nominally valid form-complex can be produced over and over, one need not follow up further on the new forms arising through the iteration of the ways of increasing complexity. You understand what is meant with an example from arithmetic. If it is at one point seen that, in the form a + b, an arbitrary number a and an arbitrary number b produce yet another number, then that means that every number in a sum is to combine again with an arbitrary number to make a sum, and consequently from the one form a + b, one can derive infinitely many forms: (a + b) + c, (a + b) + (c + d) etc. That yields a definite system of forms. And precisely the same in the realm of meaning. If it is at one point seen that an arbitrary nominal presentation N can be determinatively combined with an arbitrary adjectival presentation α to form 127
"N, which is α"; "α-being N", which is again a nominal presentation, then this one form $N\alpha$ yields infinitely many ones: $(N\alpha)\beta$, $((N\alpha)\beta)\gamma$, and so forth. Everywhere that the situation is the same, therefore an iterable operation, the only thing needed is the formulation of

the primitive complex and the statement that such a complex can be iterated.

If the possibilities of change in the primitive judgment-forms are exhausted, then one can make the transition to compound judgment-forms, <the> judgment-forms of propositionally complicated syntax and immediately to the genuinely compound judgments consisting of components from which one can detach independent judgments. Since there are now generally valid qualitative and syntactical modifications that all meanings with propositional content can undergo without altering their syntagmatic content, the possibility arises of carrying everything acquired in terms of form-formations on the basis of simple judgment-forms over to the propositionally-compounded judgments. And here too the form-formation of iterable operations is accomplished in ways that can be seen at a glance. That there is also actually iteration there is easily seen. For example, for any two independent propositional syntagmas, let us say M and N, one can conceive of a hypothetical judgment "If M, then N" formed. Through transformation of the syntactical form, we can however turn this judgment into an antecedent. We then appeal to some new propositional syntagma P and can give it the form of the consequent. Therefore, we have the form "If 'if M, then N' is, then P is" and so on *in infinitum*. Likewise, "M is and/or N is", "M and/or N and/or P", etc.

In this way, we therefore see at a glance from the start that both in the exploration of laws of meaning-forms and in ordinary mathematics, one can and must proceed operationally. Naturally, proceeding thus is not, for instance, to be addressed as an empirical grounding of laws of meaning. The operating involved, the actual proceeding, is a subjective fact. But the laws of iteration are absolutely ideal, therefore, atemporal, supraempirical in nature. They concern Ideas and pure relations of Ideas, just as in all mathesis.

<§ 28. Going Back to the Basic Form of Propositionally Simple Judgments> 128

<a) The Classification of Judgments into Propositionally Simple and Propositionally Complex Judgments>

Pertinent to the implementation of the just conceived plan is the classification: (1) definite judgments and functional-judgments; (2) the classification of judgments into propositionally simple and propositionally complex comes into consideration. I have already spoken about the fact that judgments are structured wholes and that, generally speaking, we want to distinguish among components of the first, second, third level, and so forth. If we wish then to go back to the simplest forms, we have to look for those in which the primary components no longer have secondary ones, or have forms that are also to be found in completely primitive components. Of course, we therefore have to go back to one-level judgments. Only in them are the components possibly simple presentations. Now, examples like hypothetical and disjunctive judgments teach that there are judgment-forms whose direct components are from propositional syntagmas. Therefore, these will not initially come into question. One preferably begins precisely with this division of judgments into propositionally complex judgments–into judgments whose direct components are propositional meanings, i.e. more properly stated, are from propositional syntagmas, and into propositionally simple judgments, for which that is not the case. These propositionally simple judgments are then to serve as the point of departure, and only then <is the transition to be made> to propositionally complex ones, and then again first to the simple formations. I note again that it is proper to understand as propositionally complex judgments ones that, like hypothetical or causal ones, are composed of primary components that are propositional (of really propositional syntagmas). For, since according to the law of nominalizing modification, just as any non-nominal meaning can be nominalized, so can a propositional meaning, there are therefore forms of nominal presentations that–having proceeded from nominalization of propositions–contain propositionals, therefore, <for instance>, “That A is b”, and since,

again according to the law that where a nominal form figures
129 within a complex meaning, every other nominal form can be applicable, such a form can always be formed out of a simple form "N is a" that contains a nominalized proposition for the nominal presentation-form N, then in the modification of simple propositional forms, such forms that contain nominalized propositionals in an individual component at once occur and are consequently naturally dealt with in the group of propositionally simple judgments.

<b) The Components of Propositionally Simple Judgments: Dependent–Independent Syntagmas>

We would therefore have to begin with propositionally simple judgments. What is then possible for their components? A natural path to follow is the one previously indicated according to which one looks for the conceivably simplest non-propositional forms of presentations and asks how such forms may unite in the simplest way into a judgment-unit, and how one could then step-by-step make the completely simple presentations more complex while preserving the overall form as much as possible. In my lectures 4 years ago, I pursued a somewhat different path, which also appeared to me very naturally and about which I wish to lecture.

We distinguish between complete (or independent) and dependent syntagmas. A complete syntagma is one that can act as a component on its own by taking on corresponding syntactical forms, or possibly also as a free judgment. In the latter case, it is absolutely complete, otherwise relatively complete. In any case, from the standpoint of composition, it is a completely adequate syntactical stuff for a syntactical form. That holds of every nominal presentation, as well as of every propositional one, of every adjectival one. In contrast, an incomplete syntagma is "similar", in short, any relation-syntagma. If a complete component, for example, a complete predicate, is to come out of it, then we have to say, for instance, "similar to the A", "similar to a horse", and so on. As a whole, this is an adjectival predicate, namely, a complete relation-predicate. Provided the full relation-syntagma is complex, such a thing always allows a nominal presentation to break off from itself.

On the other hand, it is not compound in the sense that it is pieced together. The nominal presentation is one-sidedly detachable. The complementary relat<ion-syntagma> is not also detachable. Now, 130
I call an independent syntagma simply-independent when it is not decomposable into several syntagmas, though an independent syntagma may also be one-sidedly detachable.

We must remain with relatively complete syntagmas for the time being. And if we inquire into their categories,[1] they are as follows: the nominal syntagma and the adjectival syntagma (with which we must probably also associate subsuming having the form "an A" as something proper to it). In the corresponding sense, compounded-independent syntagmas are, for example, "Socrates and Plato"; i.e. every conjunction of nouns, and similarly "red and green"; every conjunction of adjectives. Such conjunctions are not themselves nouns and not themselves adjectives (still speaking meaning-theoretically). We shall later hear about the fact that such conjunctions cannot occur in simple judgments. Naturally, the adjectival syntagmas then decompose into property-syntagmas and relation-syntagmas. Only the former can be simple in every sense, can really be primitive–like "red"–, while the latter precisely always contain a noun in themselves.

[1] Subsequent note by *Husserl*, "In the 1914 lecture I was unsure as to how matters stood with the categories of simple-independent meanings. It is, however, to be said that many kinds of things fall under the heading 'nominal meanings': 'this', 'PN' (proper noun) and 'a certain', 'an A', 'A man' was present. Therefore, the 'an A' (that is not particular) belongs to the nominal sphere. Now, however, on the other side, we have further simple independent meanings: the non-relative adjectival <predicate>, the relation-predicate 'similar to A', and the 'an A', whereby, however, the question remains as to how it stands to the nominal 'an A'. Obviously, it is nothing of the kind. Furthermore, the 'is identical' parts ways with the 'is similar' etc. Therefore, be careful with supposedly complete general affirmations about categories". (Editor's note)

<c) Propositionally Simple Judgments Must Be Formed Out of At Least Two Simple-Independent Meanings in Order to Be Full Thought-Meanings>

Instead of now immediately choosing the most primitive syntagmas as our point of departure, we begin with simple-independent ones. Since as simple-independent component-forms, we only have the nominal and <the> adjectival components, we therefore conceive of any nominal syntagmas or, as I also said, nominal presentations N, N′, ... (leaving it open as to whether they otherwise contain intrinsic syntaxes, detachable parts, and so forth) and again any adjectival syntagmas a, a′,... (without asking whether
131 they are property-presentations or relation-presentations and furthermore are of intrinsically rich or poor structure) and now ask: How do we obtain the propositional-simple judgments that are most primitive in terms of form? Therefore, how do we obtain those with the simplest composition? Actually–if we inquire into the latter, then the only thing that matters is the form, the syntactical category of the components, and not their content, the intrinsic structure of the syntagmas apart from the syntactical stuffs. We leave this content completely undetermined. We therefore try to form a judgment with as few nominal and as few adjectival presentations as possible.

We then first realize that a nominal presentation alone–and even an adjectival one alone–does not yield a judgment. Of course, that is vigorously contested in the case of nominal presentations, by *Marty*, for example, and I myself have wavered on occasion. For years, though, I have espoused the just stated thesis in my lectures. I can now confirm my old view for deeper reasons. It is a matter of judgments that are in fact expressed by a mere noun, for example, "the weather"; "a dog". It seems to me, namely, that the following distinction can be made: Completely generally speaking, a meaning is dependent if by essence it is in need of supplementation by additional meanings. If it is a thought-meaning, it is however not excluded that its need of supplementation is satisfied by meanings that are not thought-meanings. You recall similar discussions regarding This. There are, then, independent logical thought-meanings. Within their logical unit, the need of supplementation of the dependent components is satisfied purely in the logical meaning realm. That does not rule out that–within the specific thought-sphere satisfied by heterological meanings, therefore supplemented into independent things–such dependent components can also occur extra-logically. (The matter would become more complex if, for instance, it became apparent that every thought-meaning, as it were, presupposed an extra-logical meaning as foundation. But then we had only to eliminate this foundation and indicate different kinds of supplementation).

I need only show what I mean here for nominal ideas. From the standpoint of the logic of thought-meanings, nominal presen- 132
tations are dependent meanings, namely, because as thought-meanings they are not completed. Counterexamples are judgments such as "the first snowfall!" or, when looking out the window I say, "Mister Müller and Schulz". Here the nominal thought-presentations are based upon the sensory intuitions of the objects concerned. However, they are logically dependent, because what I mean, but do not actually express is, for instance, "The first snow of the year fell just now", or, "Messieurs Müller and Schulz are walking by!" Undoubtedly, the supplementing presentations and the thought-presentations just now expressed with the supplementary words are not realized. Other, non-logical, ideas take their place. But only when we "express", as people are in the habit of saying, these extra-logical presentations, i.e., provide suitable thought-presentations for them and coin statement-propositions in the unit-form of non-complete ones, only then do we have complete thought-meanings, only then have we expressed, namely, in thought, and expressed *everything*, that we meant.

In this sense, therefore, isolated, or strung together, conjunctively combined nouns are not independent meanings and, despite the affirmative quality, not logical judgments, by which we precisely understand completed thought-affirmations. And with this the matter of *Marty*'s reference to *Uhland*'s well-known poem is also settled[1]:

---

[1] Compare Anton Marty, *Untersuchungen zur Grundlegung der allgemeinen Grammatik und Sprachphilosophie*, vol. 1, Halle a. S. 1908, p. 279. (Editor's note)

The green of crops, violet’s fragrance,
Lark’s warble, blackbird’s song,
Sunny rain, gentle breeze!
If I such words sing,
Need there then be yet greater things
To praise thee, Spring day?

In like manner, it would be natural if, overwhelmed by the sight of a magnificent work of art, we cried out “Splendid! Wonderful!” What is meant is that the work of art is magnificent, or one thing or another about it is magnificent. It is meant, but not meant with this completeness in thought. Much is lacking for the completeness of a thought-affirmation, of a logical judgment.

133 Therefore, we stick to this: neither a nominal presentation separately, nor an adjectival one separately, yields a logical judgment. If *Brentano*’s school concluded otherwise that was owing to the false interpretation of existential propositions and to the mixing of simple existential affirmations, of “existential judgments”, with nominal positing. *Brentano* meant that a nominal presentation separately (which he thought of without any affirmation-quality) sufficed for judgments, while the nominal presentation became the foundation of a judging approving, of a belief, but that was expressed in propositional form as “A is”, for example, “God is”. And precisely this judgment, only functioning differently, is expressed when we begin, for instance, in a predicating statement, “The good God–”. I shall still have to talk about existential judgments. One thing is certain, the fact that when it comes to their matter they have two components and that “exists” is a predicate. However, I must postpone everything else for the later analyses.

<Chapter 6

Propositionally Simple Judgments> 133

<§ 29. The Basic Form of Propositionally Simple Judgments and Iterable Operations of Negation and Affirmation

<a) The Traditional Understanding of the Negative and Affirmative Quality of Categorical Judgments>

Moving on now, we realize that in any case a nominal and an adjectival presentation suffice for the unity of a judgment. Individual simple-independent meanings are not enough for any actual independent meaning (which is a complete thought-meaning). But, two suffice for this purpose, namely, in the form "N is A"–in the subject position, a nominal meaning, in the "predicate position," an arbitrary nominal or non-nominal meaning. In this way, we acquire the basic form of a propositionally simple judgment.[1] It is, as far as that 134
goes, the traditional understanding, as actually passed down, to treat the "categorical judgment" as basic form. However, the tradition goes further and immediately puts two basic forms here, "S is P," "S is not P." These are forms designated by the class-name "categorical judgments," which, however, also designates other forms related in terms of general ways of composition, especially the plural judgments to be discussed later. What is characteristic of all categorical judgments is that, grounded in a compound nominal positing or a multifold (pieced together) nominal positing as basic positing, is a positing–single or multifold–built upon that, namely,

---

[1] Subsequent note by *Husserl*, "To be placed at the beginning of the theory of propositions is what is to be shown regarding categorical judgments: all judging is identifying. Compare the unnumbered pages in the theory of existential propositions in connection with *Bolzano*, before 155." Compare §40 f-g below. (Editor's note)

E. Husserl, *Logic and General Theory of Science*,
Husserliana: Edmund Husserl – Collected Works 15,
https://doi.org/10.1007/978-3-030-14529-3_6

in such a way that the positings are positive or affirmative or negative. The first component is the subject component. It posits the subject-object, which is conscious noematically as the substrate of the subject-positing. And the predicate-positing is built upon that–whence the break. The noun that functions in the subject function–through which the subject-positing therefore goes as *positing-under*–functions for precisely that reason differently than the same noun in a different function, for instance, an object function, whereby, the positing performed there is precisely not *positing-under*. The noun is not "*subjectum*" (ὑποκείμενον). Do not confuse subject-presentation and subject-object!

The traditional understanding is such that in our language one would have to say: Out of two syntagmas S and P, a judgment can only be formed in the categorical forms indicated, "S is P" and "S is not P";–and that these forms are thereby coordinated with one another, are to be treated as equiprimordial forms. However, I am certain that that is not right. The difficulty or fallacy lies in the form of negation, and in general it creates an abundance of problems.

Negation seems to be a peculiar operation, namely, one that can, so to speak, take hold in different ways. It is a form that is directed toward, or rather against, something, more precisely, against a positing. If somebody judges, "S is P", then I can oppose that and say, "S is not P". One is at first inclined to express that in this way: I oppose P, namely, I am against the S's positing-as-P, therefore, against the predicate-positing. Meaning-theoretically speaking, I am not objecting to positing by somebody, of whom
135 there is really no question in the statement, but to something posited as such, to a positing of something. –If I think the S as P to myself, I am positing this in relation to the S–it is P. Thus, I say to myself "no", and that no holds just for that positing, which lies on the predicate side. Accordingly, the thought is not simple, not single-layered as in "S is P". The thought of negative predication <is> more complex than that of what is called affirmative (or, better, positive) predication. One must indeed not, as has so often happened, treat them as equal, nor must one in the case of a judgment "S is P" take as a basis a positing and a yes directed to it, any

more than a no directed against it in the other case. If one does that, then a new form arises: "S is however P". And one must bear in mind that, say, the "not" does not counter the entire statement, as if this thought had been "thrown away" in the negative categorical judgment and in a similar fashion "approved" in the affirmative one. The "not" lies completely exclusively on the side of the second main component of the judgment, of the predicate component. The first component is only affected by that in just the way every occurrence in the second component must affect the first. Consequently, it is completely appropriate when with the not, language, as it were, introduces a negation sign before the P, but does not use any sign of affirmation in the mere positive judgment "S is P". On the other hand, it however seems apparent to me that negation can occur not only in the second, but also in the first component. Every component of the categorical judgment has its positing, is positing of its matter, and every positing (being posited as such) can be negated. Moreover, the not can then occur in the unit of the component as form, for example, "Not Plato was the teacher of Alexander". This judgment may be equivalent to the judgment "Plato 'was not the teacher' of Alexander". But it has a different meaning. It therefore seems that we acquire two more forms, "Not S is P", "Not S is not P"–the subjective "not" also affects the remaining predication as well. Like every "not", this one also indicates a layering. All negating judgments are layered when it comes to their meaning, therefore, are not, like positive judgments, one-layered judgments.

<b) Plain Assertion as Opposed to Negation and Affirmation>

While the traditional understanding says that every categorical judgment has a twofold affirmative and negative quality, approv- 136
ing and rejecting, ascribing and denying, then for various reasons I would, therefore, have to counter it.

Every logical judgment is an affirmation, and in this sense there is only one quality for judgments. Of course, subjectively speaking, a judgment can turn into a negating of the same propositional matter. I can, initially wavering somewhat, become doubtful, then resolutely say no. I engage in negating that is directed against the only just posited "S is P"! But if I express this negation, then I affirmatively predicate, for instance, "The proposition 'S is P' is not true" or, "That S is P is not so". And these are again affirmations, but with altered matter. The consciousness of saying no only occurs in the logical realm when its no has become expression, meaningful sense. But that is only possible in the form of an affirmation. We become aware that everything that is not thought-meaning can be expressed by thought-meaning. Logic, though, reduces everything to affirmative expression. And it only deals with affirmations in general. What I call "judgment" in it is therefore always an affirmation. Now, it can be that we act negatingly as well as affirmingly toward a proposed judgment (toward any full positing). If, though, I express the affirmation, then I again have something different from the plain affirmative proposition. It is for its part expressed in an affirmation of the content "That S is P is actually so". Again we have affirmations of complex, layered matter that are not to be mixed up with the plain affirmations "S is P" or "S is not P". And again one must not confuse equivalence and identity.

The qualities of the components are to be differentiated from the quality of the affirmation that belongs to every apophantic judgment taken as overall meaning. And there too, one must take care to distinguish between the quality, which really enters into the judgment meaning, more precisely, into the matter, <and the> not-quality, which one is inclined to construe into it in reflection. If we say judgingly, "The teacher of Alexander…", then no affirming "approval" is expressed in that. Plain positing, plain position, belongs to the meaning, though. On the other hand, if we state "… was not the teacher of Alexander", the "not" belongs to the matter
137 as expression of a negating nullification and, of course, to the matter exclusive of this component. Thereby, the expression "negating nullification" obviously does not mean nullifying, but a meaningful correlate of something belonging to the meaning itself. It is wrong and incorrect when it is said (by various representatives of *Brentano*'s school) that every categorical judgment is either granting (ascribing) or confuting (disputing). These expressions are only suitable for categorical judgments from the point of view of

the predicate component, not however for the Not, which is directed against the subject-positing.

To consider would also be the following. Just as a negation is directed against a positing, so an "affirmation," in the genuine sense of the word, "consent," is directed toward a positing. It says yes, as the former says no. If negation is then expressed in the matter as "not," why should the yes not also have a correlate in the matter? Nevertheless, we actually judge, for example, in the dispute, "S is however P" or "S is in fact P, it is indeed really P," the latter as form of the confirming predication. "Aristotle was however Alexander's teacher." But also in the subject, "However, Aristotle was the teacher etc.," where the word "Aristotle" is still stressed.

The forms grow more complex when we bear in mind that we can also place the form "S is P" alongside the form "S is however not P," and again, "S is not not P." In short, we see that actually, according to the tradition, "S is P" and "S is not P" are not to be placed alongside one another as the presumably coordinated two "qualities" of the judgment, but that initially we have the one original form, which does not yet contain anything about yes and no: "S is P," the originally positive judgment, and in addition two iterable operations, the operation of negation and that of affirmation. Each of them can attach to every complete judgment, be directed against or to the overall affirmation of it. And again, it can attach to each of the two components, provided each has its own quality. And every newly formed component can again be subjected to these operations.

When we say that negation is directed against a proposition, against a component-thesis and its affirmation-quality, it is evident that this operation does not add a new one to the affirmation-quality, but it is modified in a remarkable way. It "strikes through it," but
in such a way that the striking through still (but in the modifica- 138
tion, to be precisely striking "through") is discernible. Similarly, we can say figuratively that the affirmation underscores, brings something in, even a new intermingling, but one that at the same time retains the old quality.

Naturally, I also reject the misleading talk about the copula. Since time immemorial, the "is" and <the> "is not" have been called copula of the categorical judgment, the S and <the> P are joined by a copula, or they are conjoined by the "is", separated by the "is not" (as *Aristotle* expressed it). However, it is to be observed that the situation here is nevertheless completely different than in cases of actual conjunction, say "S and S′". The syntax characteristic of every categorical judgment consists in the fact that on one basic positing another is built. This remarkable two-tieredness, in which one component rests upon the other, is retained through the iterating operations of consent and refusal. Therefore, in genuine categorical judgments an original positing or, more specifically, a plain positing thereby occurs in every component. A cancelation, or a consenting positing, or its cancelation, and so forth, can be its lot. Owing to the fact that these operations follow the original syntax, the two-tiered form is retained.

<§ 30. The Plain Is Between Two Nouns as Distinguished from the Relational "Is Identical" and from the Adjectival Is>

I make a broader differentiation within the single-layered, positive categorical judgment. The P, the predicate-presentation, can have the form m (property-presentation) or the form ςM (relation-presentation). Therefore, "N is m", "N is ςM". What presentations can then in addition figure on the predicate side apart from that? In any case, a presentation of the form "an A". But would that not be a nominal presentation, since we can have such a thing on the subject side? We shall come to doubt that. Does not the empty nucleus something occur there? To begin with, let us think about whether in general a nominal presentation can function here and leave the empty nucleus out.

Up to this point, I have made a synthesis of a nominal and an adjectival presentation. Let us now try this with two nominal pre-
139 sentations, however, in such a manner that one does not act as a constituent of a more comprehensive one, even if simple-independent predicate. Therefore, we are asking for a judgment-synthesis that merely combines both nouns through mere synthesis. Language provides us with examples here, such as "Alexander is Paris", "Plato is the son of Ariston". Here two nominal Ideas are simply joined by "is", and it is immediately obvious to say that this "is" means "is identical". We would therefore have the form "N is

identical with N_1", usually interpreted in such a way, namely, that <the> identity-judgment is understood as a special case of the relation-judgment.

I am not completely convinced that that understanding is right. It seems apparent to me that the identifying "is", which in fact has an obviously different meaning from the adjectival is, is not always short for "is the same", "is identical with". It seems apparent to me <that> substitution of the complex expression also produces a more complex thought. If, with regard to familiar shapes of analytic geometry, we say, "This is the midpoint equation of an ellipse", then we do not mean, This is identical with the midpoint equation. The statement "This is the newly named mayor" does not say, "This is one and the same person as the newly named mayor". On the other hand, we shall say, "The victor of Jena is the same person as the loser of Waterloo".

I can only point out here that the identity-thought is originally a form of the identifying joining of propositions, for example "A is α and the same thing is also β".[1] And the relational "the same" refers to such an identity-conjunction in the conjunction of names. If someone makes an affirmation about Alexander and again an affirmation about Paris without noticing that he or she is stating something about the same person, then we shout out, "Alexander $\equiv$ Paris". If, however, we are looking for the midpoint equation and find it, then we say, "This is the midpoint equation". In any case, the plain Is between two nouns seems to me to be a form of its own. It is indeed an identifying Is, but despite the sameness of expression is, on the one hand, an expression of difference in meaning in comparison with the adjectival one and is, on the other hand, also to be 140
differentiated from the relational form "is identical".[2] There is equivalence here too, though, inasmuch as from the standpoint of validity we can take the plain "is" for the "is identical".

Should we then ban the form with "is identical" from the theory of forms, perhaps, because it is a special case of the relation-form.

---

[1] Compare Appendix IX, *Plain Predicating and Actually Identifying Predicating*. (Editor's note)

[2] Subsequent note by *Husserl*, "For the theory of identity judgments! Cite negation, 'not-identical' – origin of the basic concept 'different'" (Editor's note)

No. We must include the form "N is identical with M" in the formal theory of meaning alongside the general form "N is ςM", insofar as the relation-nucleus "identical" is a concept having specifically logical character. It is a form-concept and, for this reason we could also call it a formal, non-material relation-nucleus, insofar as here too, just as with the empty something, there is no fullness–nothing that lends this relation-concept materiality in that specific sense.

<§ 31. The Iterable Operation of Attribution>

I now introduce the operation of attribution. Every categorical judgment "A is b" can be nominalized so as to form "A, which is B". Obviously, the identity-thought is thereby hiding in the "which". We can also express the nominal presentation with the words, "S, is the same as P". Hiding in the attribution A, therefore, is a proposition that has become dependent by virtue of the identifying connection. But, of course, the paraphrase we have given is precisely a paraphrase.

Now, if S is a completely arbitrary nominal presentation, and P, Q... are arbitrary predicates, then we can use all the categorical propositions to be formed here attributively and thus form the forms $S_{p'}$ $(S_p)_q$, and so forth, after the fashion of an iterable operation. In traditional logic, this kind of complexifying connection of predicate-presentations is called determinative. The nominal object is determined in stages. In so doing, one usually has in mind general conceptual predicates. However, both the attributions and the predications can also be identifying–and in general completely arbitrary. The reverse of this operation, doing away with determin-
141 ing predicates, is traditionally called abstraction, although this meaning of "abstraction" is not to be mixed up with other meanings of this word. The negative judgment can also be used determinatively: out of "S is not p" comes "S, which is not p".

<§ 32. The Introduction of the Empty Presentation “Something” into the Judgment-Matter>

<a) The Something as *quidam* and as “Something in General”>

New forms arise out of the earlier forms when we introduce the empty presentation something into the matter of judgments, when we replace the S, the form of nominal presentation in general, by the nominal empty presentation something, which, however, owing to the nature of its form itself counts as form. We, therefore, form “Something is P!”, likewise, “S is something”, so “Something is something”.

In the theory of elements, presentations, and their nuclei, I already spoke of the fact that the something figures in different functions and can undergo remarkable modifications in its meaning. I am speaking about the difference between the *quidam* and the “something in general”, where the latter can again be meant as particular and universal something in general. It will become apparent that the universal and <the> particular proposition-form make up an entirely different dimension of propositions and, despite all appearances, do not have the plain, one-level structure that the earlier forms displayed.

Of particular importance is the construction “something, that is a” or “an a”—where a is a general term—naturally supported by the negative form “something that is not a”, “a not-a”. This “not-a” is what is called the negative concept—a not-red-thing, something-being-not-greater-than-N. Naturally, these forms are also iterable: an ab, (an ab)c, ... That then yields various judgment-forms, the simplest of them “An a is P” and “S is an a”.

To begin with, let us consider the first <form>. The “a” can thereby have different empty meanings, and it is obvious that they essentially determine the whole meaning of the judgment and in it again the meaning of the copula “is”. (1) “A certain a is P”, for example, “A man is ringing the bell”, “Something is running there”. We now exclude the suggestive “a certain something”, which expressed completely logically signifies, “a certain s, namely, that one there”. Upon closer inspection, we see that after excluding the 142
concealed deixis, an empty term still remains left over as a “certain

something". (2) "An a in general is P", namely, in the sense that is to be equivalently expressed by "An a occurs that is P", and in contrast to this particular form, the "universal" form, "An a in universality (every a) is P".

Here is an example of particular judgments as opposed to *quidam*-judgments. We hear for the first time, for example, that a planet may have been found between Mercury and the Sun and, astonished, we then judge, "A planet is within Mercury's orbit!" Similarly, if we were to see a white raven and say, "A raven is white!" In such cases, we shall also use other expressions, "There are ravens that are white", "That a raven is white occurs", "There is a planet in Mercury's orbit", "That a planet is in Mercury's orbit holds". It is debatable whether these forms with "there is", which people include among existential statements, are identical with the particular statements or are merely equivalent to them. I shall later speak about this at length, since this is a subject of the most important disputes of the science of logic. In any case, the connection with the forms with "there is" shows that particular judgments are to be distinguished from *quoddam*-judgments, since the "there is"-form is not applicable in the case of the latter. Indeed, for "Someone is outside", I shall not say, "There is a man who is ringing the bell outside". The way of judging is obviously essentially different in each case. If we initially have a judgment with nothing but definite components and we replace one of them with a *quoddam*-component, then nothing else changes in the mode of judgment. Thus, for example, it does not, in this respect, make any essential difference whether we say "Someone was there", "This person was there", and "Mister Schulz was there". However, giving the matter closer consideration, we shall see that the introduction of generality-components fundamentally modifies the mode of judgment as universal and particular. Therefore, one must say that the indefinite, universal mode of judgment contrasts with the definite, particular one.

<b) The Subsuming Predicate “an a”>

Let us now turn to the form “N is an a.” It is just that it is now very important to focus more closely on the indefinite article here and its equivalent in a relative clause. In addition, there is also a 143
free something as freestanding predicate. Therefore, “N is an a,” “N is something,” “N, which is an a, is b.” Is that a particular or universal an? Or, does it have the sense of “a certain something”? There we realize we are facing a new sense of the “an” or “something.”

If we say “A raven is white,” “Some people are wise,” “A triangle has as the sum of its angles, etc.,” then we have general-judgments. The A in “a raven,” “a triangle” is an empty place of a function. If we say “This is a raven,” then we have apparently the same meaning figuring in the subject, then in the predicate position. But, in all similarity an essential difference nevertheless obtains. The form of particularity enters into the first meaning. In the second, it is missing. Indeed, the “a raven” functioning as predicate cannot at all take on the form of particularity any more than that of universality, even though the empty form “something” figures in “a raven.” In any case, nothing of a layering of the whole judgment, which nevertheless belongs to the indefinitely general judgment, is to be found here. But, such a predicate cannot take on the form of the *quidam* either. We can admittedly say, “This is a certain raven,” but we realize the essential difference. Then figuring in the predicate position is a nominal presentation that could be transplanted precisely as it is into a subject position. The “is” in the proposition “This is a certain raven” is in fact the Is of identification. We can say equivalently, “This is identical with a certain raven.” What is characteristic about identification is that nominal presentations figure in the subject and predicate positions, which precisely as nominal presentations–holding on to the other judgment-form–are interchangeable. Here, however, in the example “This is a raven,” we cannot put “a raven” into the subject position without changing the whole judgment-form. It consequently becomes clear that, strictly speaking, the predicate “an A” is not at all a nominal presentation, but a presentation-type of its own. And consequently, the An also functions in a special way here–and with the An the “is” again standing, so to speak, in “agreement” (*Einverständnis*)

with it. The Is is not adjectival, but a classifying, subsuming Is, and we call the predicate itself a subsuming predicate. It is part of its
144 essence to call for a property-nucleus or relation-nucleus.

<c) Propositional Functions and Their Quantification as Particular or Universal Judgment-Matter>

Now, on the other hand, it should not however, for instance, be affirmed that no particular and universal terms can figure in a predicate. Naturally, generality-positions, or as I have already called them, arguments, terms of universality and particularity occur in the predication as a result of the fact that (namely, in the sense of the form-formations realized up to this point), nominal presentations can occur in it in multiple ways. We can state the principle: Every nominal position in a judgment-form can become an argument place and take on the generality-forms in relation to it. Argument places are specifically nominal forms.

Let us clarify this for ourselves with the form of the indefinite-general judgment "An a is P", which ought to mean "an a in general" and not "a certain a". For example, "A triangle in general has as the sum of its angles two right <angles>". This "in general" can however be understood in two ways. In the example, we understand it universally. We shall then readily say: Every triangle, completely generally–or plurally, all triangles. What is intended in propositions of such a form can, though, also be different. For example, we know from general theorems that in application to a given figure a vertex moves to infinity. There are, however, several in the figure. And it is also not at all ruled out that several actually move to infinity. But, we only know certainly: "in general one" moves to infinity. If we then state this, we again have the verbal form of generality–the fact that, in general, a vertex moves to infinity–, but the in general does not then have the meaning of universal generality. We cannot now instead say: every, all. Rather, we can only say in paraphrase, "There is one that moves to infinity" or "The fact that in general a vertex..., is certain, holds". In so doing, it is clear that the "one" does not here have the sense of "a certain", that on the contrary <the latter> can never be affected in an equivalent way by such paraphrasing. Therefore, we have to make a distinction between two kinds of generality: besides the universal one, "particular" generality.

Where, then, one or the other occurs, it is evident that the mode of judging is completely unique. The generality obviously affects the Is and the whole mode of predication, the way in which the 145 predicate is attributed. For example, "An equilateral triangle is generally equiangular." The "generally" pertains to the "is," but the "is generally" is not merely another form of Is. Rather, it expresses the fact that equiangularity is generally attributable to an equilateral triangle. Talk of an equilateral triangle has, however, thereby been modified in a unique way. No definite subject is indeed there, and what figures there verbally in the subject position–"an equilateral triangle"–obviously functions completely differently than a nominal presentation usually does, for example, as when "this...," or "a certain..." was said.

At the same time, it occurs to us that we are so very inclined to paraphrase, "The fact that an equilateral triangle is equiangular is in general the case, holds generally," and in the particular case as well. "That a triangle is obtuse-angled in general occurs." Now, one is certainly persuaded of the fact that these paraphrases are precisely paraphrases, and every paraphrase changes not only the words, but it changes the meaning. In the new forms, the subject is, "that a triangle is obtuse-angled," "that an a is a P," and as predicate we have, "that holds generally," "holds universally and in the particular case." One will not be able to say that this organization of the meaning is really present in the original judgment, "An a generally is b." Therefore, only equivalence obtains, not identity. On the other hand, the paraphrasing in connection with the understanding of the remarkable manner in which the generality affects the entire makeup of the judgment alerts us to a certain remarkable layering in the meaning, in the syntagma, which runs through the whole syntax. This layering is, so to speak, predicatively unfolded in paraphrasing forms of speech. If we judge, "S is p," "Socrates is a philosopher," then we are judging on one layer. We have a nominal basic positing, "Socrates!," and the predicate is posited on it. If, however, we judge "Humans generally–or a human generally–rank(s) among mammals," then the double layering is expressed in the paraphrase. The fact that a human is a mammal

holds completely universally, is so generally. The fact that a triangle is obtuse-angled holds in particular cases, happens. The propositional thought "An a is b" is lower layer in a curious way, and this layer is characterized as a whole in a twofold manner by the "being
146 valid generally."

If we now ask what is figuring in the subject position here, what therefore makes up the lower layer, which in a curious way, at one time, bears the universal, at another, the particular generality-trait, then it is apparently a complete proposition, "'An a is b'–that holds generally. But, upon closer inspection, that cannot hold. In its meaning, an actual proposition is complete in such a way that, with respect to its empty places, it is either a particular or a universal proposition or even, when the empty places are *quidam*-places, a <proposition> determined in this respect. In our case, though, whether the form "An a is P" ("An equilateral triangle is equiangular") that functions as lower layer is to have particular or universal validity is first decided by the characterizing upper layer. Therefore, a something that looks like a complete proposition, and is stated grammatically as a complete proposition that is not, though, in actual fact a proposition, lies hidden in every universal or particular judgment.

We convince ourselves beyond doubt that it lies hidden there in the following way. It is after all evident that to be compared with every universal judgment–for example, "An A generally is p"–is an exactly corresponding particular judgment, only differing from it by the fact that the generality is changed into the particular meaning. If we then compare both corresponding judgments, it is evident that they have a content in common, a propositional "An A is p" that precisely usually leaves open whether it is the content of a universal or a particular generality-validity. The paraphrasing manner of speaking, "That an A is p holds generally," "...holds in special cases," just brings this content and the different characterizations more distinctly to the fore. We thereby also see that, by their essence, the predicates arising here, "holds generally" ("holds universally," "holds in individual cases") only make sense for such specific contents and not for actual, complete propositions. (One must not allow oneself to be deceived by talk such as, for example, "The theorem of the sum of the angles of a triangle holds universally". When the content indicated is not meant, that can only signify here that the proposition belongs in the class of universal validity of predicative propositions, or expressed differently, it belongs to universal judgments.)

It is now to be further observed that the curious constructions that we have been able to extract from the content of particular 147
and universal judgments are in one respect not at all anything new for us. Talk of judgment-forms or judgment-configurations, meaning-forms has really been steady. They are simply in truth nothing at all different from such substrata of particular or universal judgments (or parts of such substrata). What is new is only the analysis of these judgments themselves, the knowledge that figuring in them are forms of this sort characterized by the mode of validity of particularity, in one case, and of universality in the other.

Now, borrowing from the language of mathematics, I call these curious formations propositional functions. Those propositional functions that arise out of categorical judgments because full terms were replaced by empty terms in them are called categorical functions. It will in particular become apparent that there are as many simple functions as there are in general forms of complete propositions and syntagmas. The concept of function is in fact not just to be understood in relationship to propositions. So it is that we speak, for example, of nominal functions, "an a", and likewise, "an a and a b". The term function derives from mathematics. In particular, it became apparent, namely, that when mathematicians speak of "functions", for example of the function x + 3, they have precisely such meaning-forms in mind. x + 3 as a function of whole numbers says nothing other than "any whole number + 3", but in such a way that the "any" still leaves it open as to whether a particular or universal mode of judgment is to be made with the expression. The empty places are what mathematicians call arguments, a very convenient word, which I accept.

In my manner of expressing myself, propositional functions can then be quantified in different ways, namely, in a twofold manner. Taken in and with quantification, functions make up a complete judgment matter or a complete propositional matter (syntagma), which in terms of affirmation can be qualified as a judgment, or in the manner of a seeming, question, etc.

<d) Function-Judgments and Definite Judgments with Empty Places>

Of importance is yet another characteristic difference between judgments that are quantified and those that are not, therefore, lack function-layering–definite judgments. If we look at definite
148 judgments, we find something posited in each nominal term. "The Kaiser visited Prince Heinrich"; "Berlin is a big city". If we then look at judgments with empty places, there are two possibilities. The empty places may have positing-quality, then together with its possible attribution, the something has an "actual" nominal function, i.e., we have an actual, though incomplete, proposition, an actual judgment-component that stands out syntagmatically, for example, "A man rang the bell". There, the indefinite element is posited, and "a man" is an actual proposition-component, partial proposition, actual bearer of a positing. As against this, if we have a universal or particular judgment, it may analogously grammatically completely read there, "An equilateral triangle is equiangular"; but here the grammatical subject "an equilateral triangle" is not a component of the judgment, not a partial judgment of the overall judgment, not something posited as a foundation to be posited upon. The "an" and the attributive whole "an equilateral triangle" do not bear any positing here, but taken as a whole, the posit-less function "An equilateral triangle is equiangular" bears an overall characterization in the generality, and the affirmative positing is attached to this characterization exclusively, and without organizing the elements to form a whole. However, the place of attachment of this characterization or quantification is in the function, namely, in the "an", which acts as an argument. Once we have made this clear to ourselves, we can describe the difference between the *quidam* of other empty places in a clear-cut way. Every empty place functioning as the bearer of a positing is a *quidam*-component. Every empty place that does not and cannot bear any positing is an argument and is the bearer of a quantification in the generality-judging.

<e) Criticism of the Traditional Classification of Judgments from the Point of View of Quantity>

You understand here how wrong it was for traditional logic to say that all judgments are determined from the point of view of quantity and are to be classified into three similarly arranged classes: universal, particular and individual (singular). The expression quantity came into traditional logic because it completely unjustifiably interpreted universal generality as set-generality and in parallel fashion attributed the thought of a closed set to the thought "all A." Since I shall speak about totality at length later on, 149
I do not wish to go any further here into what is called extensional logic. I came to the conclusion that not all judgments are quantitative, or what amounts to the same thing, not every judgment has arguments, therefore, a function implicitly in itself, which, as it were, is assessed in terms of universal validity or particular validity. There are judgments without arguments. I call these definite judgments, for example, "Bismarck was the greatest statesman of the 19th century." Judgments that include a function are called function-judgments. And now we have a clear-cut division of all judgments into function-judgments and definite judgments. Harkening back to extensional logic, the former can also be called quantifying, the latter non-quantifying. However, I do not accord these terms any particular value. (If our language had simple words for clearly distinguishing between universal and particular generality, we could also call all function-judgments general, judgments that judge "in generality," that contain the curious form of generality.) A further classification of the judgments with respect to universality and particularity naturally occurs on the side of function-judgments, on the side of definite judgments depending on the terms, provided it is possible that they either include only terms with full-nuclei or even the This or finally an empty place, but with positing, a *quidam*.

<§ 33. Steps of Form-Formation>

You see what an infinite wealth of forms is included in the previous ones from the fact that every possible new form yields new nominal forms when we transform predication into attribution, furthermore, that every nominal form having the form of the relation-predicate ϛN can be put in the place of the N, that we can then bring the different modes of generality into every such N-place, and so forth. A generality-term (argument) can already be brought into one and the same judgment-form an arbitrary number of times in this way, with the result that we see that one and the same judg-
150 ment can at once be universally and particularly indefinite[1] and that it is consequently fundamentally wrong to divide up judgments in this respect. We can only divide judgments up into ones having generality-terms and ones not having them, for example: A straight line intersects a circle at two points, at one point or no point.

In the multiplicity of established forms, the indefinite N always occurs as mere general form of a nominal presentation. This N could then itself already have all the forms we have found to be nominal up to now. We now take yet another step in formation of forms when we proceed in the opposite direction and stipulate that the N is to have undergone the reversals of the operations concerned. The N can contain attributions. We now perform the so-called abstractions in reverse. Then we ultimately come to an N that no longer has the form "x, which is a" or "ax". It is then necessarily merely a free something or This. Or, it is a non-attributive full-presentation. The nominal presentations therefore generally fall into attributive and non-attributive ones. The non-attributive nominal full-presentations are either original nominal full-presentations, i.e. individual proper-presentations, or they are nominalizations of other presentations (perhaps, complete propositions), so, for example, depending on the kind of full-nucleus, "redness", "similarity", "being a house", "being red", being similar". However,

[1] Subsequent note by *Husserl*, "Compare this to 119, 2nd page <§36a>. One and the same judgment can surely contain terms of universality and particularity, but are these arranged in a similar fashion? Do not the terms of particularity fall under those of universality and perhaps the other way around? That is important and to be studied". (Editor's note)

there are yet other nominalization-forms that we have not yet been able to encounter up to now.

<§ 34. In Terms of Meaning, Plural Judgments Are Propositionally Simple Judgments with Multifold Subjects or Predicates>

I now proceed to a series of new judgment-configurations, in particular, to begin with, <to> configurations of genuine categorical judgments in which, along with the new formations that occur on the subject and predicate sides in the categorical basic-form, the form of the so-called copula is also changed. Out of the singular Is, what is plural in general becomes what is multifold. We can deal with all the forms to be discussed now under the somewhat broadened heading of the plural.[1] 151

The characteristic basic type of categorical predication, the double layering discussed earlier, is thereby retained, and the multifold nature of copulation spoken of here does not involve making compound propositions. We are truly in the realm of forms of propositionally simple judgments. Naturally, not meant either is that multipleness of copulation that occurs in the judgment-form through each repeated attribution inasmuch as a relative clause and its own "is" truly figures in each one as distinct from the main copula of the overall proposition, whose categorical form is "S is p." Rather, what is now remarkable is that this main copula itself–for example, by being expressed grammatically in the plural form–harbors in itself a certain multiplicity, namely, as a peculiar basic form that consequently gives rise to the establishment of a basic type of judgments as plural.

I am speaking here of plural judgments in the extended sense. Language does not in fact always express the theoretically ubiquitous analogous multiplicity of copulation by the plural form of the auxiliary verb. We say, "M and N are p," but not, "M are p and q."

[1] Subsequent note by *Husserl*, "Since plural judgments refer to conjunctive or disjunctive connections of judgments, it would however probably be consistent to deal with them first after judgment-connections or the propositionally complex judgments as 'concentrations' made of them." (Editor's note)

You see that here, in one case, a conjunction occurs in the subject-component, a conjunction of two subjects, in the other case, on the predicate side, as conjunction of two predicates. However, <the> disjunction runs parallel to the conjunction, therefore, for example, “M or N is p” and again “M is p or q.” And even there this multiplicity obtains. To be noted as well is the fact that entirely arbitrary nouns can be combined in the form of conjunction or disjunction on the subject side, and completely arbitrary full predicates on the predicate side.

We first look at the predicate forms. The propositions “The Platonic dialogue the *Protagorus* is genuine and dates back to the philosopher's early period,” “π is an algebraic or transcendental number,” “M is p and q,” “M is p or q” may serve as examples for this. It immediately suggests itself to say that those are compound propositions, that the judgments are propositionally compounded. What is said is that the *Protagorus* is genuine and that it, namely the *Protagorus* again, dates back to the early period. The other example says that either π is an algebraic number or π is a transcendental number. By delving into the judgment-thought as it is realized in the stating of the original statement, we assure ourselves, however, that, *de facto*, the subject does not appear in it
152 twice. Now, one would like to say that that is merely a grammatical abbreviation, that there are really two predications, but not with the same subject in both cases. Rather, in the second case, the identity connective occurs instead: π is either an algebraic number or the same thing is a transcendental number. But even this interpretation falsely attributes to the original thought another thought, which in a sense unfolds the former, which is precisely not in itself unfolded. The two propositions lie side by side in the explication. In the implicit form, they are not side by side, but they, so to speak, pervade and coincide with one another. In a plural proposition, in a dual judgment-ray going out from the same subject, the genuineness and the originating-in-the-philosopher's-early-period are stated at the same time, and it <is> likewise stated in a dual judgment-ray of the π that it is algebraic or transcendent. In both cases, the oneness of the dually composed predicate is apparent in the fact that, like every coherent predicate, we can coherently nominalize it, “the algebraicity or transcendentality” or, “the A-ness

and B-ness". And so we do not actually have two propositions lying next to each other merely linked by an identity-bond, but in each case we have a propositionally simple judgment with a subject and a coherent predicate built out of two implicit predicates. Precisely the same thing can be easily undertaken for plural judgments in the ordinary sense, where–for instance, conformable to the form "M and N are p"–the dual ray of predication goes out from the M and N and is united in the single predicate p.

The fact that traditional logic has consistently overlooked these syntaxes is again due to the fact that it has not understood about separating theory of forms and theory of validity, meaning analysis and validity analysis. Instead of analyzing the judgments as meanings with respect to the judgment's form, it decomposed the state-of-affairs posited in the validity-consciousness. In terms of validity, the propositions subsequently retrieved lie in the plural judgment, and those propositionally complex judgments are equivalent to the simpler plural judgments. But, over and over again, it is maintained that equivalence is a matter of validity and not a matter of meaning in itself. However, the situation is a little different here than in the case of other confusions of equivalence and identity of meanings. In the meaning unit of plural categorical predications, the predications to be stated in particular propositions (and following the different subjects or predicates), however, also already 153
lie implied in terms of meaning in certain ways. This way of increasing complexity is, though, quite peculiar, and is so in such a way that one cannot, however, speak of a genuine conjunction of two or several propositions–and even of one, since it just implies the unfolding.

As concerns the number of components that can be combined with a conjunctive and disjunctive subject, or with a conjunctive and disjunctive predicate, it is thus obviously arbitrarily large, so that as a result we already have an immense number of possible forms in every direction. Every singular subject in general possible can function as a component on the subject side, every singular predicate in general as a rule possible as a predicate component. In this respect, we could generally contrast the predicative presentation with the subject presentation–you understand the terms, every presentation completely adequate for the subject function, each one for the predicate function. Each nominal presentation, and only it, is singly-subject. Partly nominal, partly non-nominal presentations <are>, however, singularly-predicative. In the identifying judgment, nouns can also act multiply on the predicate side, for example, “Aristotle was the student of Plato and the teacher of Theophrastus.”

I could furthermore explain that in form-formation, multiple subjects and predicates function completely analogously to single ones, to begin with, in such a way that secondary combinations can also occur in conjunctions that function analogously to simple components. For example, “Aristippus and Euclid–for that matter Plato–were personal students of Socrates.” There, we therefore have a combination analogous to the form (a + b) + c. Combinations function as simple components in the formation of forms in other ways as well, for example, “an (a and b)–(something red and something round),” “an a that is b and c,” “an (a or b),” and so forth. Obviously the forms “an a, which is b” and “something, which is a and b = an (a and b)” are thereby differentiated in a clear-cut way when it comes to meaning. Since the conjunctions occur as complete components and, in their manner, bear the basic positing upon which the predicate clause is built in a multi-rayed way, we should actually call the plural judgments categorical–and call the combinations in the subject, pluralistic subject, likewise, pluralistic predicate.

154 Very different conjunctive and disjunctive forms now easily arise that in terms of their presentation-constituents can have multiples, partly on the subject side, partly on the predicate side, and combined on both sides. I leave closer study to your industriousness at home. The negative forms, and likewise the affirmatively-consenting forms, are particularly worth noting, because they make it evident that the usual treatment of negation is not adequate and confirm my theory <of> negation. Rays of affirmation, rays of negation, and perhaps <of the> plain position are layered upon one another in there. For the form “M and N are p,” we have the negation “M and N are not p,” but also the very common forms “Neither M nor N is p” or “not p.” Negation is thereby directed *expressis <verbis>* against M and against N, “Not M and not N,” furthermore, “M and not N is p,” “Socrates and only Socrates,” but

also, "M and only M" = "M and nothing other than M". To be noted well is that what is negated are not nouns, but negations that are precisely directed against the nominal positing. Things are different when we have, "a not-a", "a not-rich person", "an a, which is not b", and so on.

I must not neglect to mention here yet another restriction to which the free variability of conjunctive and disjunctive variables is subject in the forms. Every such form, for example "M and N", loses its meaning when M and N become identical in meaning. It is part of the essence of every and-connective and or-connective in general that what is connected must not be identical, *nota bene*, identical in meaning. ("Socrates and Socrates".)[1]

<§ 35. The Nominalization of Pluralities–The Origin of the Concept of Cardinal Number>

Of great importance are certain new nominal presentations that originate in the sphere of the various modes of judgment, namely, through nominalization. The unitary subject of a plural categorical predication is a conjunctive or disjunctive combination of nominal presentations, but as a whole not itself a nominal presentation. And the situation is the same with the various predicate 155
formations, where a conjunctive predicate formed out of adjectival presentations is not perhaps itself again an adjectival presentation–or formed out of subsuming presentations, not itself again a subsuming presentation. In contrast, every such unit can be nominalized. And above all else, we turn our attention to the conjunctions of nominal presentations. "Jupiter, Mercury, Neptune were Greek gods". In plural predication, nominal presentations occur as subject-unit, therefore, in certain ways present a plurality. But note well that this plurality is only presentational in them in an inauthentic sense. From the multiple ray, <the> predication ray, one implicit ray each indeed goes to each component, and only the bundling of rays, their coming together in the one ray going into the predicate, creates the unit. But a plurality can also become

[1] Compare Appendix X, *The Conjunction of a Nominal Presentation with What is Negated in Another Nominal Presentation*. (Editor's note)

objective in the authentic sense provided that it turns into the singular object of a single nominal presentation and that that is achieved through the nominalization of the conjunctive subject. We accomplish this nominalization when, instead of distributively, therefore plurally, judging about the god Jupiter and the god Mercury, we rather make a singular statement collectively about them together, about the pair of gods, the aggregate, or when, instead of stating of a and b that they stand in a relationship of friendship, are similarity-components etc., we speak of the pair of friends, of the similarity-group, and so on. We can call nominal presentations like "Socrates and Plato" etc. proper-presentations of multiplicities.

Now, these nominalizations take on particular importance when they take place on conjunctive nominal forms with indefinite attributions. If we form such forms first, then we would perhaps have "an a and a b";"an a and a b and a c"; and so forth. As a special case of that: "an a and an a";"an a and an a and an a"; and so forth. In addition, "something and something"; "something and something and something" etc., for which we can also say "one and one"; "one and one and one"... All these nominal constructions have their definite meaning as subjects of possible plural judgments. Thus cardinal number-presentations, namely, in the last sequence pure abstract number-presentations, arise for us as plural subject-forms, therefore, as purely logical meaning-constructions; the presentations of concrete numbers do so likewise.
156

Like every plural judgment, that with the dual subject "one and one" can also apply attributively and consequently determine the plural subject. Therefore, in plural determination, we can form "(one and one), that are a";"(one and one and one), that are a"; and so forth–i.e. "two a";"three a"; and so forth. A second pure number concept arises here at the same time, pure, provided that the units are not definitely concrete or entirely concrete. "Two" can mean a (one and one) of an arbitrary determination a, likewise "three" of an arbitrary determination a, and so forth, whereby in indefinitely-general thinking, one can restrict oneself to units having one and the same denomination to be adopted, moreover, arbitrarily in every domain.

In any case, concrete cardinal numbers are attributive presentations, in which an empty cardinal number—a conjunctive presentation, made up of empty units—obtains a plural determination. (Or rather, presentations of concrete numbers.)

What must fill us with astonishment here is the circumstance that the basic concepts of a science that is commonly considered to seem far-removed from logic and that we were also not accustomed to thinking of as having any especially close relationship to it are naturally located on the lower level of logic, in the theory of forms of meanings. That is actually a very momentous insight. However, we do not have the time now to embark on general reflections, and still have some duties to fulfill regarding the clarification of the specific problems now of concern to us.

It strikes me that plural subjects that plainly contain the same components are present in these remarkable constructions, whereas I have nevertheless stated the principle—and in the belief of the certainty of its Evidenz—, that subjects that are the same are not permitted. "Socrates and Socrates were great philosophers" makes no sense if in both places we thereby continue to associate the same proper-meaning with the same word "Socrates", and do so everywhere.

My principle is obviously universally valid. On the other hand, we shall be able to ask whether the repetition of the expression in the forms "an a and an a", "one and one" actually amounts to a repetition of the identical meaning. We know that This changes its 157
meaning from case to case, *quidam* does likewise. Ought not something similar also hold for the "one" that acts as a number-forming moment here? What kind of "one" or "something" is it? Obviously, the "something in general", or the *quidam* and this indefinite thing do not have a determinate meaning in the sense of a complete meaning. Lying right in the indefiniteness, which is an arbitrary determinability, is a source of the shift in meaning amounting to no less than ambiguity. As soon as "one" enters into a conjunction with "one", the indeterminacies are delimited and with that a certain change of meaning occurs. We also say "one and another one" or "an a and another a". However, we need not say that at all. The meaning of the And forces otherness provided the whole in general is to be understood as a meaning unit. And the situation is also

the same if we extend the sequence of ones in the conjunction. The meaning of the whole requires that every "one" is meant as different from every other, and that it is not a matter of particular attributions to be added on, but of delimiting meaning, which is accomplished by the conjunctive connection itself.

It is, therefore, apparent to me that it is not a matter here of additions to the concept of the meaning of cardinal number that aim to make it into a validity-concept there. It is not as if the "one and one" first had to be related to something different by constraint, and this constraint was required to yield valid twoness. It is the other way around. A constraint is first needed in order to require that identically the same "one" is posited in the "one and one", and that then yields a complex meaning that is *widersinnig* from the standpoint of validity. In contrast, in every case of plural presentations, a question of validity is that as to whether the different meanings of conjunctively connected presentations are of such a nature that different objects also truly correspond to them. Only then can numbers truly exist. From the realm of meaning originate, not valid number-meanings, but number-meanings in general.

Precisely for that reason, one must be careful not to identify what arises differently when it comes to meaning, therefore, for example, the "an a and an a", on the one side, and the "two A" on the other side. Both these constructions originate in different ways and are also of two different kinds when it comes to meaning. The
158 fact that an a and an a equals to two A is a principle of the theory of validity of number-meanings, but not one of the theory of forms. Both sides of the equation are equally valid, but do not have the same meaning.

<§ 36. Arithmetic Number-Presentations>

<a) Cardinal Numbers as a Basic Kind of Pure-Logical General-Nuclei. (Original Cardinal Numbers and Associative Constructions)>

But, it is now of particular importance to emphasize that the number-presentations that we have obtained up to this point are still not at all arithmetic number-presentations. Naturally, in the plural sphere we encounter precisely plural constructions, more specifically, empty plural formations out of units. But, no more than the nominal construction "Socrates and Plato" is a nominal Idea, and hence, no more than it is directed toward *one* object proper to it, no more does that hold of the construction "one and one" or "two A","two humans", and so on. Only when we engage in a nominalization do we obtain out of the plural something singular, –i.e. a nominal presentation directed toward a collective as an object. Only thus, do we therefore obtain the nominal presentations "a pair", "a two", "a three", and so forth. These latter expressions are not to be understood as subsumption-predicates, as we can of course also though do. There is therefore a remarkable linguistic ambiguity here that we must follow up on. We wish first to familiarize ourselves with the fact that there are also genuine subsuming number predicates.

Just as the nominative "an A", which functions as a subject, has its counterpart in the subsuming, non-nominal predicate "an A"– just as the nominative "something" has its counterpart in the subsuming predicate "something"–, so the "an A and an A" (the nominalized one as well) functioning as a subject has its counterpart in a subsuming "an A and an A." In like manner, on one side, the "one and one" has as its counterpart as subsuming "two" or "a two." Finally, and in like manner, we also have the subsuming predicative "two A", "three A", and so forth. Thus, we can, for example, state of the nominally presented aggregate "A and B" that it is "a two", that paraphrased, it is of the "form" "one and one." And this 159
predicate is of the nature of an indefinitely-general predicate. Namely, in the way we state of one specific house or another that it is a house and this is a general predicate, we can state of the aggregate "A and B", of the other aggregate "A′ and B′", and of every such aggregate generally of the form "one and one." That is also therefore a general predicate. And it is also possible for us to form the general predicates "two A", "three A", and so forth. If, for example, A and B are philosophers, then we can state of this aggregate of the two that they are two philosophers or a pair of philosophers. Likewise, in glancing at the set of balls standing before our eyes, we can say, "This is a three, four, etc."., although we will usually say verbally, "There are three of them"–although language

does not express <that> correctly, since the "three" is not a plural predicate here, but a singular predicate of the subject-aggregate. Corresponding to the predicative, subsuming "three" is then the "a three" as subject. The number then figures in the subject position and then functions completely analogously to an isolated "something" or "one"; only that it is a nominal construct out of several "somethings"; namely, the nominal presentation "one and one and one"; and an arbitrarily different number–the "a two"; "a three"; "a four"; and so forth–can likewise <figure> in the subject position.

(Our way was therefore this: First, the constructions "one and one"; "an a and an a"; "two A" occurred as plural subjects and also as predicates of such plural <subjects>, which perhaps could also be nominalized. These plural subjects and the plural predicates are not the only specifically arithmetic number-presentations, as whichever nominal presentations and the coherent classification-presentations corresponding to them are. But, corresponding (and in equivalence validity-theoretically) to the plural constructions are singular ones through nominalization or adjectivization–therefore, "a two"; "a three"; and so forth.)

These constructions now function in precisely the same way as individual empty places do. They therefore function everywhere where they figure nominally, either in the forms "thisness" and "certainness"; or in the forms of generality, as numerical terms of particularity and universality. For example, "These three men (or three men) are coming as a delegation." In addition, particularly, "An ellipse has two foci." Figuring here with "an ellipse" is a term
160 of universality. However, the two is, so to speak, a double point of particularity, of a particularity that, so to speak, falls under the universality of the "an ellipse has in general." (As I must then note the fact that it is generally speaking an important task to study the sub- and superordination of the generality-function, which is really in no way a mere form in the isolated position). Finally, mathematics offers examples of universal number terms in most propositions, "Two straight lines intersect in one point"; "2 + 3 = 5"; also for concrete numbers, "Two apples and three apples are five apples."

With the essentially new kinds of nominal and subsuming presentations that have arisen here for us, we also encounter new kinds of logical "concepts"; namely, new kinds of nuclei. The at times universally, at times particularly, functioning number-presentations and, in turn, in the changing of their meaning-forms, the parallel nominal and subsuming number-presentations, have a common nucleus, and these are the cardinal number concepts in the concise meaning of the word I have established. They belong to the general concepts and like every general concept, corresponding to every cardinal number concept is an Idea as universal object. The series of cardinal numbers can be viewed as a series of numbers as Ideas. It is important for us that new basic kinds of general-nuclei, of general concepts arise here that are themselves purely logical and do not fall into pure logic merely by virtue of their form and category. For, although nuclei, they are nevertheless at the same time logical forms, precisely as the same thing was realized out of the singular something, out of "identity" also.

Finally, it is yet to be noted that to be distinguished from the original cardinal numbers as they have come to us as constructions of the form "one and one"; "one and one and one"; are the associative constructions that arithmetic favors, "one and one"; "(one and one) and one"; and so forth, i.e. in line with the chain of definitions, one builds: 2, 2 + 1 = 3, 3 + 1 = 4, and so forth. These associative formations are equivalent to the non-associative ones and hence are not to be distinguished from them in the arithmetic theory of validity. Nevertheless, one will have to say that indirect presentations, indirectly built one upon the other, at the same time arise with each new denomination, so that the concept-formations corresponding to the chain of definitions are genuinely higher-level presentations, as I already remarked every so often in the 161
Logical Investigations. For sake of simplicity, I am not going into these higher-level Ideas, however important they are meaning-theoretically.

<b) The Difference Between Cardinal Numbers and Numbers>

So as not to allow misinterpretations to arise from the outset, I stress that we have built cardinal number presentations and cardinal number concepts here in our systematic system, which is not to say that these are the only presentations legitimately called number presentations. Number and cardinal number are not the same thing, even though one certainly has a lot to do with the other. Thus, for example, ordinal numbers are not cardinal numbers, and again the different numbers in the realm of the continuous magnitudes are not cardinal numbers, nor are they ordinal numbers. Since, although they all belong in pure logic, I cannot go into the other number formations in these lectures, I just want to note with regard to ordinal numbers that they are related to the sphere of the forms of relation-formations, and more precisely to those purely logical forms of relation-formations that are called orders, more precisely simple and open orders, sequences.

The difference in comparison with cardinal numbers becomes evident if you think about serial constructions like aςbςcςd etc., where ς is thought of as an irreversible relation. If aςb is, then bςa is not, but rather in a reversed relation bς′a. If we then have a group of objects, of arbitrary somethings, and along with it in indefinitely-universal presentation an irreversible relation-concept ς, which is therefore genuinely also thought of as something, as something relational, then with respect to such a chain of relations, where any two adjacent components are linked by the same relation-moment, the something next linked to a something–which is the last component of the relation ς—the first one in relation to this initial-something, the first one relating to that <is called> first, the one in relation to the initial-something second, and so forth. You see, it is again a matter of formations out of indefinites, only that here something indefinite of a new category, namely, a relational indefinite is also involved. One can also now purely logically form the Idea of double sequences, triple sequences, and so forth, depending on whether two, three relat<ional> indefinites come into play at the same time. Accordingly, <one can> form ordinal
162 numbers of the second, third dimension, and so forth. Naturally, a distinction is to be made here between the new kinds of meaning formations and the new kinds of objects, and the conditions of the validity of meanings or of the possible existence of such objects <are> to be set in a corresponding <theory of> validity. In a logic of validity for these forms, i.e. in a theory of ordinal numbers, as well as with the transition from plural presentations, one then arrives at a theory of cardinal numbers as a logic of validity of meanings of cardinal numbers.

You understand the immense importance of such considerations. For, they already provide Evidenz that the entire theory of cardinal and ordinal numbers is no more than a piece of formal logic. And the same holds, as becomes further apparent, for all of pure formal mathematics, for everything mathematicians deal with under the heading of analysis, number theory, function theory, and so forth. However, let us return to our subject after this digression.

<§ 37. The Indefiniteness of the Left-Open Continuation of Connections Lying in the "And So Forth". The Origin of the Concept of Set>

The nuclei of cardinal numbers function as all general-nuclei do. All the judgment-forms in the general theory of forms that we have differentiated are unaffected by the kind of general-nuclei. Related to that is the fact that judgments can be made precisely about cardinal numbers in the same forms as about other objects, among them also universal and particular ones. The constructions "a two", "a three", and so forth stand opposite "a cardinal number". So, the cardinal number nuclei can also be replaced by an indefinite something, and this something can then generally be determined as cardinal number. So universal judgments like $2 + 3 = 5$ and $a + b = b + a$ differ. The former is universal as regards the empty places, which figure in the "2", "3", "5" as so-called units. Cardinal numbers are really nothing more than certain nominalized complexes of empty places that are called units. In contrast, "$a + b = b + a$" is universal as regards the concept cardinal number– the explicit meaning is: The sum of a cardinal number and another cardinal number remains unaltered when these cardinal numbers change places in addition.

When we speak indefinitely of "cardinal numbers" and of "a 163
number" in arithmetic, this general designation refers to the previously terminologically and conceptually firmly fixed 2, $3 = 2 + 1$, and so forth. (Every sum-form of "one" is distinguished from every other one and identified in its fixed distinctiveness.) Indefinite multiplicity-presentations are, however, possible that do not require this firm fixating, but rather precede it. I am thinking of the

indefinite "a multiplicity", "a multiplicity of A's". If we go back to the plural judgment, then this nominalized form goes before the not yet nominalized one that occurs in the form of the plural judgment–"A's are b" (or "Some a are b"), "A's are generally b", "Certain a are b".[1] To be noted well is that therefore in the plural judgment "Some A are b", a multiplicity does not, for instance, occur in the subject position, "A multiplicity of a are b"; "A's"; i.e. "an a and an a, and so forth". This "and so forth" is one of the most remarkable presentations, something original and in no way to be divorced from the meaning realm. The "and so forth" signifies that another a may come and another a, but it need not come. The "and so forth" can also occur within the context of constructs out of proper-presentations, for example, "Socrates, Plato, Aristotle, and so forth, were famous philosophers of Antiquity". Moreover, the "and so forth" can occur everywhere a conjunction is staged. It is a new, distinctive sort of indefiniteness, which plays no small role in thinking. This indefiniteness of the left-open continuation of concatenation still precedes the nominalization of the plural "some" and then makes the transition to nominalization, where a multiplicity, a set, an aggregate results–all, properly understood, synonymous words. This is the basic concept of set theory. Corresponding to the set-presentation is an objectivity proper to it, the set, the multiplicity, and valid statements can then be made about sets, such as that they are contained in one another, that they intersect, that they are identical, or completely different, and so on. These set
164 predications are not about cardinal number predications, because it is just a matter of indefinite sets and not of numbers.

Naturally, running parallel to the plural form "Some a are b" are the other plural forms, "These a are b" and "Certain a are b", furthermore particular and universal forms "There are A's that are b", "A's in general are b". The copula affects all the indefiniteness of the "and so forth" in them in remarkable ways, insofar as a multi-

---

[1] Subsequent note by *Husserl*, "Mediation forms the form 'A are generally b'.

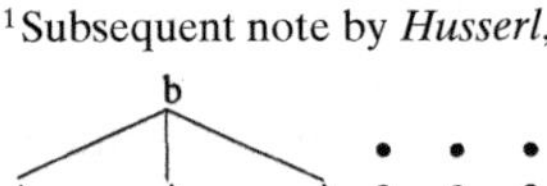

The generality relates to the points, to the continuation of the plural". (Editor's note).

plicity indeed meaningfully belongs to it, but precisely an indeterminate one corresponding to the "and so forth". (There are probably at least two pluralisms there–as becomes apparent when we make the meaning of this plural predication clear to ourselves–but not exactly two or three pluralisms are meant. The "and so forth" therefore also figures in the copula. The judgment's mode of synthesis therefore has forms of its own and also a specific form here. You see how little what people call the copula of the categorical judgment is everywhere the same kind of bond, and how many kinds of connection of predicates to subjects there are that hinge on the forms of the subjects and predicates throughout and are themselves component parts of the judgment–form.)

<§ 38. The Thought-Form of "Again and Again">

As we have heard, running precisely parallel to the conjunctive forms of the categorical judgments are the disjunctive ones, and in the case of the introduction of indefinites into the disjunctive realm, there must also be number formations regarding which reflections would then have to be able to be engaged in similar to those for cardinal numbers, therefore, formations like "something or something", "something or something or something", and so forth. These formations do not, however, therefore play any special role in scientific thinking because, as we shall come to understand even more precisely, immediately equivalent, only somewhat more complex, constructs slide under that contain indefinites altogether (in collective form) collectedly as cardinal numbers. It is here a matter of equivalent expressions that can occur in all plural forms and are very important, because a new, very remarkable thought-form occurs in them again, namely, that of the "again and again"–or more simply of "every".

To begin with, in the case of the And-forms we have as equivalents: (1) the form "A and B are p"; (2) the forms essentially distinguished from them in terms of content and syntax, "Of A and B, the A and as well as <the> B, is a p"–or more clearly, "Of A and B, 165
A as well as B, is again and again a p". An equivalent, new form is again, "Of A and B, each one is, one by one, p, the one like the other is p". Here the A and B are not themselves repeated, but

thought indefinitely in the repeating. In such predications, the plural presentation occurs twice in remarkable ways, namely, in different forms, the first time as nominal presentation of the aggregate of “A and B”; the second time non-nominally, as plural subject of the predicate p. And, a certain sliding-over coinciding of both whole presentations thereby takes place through identifying positing-in-one of the nominal components of these whole presentations corresponding to one another. What has not yet been determined as being-p is always brought to coincide with what has been determined thus in the plural predication. The “every” expresses this overall coincidence in the step by step coinciding of components and consequently pertains both to the subject and to the “is”. This remarkable “again and again” thought-form then also occurs in the forms of indefiniteness, co-determining their meaning, itself becoming indefinite in the process. “Of these A each one is p”; likewise, “Of certain A–” and again “Of several A, of some A each one is p”; then “of these two, three A, and so forth”. Judgments like “Each one of the roses of this flowerbed is a carnation”; “Each A is B”–namely, each of the A, each of the aggregate of roses of the flowerbed also therefore belongs right here. I shall still have some things to discuss regarding this.

Before we enter into that, let me talk more about the parallel disjunctive forms “M or N are p” is equivalent to the new form, “Of M and N, M or N is again and again p”–or again equivalent and without repetition, “Of M and N, one or the other is p” = “–*one* is a p” (disjunctive one). An “again and again” also lies in this “one”; one out of the aggregate “M and N”; only one modified in certain ways. Here too, there is a running-through of the components and a sliding-over coincidence of the aggregate-presentation “A and B” with the non-nominal presentation of the same A or B, but as disjunctive presentation. The indefinite forms of the disjunctively-multifold subjects are now obviously reduced to forms with cardinal numbers. “This or that a” becomes “one of the two of these A”; “one of these two A”; “anything or anything else” becomes “any of two”; likewise “any of three”; and so forth. No new number formations are therefore needed. One has equivalent formations
166 from cardinal numbers.

Moreover, the “one of the two”; “one of A and B”; is ambiguous within the equivalence. “Of A and B one is red”; it need not be said,

"–one or the other is red". If we start with the proposition "Of A and B, A is red", then there is no disjunction there. Likewise, "Of A and B this is red there", and then indefinitely, "–one is red". The indefinitely remaining coinciding of the "one" that is red there with a component of A and B belongs to the meaning, only that the thought of disjunction is missing. In the same way, it can read, "Of A, B, C,... A and B are red", "–are two, some are red". Therefore, formally, "Of A, B, C,... some are p". (When it comes to logical validity, the indefiniteness of the "some", and also of the "a", does not rule out that each one of the subject constituents is p, but this is of no concern to us when it comes to meaning.) Naturally, we could also allow indefiniteness to prevail in the subject, "Among the α one is, some are p". ("Of the multiplicity of α's is a component, is a plurality p"). Thus, one notes the difference of the meaning for the formal "Some S are p", which can shift somewhat. "Several S are p" as plain plural form is different from the form, "Among the S are some P" (= "Some P are among the S"), and again from the form "Some among the S are p" (= "some among the S being P").

<§ 39. The Totality-Thought–It Does Not Contain the Thought of Generality, of the Universal in General. Surrogative Thinking>

An infinite number of disputes and difficulties are related to the ordinary formulation of what is called the general judgment, "Every A is B", provided that "All A are B" and "The As are B" are said with the same meaning. They are forms that contain negations buried in their meaning. If we produce some of the negations belonging to our positive forms, then we shall have everything gathered together also to deal with those common verbal complex-forms that the tradition treated as simple. Such negative forms from our sphere are, for example, "Among A and B and C is one that is not p", for which we can also say, among them some p is missing. This predicate can also be negated again, and so we have, "Among A and B and so forth, no p is missing". There the negation is directed against the particular in general.

When instead of proper-presentations of aggregates, we choose 167
indefinite presentations of aggregates, we can find nominal presen-

tations of the following form in this sphere, "a multiplicity in which a p is missing", and again, "a multiplicity in which no p is missing". A multiplicity can be presentational as an aggregate of A's through attributive determination, therefore, with the content "an A and an A, and so forth", and this aggregate can then be further determined by the fact that no p is missing in it.

Presentations of totalities then pertain to these important presentations. The totality of p, i.e., a multiplicity of p in which no p is missing. Two negations are therefore in the determination. (The equivalent of this totality-thought is in the thought "a multiplicity in which there is no component that is not p and in which no p is missing". We would have three negations there. But it is obvious that this is a complex thought and that we in general can frequently form equivalent complexes in this manner.)

We then encounter the totality-thought in the judgment-form, "All A are B". Obviously, "all A" is not thereby, as it might appear from the verbal form, a plural subject, and the "all" perhaps entirely an adjectival meaning like "red". Rather, it is clear that it is stated of the totality of A's that every component of it is B. Every <component> and the entirety of A's is B. Precisely this same thought is the meaning of the other common forms "Every A is B" and "The A are B", although the former is grammatically expressed in the singular and, through the grammatical plural, the latter indeed indicates a distribution, but not the "every", and not the entirety. But both the totality-thought and the thought of "every" belong together in there as well. The meaning of many, really most, of the propositions about totality of everyday life, is undoubtedly so, for example, "All trees in this forest are pines", "All the flowers in my garden are roses", and so forth. However, it is completely different when we state, "Every triangle has (or all triangles have) as the sum of their angles two right <angles>–and so it is in every case in which in mathematics, or in some other science, we make a general statement having the nature of a law, the nature of "rigorous and unconditional generality".

Let us be more precise. We often *de facto* form the meanings thoroughly described earlier. We form, for instance, the logical
168 presentation of the totality of triangles and think that in this comprehensive aggregate of triangles each one has two right <angles>

as the sum of its angles. On the other hand, we cannot however at all begin with this meaning in the context of insightful proving, for it cannot be brought to Evidenz, as we can otherwise very well do when we speak of the totality of the roses of this garden. There, we can go into the garden, look at the roses individually and ultimately make a comprehensive aggregate of them. But an aggregate of triangles in which no triangle is missing cannot be brought to givenness. That is nonsense. It also immediately strikes us that the original, genuine totality-thought, "All A are B", that I have described precisely contains the idea of totality, but in no way that of generality–that of the universal in general–, while the latter is obviously what is most essential to the thought of every mathematical <judgment> and every law-judgment. It is furthermore also clear that the generality-thought also plays its role outside the sphere of laws and that it is also meant in the speech-form along with "all A" and "every A".

The ambiguousness of these statement-forms has become a true cross for logic to bear. Besides *Brentano*, *Sigwart* deserves credit for having subjected it to a comprehensive investigation for the first time, which, however, despite much that was worthwhile about it, was in no way conclusive. For that, he lacked much too much of the whole leading thought of a theory of forms, an understanding of the difference between equivalence and identity of meaning, and likewise the yet to be discussed <difference> between surrogate-meaning and validity meaning. It is not enough–and is also besides not correct–to contrast, as *Sigwart* does, two meanings, one of which refers to a complete, countable multiplicity in an empirical sphere and expresses the thought of admitting no exception, while the other does not seek to make a statement about such a thing, but about a relationship of necessity.

Without being able to embark on critical arguments here, I note that the totality-form in the sense fixed and normal up until now does not in itself already affirm non-admission of exceptions. Rather, the thought of not admitting exceptions adds a certainly fairly common complexity to our normal statement-form. If we say, "All A are–without exception–B", then we are initially saying 169
that all A are B, i.e. that in the entirety of A's each one is B, but we

still stress in addition, "Each one is B—none that <is> not <B> (no exception!)". We therefore introduce another double negation, so that besides the two negations figuring in the Idea of entirety, now two more negations also figure on the predicate side. The complex thought is thereby indeed equivalent to the original one. (Often merely stressing "All A are B" serves to indicate the same thing, therefore, to make it tangible to us that totality is completely seriously meant, without exception).

While *Sigwart* then secondly draws in the necessity-thought in order to obtain another meaning of the statement with "all" and hypothetically grasps it in the form, "If something is A, then it is B", he does not note that, besides the hypothetical form as something new, the "something generally" occurs there, and that this very "generally", the form of universality, can also occur without hypothetical form. (It would be *widersinnig* to restrict it to this one form of the hypothetical relationship. And, it is furthermore also clear that the generally also plays its role in both the empirical sphere and in any purely conceptual sphere.) Nor need we use the triangle example at all to eliminate the mathematically incongruous thought of a countable totality of triangles. We merely need to say, "A triangle generally has as the sum of its angles two right <angles>". And obviously that is exactly what ought to be said, while, however, nothing more is then thought of either "all", or "every". In the other example of the garden, we can then indeed also form, "A rose of my garden is generally a rose", but here it is not exactly what we want to say. There we want to make a statement about the entirety.

Only one difficulty still remains left. We granted that for the most part the set-presentation, the totality-presentation, is also realized by mathematicians to begin with, while it is nonetheless of no use for argumentation, since it involves a *Widersinn*. A totality of triangles, a totality of numbers, is not, if we hold fast to the meaning of totality, to be had intuitively, therefore cannot exist either. Even in inferential thinking, we operate with such *widersinnigen* presentations to some extent. If it is said that all triangles are
170 polygons, all polygons closed figures—therefore etc., we then conceive of the triangles as a set and as inserted into the set of polygons. And if the inference is then to be made, then we make it from

the sets in accordance with the schema: If a set is contained in a second one, the latter in the third, then the first in the third. It is then to be said there: Psychologically, the totality-thought slides under the generality-thought, because our thinking generally loves to operate with intuitive and analogizing auxiliary presentations. If we inquire into the reasons for bringing set presentations into general thinking, then they are obviously as follows: We have, for example, already frequently had dealings with triangles; and running through them and comparing them in thinking, aggregates of triangles arise for us, namely, cases of triangles dealt with earlier. If we now have come to understand that every triangle has the property of being Pythagorean, then it is worth our while to carry this over to the earlier cases and new possible cases. And so we arrive at the thought, "Arbitrary triangles generally have this property– arbitrary triangles are so constituted that there is nothing in them that would not have the property". There we have the totality, but the multiplicity is generally thought of as one of triangles. We therefore have formed the thought, "In universal generality, a triangle multiplicity is a totality of those that have the property–" or, "all –are in a multiplicity of triangles, etc."

All the traditional talk of extensions of concepts, according to which the totality of objects that are to be subsumed under each valid universal-concept are to belong to it as its extension, is based on such analogizing. No pure concept in fact has anything like an extension, and saying that for every concept a distinction is to be made between intension and extension is actually *Unsinn*. Since *Euler*, extensions of concepts are visualized by means of circles, i.e. something fictitious, the totality of objects of a concept, is visualized by means of an arbitrary circle or a simple closed curve. And then the laws of syllogistics (i.e. of the theory of relationships of subsumption) are symbolically derived upon intuition of relationships of circles, and that happens by virtue of a certain analogy between circle relationships and subsumption relationships, the meaning and boundaries of which have, however, never been scientifically established. In this way, the difference of intuitively 171
analogizing thinking for actual thinking itself therefore makes its way into logic, which nevertheless *expressis <verbis>* is called to become the guardian of exactitude and to develop its theories with

the purest theoretical rigor. Naturally, as regards its essence, as regards its performance, as regards the limits of its validity, surrogative thinking itself can be accepted and methodically developed, only it must not come into use without prior regulation and investigation. It spontaneously, inadvertently forces itself upon natural thinking–prior to any study of these matters. For purely psychological reasons, an intuitive symbol for a concept slides itself in, and analogies between conceptual and intuitive relationships come into effect.

And, the great difficulties of a pure theory of meaning are also connected precisely with this. It is apparent that in meaning analysis, one must not settle for adding meanings that initially present themselves in the elucidation of the meaning of our statements. The surrogating just discussed marks a flood of equivocations. The same statement, "All A are B" as mathematical statement, for example, changes its meaning in certain ways, according as it functions in the sphere of the surrogative thinking or in the sphere of substantiating thinking. And it is nonetheless otherwise called "the same statement", "the same mathematical statement". Inasmuch as mathematicians wish to state truths with their propositions, if they arrive at an understanding of the duality of the intuitive presentability of presupposing set-statements and of the completely different generality-statements, they will naturally say that the right, scientifically valid, meaning is the one attaching to the proposition in the reasoning. The other may only hold sway as a surrogate, as a convenient tool of inventive thinking conducted by the play of analogizing. What cannot be used and redeemed in reasoning is inauthentic and false meaning, a changeling. And in this sense, from the standpoint of the theory of forms, we must also logically say that we have to inquire into all the twists and turns of meaning that a proposition undergoes within different contexts of inventive thinking, but also reasoning. We first have to distinguish purely formally between the
172 different meanings and then–with an eye to furthering thinking–point out in the theory that some slip in for others and that, where it is a matter of validity, they must be carefully dissociated.[1]

[1] Subsequent note by *Husserl* in the margin. "1917 the theory of totality with the surrogates explained differently". (Editor's note)

My remarks about universal statement-form have shown you what difficulties the surrogative confusing of plurals and forms of indefiniteness present for analysis and how hopeless attempts must be here to pierce through to clarity as long as one has not systematically developed the elementary forms and gained an understanding of their intrinsic structure. In this respect, then, the area of statements in generality-consciousness then especially still presents many not fully clarified points which become perceptible precisely in the analysis of the ambiguities of common, customary statement-forms.

<§ 40. Existential Judgments and Impersonals>[1]

<a) The Issue of Existential Judgments. In Conflict with Brentano>

The other day, I called universal statements a cross for logic to bear. This holds to an even greater degree for existential statements and for the impersonals closely related to them in terms of meaning. Examples of existential propositions are, "God exists" or "God is";"A regular decahedron does not exist, is not". Examples of impersonals, "There is not enough money"; "It is burning"; "There are regular hexahedrons";"There are not any regular decahedrons"; and so forth. In the never-ending literature about these statement-forms, it is not merely a matter of questions of grammatical interpretation. Rather, it is thought that this is the battleground upon which the fundamental questions of the descriptive psychology of judgment are decided. It is said that one must study the essence of the judgment-function in the cases in which judgments with the most primitive judgment-content (or as I would say, meaning-content) are made. If not always, then nevertheless in certain cases, an actual simple judgment is expressed in the controversial primitive 173
statement-forms. The issue around which everything turns here is that of whether every simple judgment actually is categorical, therefore, has that distinctive dual composition that allows for

[1] Compare Appendix XI, *Old Pages on the Problems of Propositionally Simple Judgments*. (Editor's note)

differentiation between subject component and predicate component–as the age-old, and even now most determinative, tradition going back to *Aristotle*, held.

If we now also take but a fleeting look at the controversial statement-forms, we notice their anomalousness in this regard. In the case of "A is" existential judgments, the predicate is missing. Considered grammatically, they are predicateless propositions. Contrariwise, in the case of impersonals, the subject seems to be missing. They are subjectless propositions. The "it" figuring in the subject position can only be allowed to count as a subject word in exceptional cases. And these cases are left out of consideration in the dispute, as, for example, everywhere where in saying "It is a", we meaningfully ask, "What is a?", therefore, ask for a more precise determination of the subject. For example, "It is moving again"–"What is moving?" Well, for instance, the little animal we are watching. One can, however, ask such questions in the case of propositions such as "It is raining", "Money is lacking", etc., especially in the case of propositions with "there are", "There are regular bodies". The problem is, then, whether these propositions are actually subjectless in terms of their meaning, or whether–even though the "it" is not functioning in the sense of "a certain something"–a subject is nevertheless not thought in the meaning of the proposition and a predicate belonging to it, which perhaps does not coincide with the grammatical predicate.

This first became a burning issue owing to *Brentano*'s *Psychology* of 1874,[1] where he defended the theory that if one goes back to the essence of judgments, the existential statement-form emerges as the only suitable one, that it is therefore the normal form to which every statement would have to be reducible and would also have to be reduced to if the true meaning of the judgment was to be established. Later, *Brentano* interpreted his theory in there that every simple judgment has to be of the form "A is" or "A is not". The "is" ought not thereby harbor within it something like a predicate, so that the primitive judgments are actually predicateless and consequently also subjectless. *Brentano* thinks that

[1] For the bibliographical information, compare the editor's introduction to the original German volume (Hua XXX), p. XLVIII. (Editor's note)

the choice of different grammatical forms for primitive judgments is explainable on merely psychological grounds, provided extra- 174
logical reasons are to be held fully accountable for the development and the choice of grammatical forms.

Every statement of categorical syntax is either logically unsuitable, or is not an expression of a simple, but rather of a composite judgment, though it may not be decomposable into independent judgments either: "This rose is a plant". It is thereby to be noted that *Brentano* sees the essence of judgments in affirmative positing (which according to him is either an affirmation or negation, either approval or rejection) and that he accepts as many partial judgments into a consistent judgment as can be differentiated in these affirmative positings.

Therefore, for *Brentano*, genuine existential propositions—those expressing a primitive judgment—are not categorical. Where, however, "existence" occurs grammatically as a predicate —"A is existing", "A is actual", and so on–it is either a matter of a pseudo-predicate, or the judgment already involves some complexity owing to which it is to be approached as being complex.

In contrast to this, other scholars uphold the old theory. They in general deny that there are simple existential propositions in *Brentano*'s sense, therefore propositions that are not composed of subject and predicate.

Related to that is the controversy surrounding the univocality or ambiguity of what is called the copula, i.e. of the little word "is". *Brentano* champions univocality. "Is" is everywhere an expression of accepting, "is not" an expression of a rejecting, in the existential proposition "A is", as well as in the categorically explicit proposition "A is b". According to the opposing view, the "copula" is ambiguous. In existential propositions, it also expresses existence. It expresses "is existing", "is actual", and so on. Besides the Is as an actual copula, we would therefore still have a predicate, although, as *Kant* proved, not a real predicate. In categorical propositions, however, the Is expresses merely the logical copula that otherwise as a rule fuses with any predicates, as in verbal forms.

To illustrate *Brentano's* view again (using examples), according to him, each judgment of the kind "Socrates is a philosopher" is already a composite judgment. Implied in this is the positing of Socrates as something actual. Therefore, as *Brentano* thinks, the

existential judgment "Socrates is!" is implied, woven in and,
175 according to him, based upon this is a dependent judgment not detachable in its own right in which Socrates' philosopherness is affirmed. We can also say: Every categorical statement with definite fixed terms, with proper names, or with places in which "this" or the *quiddam* occurs is composite, namely, is as multifold as places of that kind occur. According to him, simple judgments are only expressed in obvious categorical-forms when such terms are missing, and only genuine empty places, "generality-places" in our sense, occur. Therefore, it is then a matter of purely particular and universal judgments. According to *Brentano,* all such judgments are actually existential judgments. Thus, the form "Some S are p"; "Some men are learned" is to be reduced to, "There is an S-being p" or, quite purely interpreted, "A being-p S is"; "All S are p" = "There is no S ~~p~~"; and so forth.

As regards impersonals, as has already been said above, a distinction is made between genuine and non-genuine ones. The genuine ones, in which a subject is not to be attributed to the "it"; are existential propositions according to *Brentano* (and *Marty*, who in the main follows him).[1] It is thereby worth noting that for these philosophers all impersonals can be reduced to the expression "There is an A"; since discourse with "there is" is then in general a favorite form for expressing the thought of an existential proposition. The "there is" as the genuine impersonal "there is A" promptly turns into "A exists".

I am satisfied with this account of *Brentano*'s theory, without entering into the ideas of his opponents, which you may consult in major logical works and in specific writings. Even though I consider it erroneous, I am of the conviction that what *Brentano* is proposing here nevertheless stands far above the theoretical attempts of his followers, although in their criticism they often hit upon real deficiencies in *Brentano*'s theory. It is characteristic of genius to drag out into the open, in one-sided and often completely false theories, fundamental problems that had theretofore remained buried and in this way to serve knowledge to a much

[1] A non-genuine <impersonal> would be, for example, the following "What is rustling there in the foliage? It is a bird".–"It does not burn like the match".

greater degree than everyday theories <do>, which–harmonizing more with common sense–indeed remain closer to the truth, but remain on the surface and overlook radical problems. 176

Brentano has actually earned a great deal of credit for his attacks against traditional logic, which had completely shoved aside the existential judgment. The young *Hume,* and also *Kant* in his early writings, had already become aware of the difficulties of existential propositions, but they had no successors. In the nineteenth century, *Bolzano*'s *Wissenschaftslehre* made some worthwhile contributions, but up to our times, they have in general remained without due effect. *Brentano* was the first to have compelled us to study the complex and extremely difficult problems connected with the elucidation of existential propositions. One must, however, also see these problems and–inasmuch as they are not neatly set forth–work them out. And in this regard, people are very far from having done anything substantial when through criticism they themselves and others are led to the conviction that *Brentano*'s paradoxical theories are false. I mean, *Brentano* forces certain problems on us.

1. To begin with, never before him were the fundamental questions raised as to what to have count as a simple judgment, how to implement the differences between simplicity and complexity in radical and fundamentally clear ways in the realm of judgment. For him, a fundamentally important kind of complexity emerges, namely, that in accordance with the singular and plural <of the> the affirmative positings contained in the unity of a judgment. He has not worked his way through to the standpoint of a pure theory of forms of meaning, and so those other differences between simplicity and complexity that we have studied as differences between detachability and non-detachability, between propositional simplicity and propositional complexity, and finally between plural singleness and multifoldness do not become important for him. Even so, *Brentano* made an important start, of which his critics have unfortunately made little use.
2. I thoroughly object to *Brentano's* claim that a judgment like "Socrates is a philosopher" is complex and reads into every proper name like "Socrates"–and likewise into every (namely, affirmatively posited) name–an existential judgment of its own.

177 Unity of judgment, I say, requires two things: unity of quality and unity of matter. And if the judgment is to be an independent whole, then the matter must be independent, complete. Obviously, the concept of primitive independent judgment must count as a basic concept of the theory of judgment, and that is the ordinary concept of judgment that has been scientifically defined through my analyses. But a broader definition is still needed. It is fundamentally important to grasp judgments in the logical sense, in the sense of an independent, complete thought-unit. Bolder criticism on *Brentano*'s part could have led to those distinctions, and one also has to have grasped them in a clear-cut way and to keep them constantly in sight if one wants to find one's way amid the clutter of arguments surrounding existential judgments. Therefore, to emphasize the last point once more: A fundamental distinction is to be made between logically defective, logically incomplete meanings and logically complete ones.[1] Interjections like "There!", or mere names like "Mister Schulz!" are logically incomplete, that is to say that not every judgment-intention is logical. Rather, for apophantic logic, the judgments exclusively come into consideration as thought-meaning. An intending can also be pre-conceptual, unconceptualized, not at all, or only partially, grasped in thought. What I intend is: Mister Schulze walks over and gives greetings, but I only said "Mister Schulze!" (And there are thereby two more possibilities: either only the expression is incomplete, i.e. I have consciously formulated, but not actually said, the conceptual version; or the meaning is also incomplete, no conceptual versions at all were formulated. This incompleteness or deficiency is naturally to be essentially distinguished from that pertaining to the meaning, namely, to the logical meaning itself, namely, that expressed in the empty places, on the one hand, and that buried in the nuclei themselves, on the other hand, for example, in the proper name that names the object, but tells us nothing about what objectively determines the object as this person, and so forth).

[1] Compare Appendix XII, *Logically Incomplete Meanings*. (Editor's note)

The difference between the completeness and incompleteness of meaning essentially comes into question in the debate surrounding impersonals, insofar as the latter are incomplete for good reasons in countlessly many cases. For example, the meteorological impersonal, “It is raining” does not say, “There 178
is in general rain”, “There is something like rain” in the sense of particular judgments, as if we left it open where and when there is rain. Rather <it says> “Here and now, outside, there is rain!” Obviously that is no particular judgment at all. *Brentano* overlooked that when he regarded these meteorological judgments as existential judgments, while on the other hand he placed existential judgments on a par with indefinite, particular judgments and universal ones, as the case may be.

3. The changes in meaning in existential propositions that particular and universal propositions undergo at the hands of *Brentano* (first, in propositions with “there is”, which are for their part equivalent to existential propositions) compel one to undertake a deeper study of the intrinsic structure of particular and universal judgments than has happened in the past. The old, as well as <the> new, logic has not only lacked an analysis of existential propositions, but also a serious analysis of its favorite propositions, <namely>, of “quantifying” forms.

The following, which I did not emphasize in a clear-cut enough way earlier, should be of particular importance there. We separate the judgments with indefinites, in which only *quidam* places occur, from those in which genuine terms of particularity and universality occur. We can say that the former are bearers of their own affirmative positing, the latter not. We become aware of this due to *Brentano*'s reductions, which properly considered, precisely presuppose the fact that the empty places do not already contain positings, since only then can his interpretation, which in the case of complete judgments is a matter of simple affirmations, be correct. Places of affirmative positing can admittedly occur elsewhere in a particular and universal judgment, but it is certain that the indefiniteness-terms cannot be those places. The point is not, *nota bene*, that we have beliefs in the case of those places, but only that the indefiniteness-judgment contains affirmative qualities in its meaning.

Consequently, there is actually a radical difference in the sphere of propositionally simple judgments, that of the fact that some are also simple as affirmations, the others not. *Brentano*'s existential propositions may be very richly composed. For example, the simple universal judgment "All A are B", which has two main terms–"All triangles have two right <angles> as the sum of their angles"–already supplies him with the existential proposition of 179 correspondingly rich composition, not have as the sum of its angles two right <angles>, is not". But, no component in it bears its own affirmative positing, and the same thing obviously holds for the original judgment, which in my opinion is not in any way identical with the existential proposition, but is only equivalent to it. You notice, moreover, that the "generally" also occurs in the existential proposition and has not disappeared, say, through reduction. *Brentano* did not see this "generally" at all.

<b) Further Discussion of the Problems of Functional-Judgments. The Equivalence of Particularized (and Universal) Functions and Positive (and Negative) Existential Propositions>[1]

It now seems to me upon deeper reflection that from here on out we must go one step further and make a fundamental cut in the theory of simple judgments–and of judgments in general. (Unfortunately, I already wrote down the remark to be made now some years ago, but have forgotten it again.) We can also express a particular judgment, for example, "An A is b", "Some A are b", in the form, "That an A is b, holds in one, holds in some cases". We can also transfer the *quidam*-judgment thus: "That an A is b holds in a certain case, holds in certain cases"; similarly, every universal judgment, for example, "Every A, an A in general is b", "That an A is b holds in general, holds universally, holds in every case". That this recasting only transforms equivalence into equivalence is clear. For, should the explicated form merely be identifying eluci-

[1] Compare Appendix XIII, *Propositional Function and Proposition*. (Editor's note)

dation of the meaning of the original, we would come to an infinite regress, since the new forms themselves are again particular, universal or *quidam*-judgments, therefore need the same explication, and so forth. All the same, the recasting is not without value. It makes us–namely, for the same generality-judgments (I mean <for> the particular and universal ones)–aware of an intrinsic separation of the forms of their syntax, which in the equivalent explication leads to a contrasting of the two components, one of which is nominalized, while the other is adjectivized in it. In the explication, "the fact that an A is b" figures in the subject position, and 180
stated of that is that it has particular or universal validity. What is it that is figuring in the subject position? Apparently a proposition. But, the full and actual proposition, the judgment, is either necessarily particular or universal, when it is, as here, not to be a *quidam*-judgment. We just state, however, that in one case it is particularly valid, in the other case, universally valid.

Obviously, the subject can therefore not be a judgment. It is what the universal judgment "An A is b" and the particular judgment "An A is b" have in common and what is accentuated in nominalized form. What they have in common is then what is conscious in the generality-consciousness, in the one case, and in the particularity-consciousness in the other case, or meaning-theoretically speaking, what bears the meaning-form of universal generality in one case, that of particular generality in the other. And this higher form thereby belongs to the "an", to the empty place in what they have in common, as well as to its "is". In the explanation, we make what they have in common into the subject through nominalization and the generality-form into predicate term in its own way.

What they have in common is then nothing other than what in their field mathematicians call function, and–without otherwise having thereby arrived at the full descriptive analysis of the kinds of judgment concerned–the sharp-witted *Frege* has the merit of having recognized this in his article "Function and Concept" (which appeared in the early 1890s). Therefore, for example, "A human being is mortal" is a function or propositional function that is featured nominally when we state, "That a human being is mortal is valid generally". We cannot state general validity of the judg-

ment. The judgment "A human being in general is mortal" or "Every human being is mortal" is a general judgment and a valid judgment. But closely examined, it makes no sense to say of it that it is valid in every case. In contrast, this simple proposition-form "A human being is mortal" with the empty place "a" holds in every case. Namely, it always yields a valid proposition, where we may <also> bring particular humans to the empty place. The empty places of a propositional function are what mathematicians call arguments. We would then have to distinguish between quantified and not quantified functions. Quantified propositional functions are those making up the full logical matter of an affirmation or of some other quality, because they can only be qualified in one of
181 the two "quantities", in the quantitative mode of particularity or that of universality. To be noted thereby is also that talk of functions carries over—but, of course, in a figurative sense—from the whole functions to their components, to their nominal ones especially–provided, obviously, that one can also speak of nominal functions–precisely <to> nominal components of the form "an A" or "one". Nonetheless, these no doubt do not admit of any separate and genuine quantification, something which harmonizes splendidly with my theory that in generality-judgments the terms do not, indeed cannot, ever, take on any positing, without the whole judgment losing the character of generality-judgment.

Once we have made these distinctions, it is easy to see that, with respect to their matter, all judgments break down into two great classes. The matter is either quantified functions, or it is individuations of unquantified functions proceeding from corresponding functions as a result of filling the empty places, the arguments, with definite "values". As a result of this, definite judgments emerge that do not contain "arguments", but rather "definite terms", full terms or ones with "this" or "certain" standing parallel to them there. A positing is buried in each definite term, while the quantified functions cannot contain any partial positings in the corresponding argument places. Functions are pure functions when they do not in general contain any definite terms and positings and are mathematical functions when they exclusively contain purely logical forms and concepts. Accordingly, a layering in the matter becomes apparent in the function-judgments that again allows differentiating in it between a content proper to it, namely the function, and a form, namely the quantity. And, in addition, the function then itself has forms and stuff. Otherwise, what I said earlier about the difference between syntactical stuff and syntactical forms by all means holds good, only that, for example, a distinction is to be made between nominal function and determinate nominal presentation.

A short while ago, I said that all judgments break down into two groups. That does not therefore merely apply to propositionally simple judgments. It is actually clear that, just like judgments (and like the propositions that occur as complete judgment-components), propositional functions break down into simple and composite. 182

The theory just now outlined is particularly further confirmed through reflection upon negation. Negations can occur as component parts of the distinctive kind of "matter" that we called function. They can, however, also fasten on to generality, to the mode of quantity. Provided that it is directed against the quantity, the negation of "Some A are b" and "An A in general is b" can read, "In no case is A b" and "An A is not generally b, not in every case". That therefore gives, if we take everything into account, many kinds of forms of negatives that permit judgment-functions.

The particularized functions, those grasped in the quantitative mode of particularity, can then really be considered existential propositions in the broadest sense.[1] They are—no matter how we rationally grasp the concept of existential proposition—at least equivalent to such propositions. One can always equivalently say: There is a system of values that satisfies the function as equivalent of the particular proposition, just as one can place universal propositions on a par with equivalents of negative existential judgments. There is no system of values that does not satisfy the function, though, here, the existence-thought will remain remote.

[1] Subsequent note by *Husserl*, "I actually abandoned the theory that sets forth a form of the particular judgment 'Socrates is something' as a form of an existential sentence in 1917. It is seriously open to doubt". (Editor's note)

<c) Differentiating Between Categorical Judgment and Categoroid Function>

If the theory of functional-judgments I have outlined is correct, then it results that *Brentano*'s opposition to the fact that particular and universal judgments are categorical judgments in the genuine sense contains a valuable kernel of knowledge, or that it is necessary to secure the concept of categorical judgment scientifically. It becomes apparent that every judgment is either a functional-judgment, a judgment with arguments, like every universal or particular judgment, or a definite judgment, like "Plato is the student of Socrates". As regards its matter, we can view every definite judg-
183 ment as fulfilling, fixing a function. In other words, and inversely, we can mathematize, not-purely or purely mathematize any definite matter. In the latter case, propositional configurations arise as pure propositional functions. Even the conceptual nuclei are then mathematized. Everywhere, we have only arguments and simple syntactical and other forms.

Initially, we then find in definite judgments a broad class of cases of characteristic dual composition, where a second thing, a positing-on (for example, "The oven is warm") is built upon a singular or plural positing-under. If we understand by that categorical judgment, then it is clear that one cannot talk of categorical judgment in the sphere of the corresponding functional-judgments in the authentic sense, because a distinction is to be made between a function of a categorical judgment and a categorical judgment itself. The function "A human being is mortal" is a configuration of a possible categorical judgment, but not itself a categorical judgment. A positing-on is not carried out on a positing-under there. But the functional-judgment does not do that either. Quantifying and qualifying do not bring any positings into the judgment material.

On the other hand, one will say that it is not in vain that judgments of the kind "This A is b" and "An A generally is b" are referred to as being closely related, indeed that both are called categorical. What is categorical is, so to speak, posited on what is functional. The function has dual composition and with some modification includes positing-under and positing-upon, which is of course not real positing-under and not real positing-upon. The function is, so to speak, categoroid, not however really categorical. And the functional-judgment itself is determined by its function, which is indeed immanent to it. So, one may therefore after all take categorical judgments and quantifying judgments of categorical function together as one, but more as parallels than by virtue of actually identical syntaxes. The actual syntaxes are basically essentially different, despite a certain mirroring of the ones in the others. (One will not even be able to speak of actual dual composition in the functional-judgments, because the generality that enters into a special intimate alliance with the Is and the argument does not actually divide. It is nevertheless a thoroughly coherent form animating the whole function. The composition lies only in the function.) 184

And what holds for the categorical functional-judgments, holds likewise, as I can say in advance, of hypothetical and all remaining functional-judgments. There are propositionally simple and propositionally-composite functions parallel to the differentiation of judgments. But actually, this again signifies that the composition of the propositional functions does not make the functional-judgment itself into a propositionally-composite one, so that, in order to take the viewpoint of the greatest possible simplicity into account from the propositional point of view, in dealing with functional-judgments, it would have to be said: Functional-judgments ought first to be dealt with along with propositionally simple functions.

If then the propositional composition of the function out of the functional-judgment itself does not make a propositionally-composite one in terms of matter, then all the more not a composite judgment as judgment. For example, "If something is A, then it is B"—the whole judgment is a thoroughly simple judgment in terms of matter and quality. On the other hand, there is actual composition of functional matter and functional-judgments. The first holds for hypothetical forms of the kind, "If universally an A is B, then B is universally C." So, the corresponding causal judgment "because..." is then composite in terms of quality. It is thereby to be noted that my concept of propositionally-composite judgments is determined purely as composition of judgments with respect to their matter. Simple propositional matter is simply-independent judgment-matter, and this only assumes a composite form in the composition.

About the controversial issue as to whether every judgment–and more precisely, every simple judgment–is categorical or not, remaining left over after what has been explained is therefore only the question as to whether independent judgment-matter, a proposition with fixed content (therefore, a definite proposition), must necessarily be categorical if it is to be simple, and whether accordingly a simple propositional function must necessarily be categoroid, or whether there is yet another function form here, say the existential form. Let us therefore move on to existential
185 judgments.

<d) The Concept of Inexistence. The Impersonals (Definite Categorical or Functional) Are Existential Propositions in This Sense of Existence>

The concept of existential judgment is by no means completely unshakably defined. A fixed definition was obtained by identifying the concepts existential judgment and particular judgment, because the latter concept is fully defined. But, this identification will not as a rule be countenanced. The term existential judgment is not as a rule used in this sense.

A certain range of related instances of application of discourse about existing is traceable back to the concept of inexistence, to the concept of being-in-something. It can be said of a subject that it <may be> a part of some whole, an item of some collection (of a multiplicity) in some context (for example, in a spatiotemporal context). Related to that is also being-somewhere or -at-sometime. We often use "somewhere" broadly in a non-localized sense for every being-in-something. People then often say "let the object exist"; let it exist within the context, let it be "existent"; present, there. For events too, "they take place" and so on. There is a number in the number series. There is the book in the library, the letter in the box, etc. The letter is no more. It is "no longer existent". It has been burned.

The being-in-something, -somewhere, -at-some-time is often stated in incomplete predication, provided the context is intended, but not stated, often, however, even in indefinite, namely, functional predication. In affirming such statements, we arrive at particular judgments. If we initially still keep both cases together, the undeniable ambiguity of what is called the copula, of the little word "is" results, insofar as the "is" frequently also means "exists", and "exists" contains a relative, namely, the being-in-something. Where, however, the Is acts as pure copula of the categorical judgment, we can never instead say "exists" for it. For example, for "The castle is a Renaissance building", we cannot say "The castle exists a Renaissance building". In contrast, wherever we say "being in something", "being somewhere, at some time", "exists" can be substituted there. The thing exists here, there, now. It exists in the 186
room, as part in the whole, and so forth. For "exists" we then frequently say "there is". Obviously, all impersonals are existential statements in this sense. We can substitute the "exists" precisely in this sense everywhere in them. "It is raining = raining exists", and obviously "here" and "now" is to be added. "There is a dance, it is, it exists", namely, "at the Green Tree", "today", and so forth. "Money is lacking: No money exists in the cash register". "There are polyhedrons, polyhedrons are, exist in space, or among geometrical shapes", and so forth. Naturally, relations can also play out in the ideal sphere, and infinite aggregates, infinite extensions of concepts can act as members of relations for surrogative reasoning, as ones in which something exists.

Not all the individuations exhibiting the relations relevant here belong in a theory of forms of relation-judgments or in a pure theory of forms of logical relation-forms, but a part's being contained in the whole or in the sum of the parts of the whole, and likewise membership in a multiplicity surely do. Those are purely logical relations, while the being-somewhere and at-some-time-in-the-world no longer belong in formal logic. There are, therefore, in the theory of forms of relation-judgments that must be pursued more closely the surely to be cited forms "In A, B is (exists)", "Under the A is, B exists". We also speak here of a sphere of existence, a domain of existence. The B is posited as being in it. Whether the domain has remained concealed or "unthought" in incomplete thinking is of no concern to formal logic. Naturally, like all categorical propositions, these relation-propositions break down into two groups: that of definite categorical propositions and <that> of function-propositions. "It is raining here" ("The letter exists in the box") is a definite proposition. "There are numbers smaller than an arbitrary number n" is a function-proposition.

<e) Obtaining Existential Judgments with a Definite Term. They Are Categoroid Functional-Propositions with a Quasi-Subject Without Positing>

With this I have completely defined the original existence-concept and the domain of statements associated with it: Existence = Inexistence. However, the talk of existence shifts by itself, that is,
187 owing to the fact that originary existential judgments, judgments of inexistence, easily turn in directly equivalent ways into those in which inexistence is no longer stated. We thereby come, particularly in the sphere of function-judgments, upon some important forms, namely, <upon> those that *Brentano* exclusively took into account without having systematically comprehended and defined them.

Instead of judging, "Under the B is an A" or, what <is> the same thing, "An A is under the B (occurs in it, exists in it)"; our judging easily changes in such a way that we state "Anything under the B is an A" where the thought of inexistence is slipped into the subject, whereas the Is does not now imply existence. The relationship to B can now be indefinite. The only thing of interest to us is the fact that something (that is known to be in the group of Bs) is an A. And we can also state that. And, "Something is an A" may perhaps interest us only in general. The something always possibly occurs within a context, for example, figures always on a plane, numbers in the number series, and that may again convey talk of "it exists", "there is". But it is no longer actually intended. We intend that in general something is an A and we say "It is an A, there exists an A." Also, the other way around, "An A generally is, exists under the B", "An A generally is in some context", "An A generally is existent, occurs"–the context is still thought in functional form there, and the existence-thought therefore involved in the function. Finally, equivalent with that however is "An A is generally something", where the relationship to a context has fallen out, even in the function.

Time has come for me to <speak of> the borderline cases in inexistential judgments and categoroid functional-judgments. <To the form> "An A exists in some context", I contrast "An A is something, is a something", as well as the reverse form "Something is an A", all of which are equivalent to one another. Those are one-place existential propositions–in a certain sense, namely, provided they contain only one term, not counting the arguments. Nothing very essential changes when we bring in the plural multiplicity and say instead of "An A" "A", "some A" in the indefinite plural.

That this group of propositions is the *Brentan*ian group, namely, the group of existential propositions that he has in mind in his 188 reductions, is clear. It is an outstanding group of judgment-configurations or of functional-judgments, outstanding owing to the fact that in them only one term occurs besides the arguments.

We can finally transform this term into an empty one as well, and then, for example the form "An A is something" is transformed into "Anything is something", and the inverse form further reads similarly. In addition to that, "There is anything whatever (namely in some context)". We can take these most extreme borderline cases of functional judging together with the previous ones as a class of functional-judgments with a simple function that either contain one or no actual term. Obviously, more complex constructions can be generated if one makes the functions more complex, for example, "Anything whatever that is somewhere, is something", and so on. These are seemingly uninteresting complexities. Yet, closer reflection would show that they too have a valuable content.

The borderline cases of categoroid functional-judgments, more precisely, of one-place particular and universal judgments that we have arrived at in this way, lead to an only apparently completely clear and complete determination of a concept <of> existential judgment that is now no longer to imply inexistence. The difficulty lies above all in deciding how far the range of these judgments extends and whether we thereby really come upon the normal concept of existence that makes up a tangible unit. The judgments we have obtained are functional-judgments (genuine particular judgments when they <are> affirmative, when <they are> negative, then, universal). How do matters stand then, if for "an A" we put the definite term "Socrates" in these forms "An A is something" or "Something is an A"? How do we obtain completely ordinary existential judgments with definite terms such as "God is", "Socrates is, exists", and so forth? For example, in specified ways by means of substitution? Is "God is" = "God is something, something is God" therefore the true meaning of the existential proposition? The fact that in denying God's existence atheists will also possibly express themselves using the proposition "God is nothing" seems to agree splendidly with that. Of course, not the other way around, "Nothing is God", which points to the fact that only one of the equivalent forms can count as the explication of the existential proposition, namely, "A is something", "God is something".

189 But the difficulty precisely lies in understanding this proposition. The much debated, and never resolved problem as to whether the existential proposition "God exists" is or not categorical, now emerges in the form "God is something", "The centaur Chiron is nothing". At first, we would like to consider those categorical propositions, just like propositions with "an A", for instance, <the> propositions "A centaur is nothing", "A regular body is something", understood as illustrations of "A centaur does not exist", "A regular body exists, so there is something". But, one must carefully guard against mixing things there. I had defined the latter proposition-forms, "An A is something" and the inverse <form> as functional-propositions and not as definite categorical propositions and, in spite of the sameness of expression, that is an enormous difference. Note that in the categorical proposition, a double positing, a positing-under and a positing-upon, is given. The like can be accomplished with definite or indefinite terms, which then are not, however, arguments. And accordingly, the formula "An A is b", and consequently also "An A is something", is ambivalent. The proposition "A glass is broken" is understood as a categorical judgment. It is predicated of an indefinite glass posited as a reality that it is broken. Were "A glass is something" predicated in this sense, it would be a pure ridiculous tautology. And when "Socrates is something", "God is something" is predicated in this sense, it is empty

talk in precisely the same way. For, if with the subject-positing, Socrates is posited in the reality mode, then the empty "something" states nothing new about him at all. Naturally, every subject is something. That is already implied in the subject-positing as such. Likewise, it is clear and correlative that, if it is to be categorical, the corresponding negation would be a contradiction. "Socrates is not something"–positing "Socrates" as actual in the subject and then denying that it is "something" is a pure contradiction, namely, contra the subject-positing as such. Therefore, when we understand the propositions "Socrates is something," "Socrates is nothing" as categorical, then they roughly amount to something like "Socrates is Socrates" and "Socrates is not Socrates," in any case something equivalent to that. It is, however, completely evident that they cannot then be equivalent to the existential propositions "Socrates exists," "Socrates does not exist."

On the other hand, it is however also clear that the same modes of expression, "A is something," "A is nothing," from time to time serve–therefore, can serve–to denote existential judgments. 190
Socrates, the famous philosopher, who was in Athens of old, is something. The centaur Chiron is nothing. It is a mythical fiction. These ways of speaking are found alongside one another in all spheres of being, not only for spatiotemporal realities, but also for atemporal idealities, such as numbers, geometrical shapes, and so forth–namely, for what is definite as well as what is indefinite. "A regular decahedron is nothing," "a round square is nothing," and so forth. Now, for the forms with "an A"–therefore, "An A is something," "An A is nothing"–we have not had any reason to encounter a problem because they were functional forms. These particular or universal propositions are categoroid, but not by any means themselves categorical. We must only still always make sure in a clear-cut way not to attribute falsely something actually categorical to categoroid judgments owing to the ambiguity of the mode of expression. And we must in general make sure that, in a function, the arguments do not include any positing.

If we however now move on to those individuations of these forms that arise when–as in "An A is something," "An A is nothing"–we use a definite term for "an A," then a remarkable splitting results and a most remarkable feature on one side. Either we posit

something in the case of the definite term, we state, "Socrates is something", "God is something"–where the subject is an actually posited, and hence, genuine subject–, then the "something" on the predicate side loses the status of argument, and the judgment turns into a tautology. Similarly, in the case of negation, the categorical judgment that then arises turns into the negative counterpart of a tautology, i.e., into a contradiction, or we do not posit anything, and with the same words–"Socrates is something", "The centaur Chiron is nothing"–a totally different thought is realized. The "something" retains the status of argument. We now have remarkable borderline cases of particular judgments that only have a single argument. We have a judgment that looks exactly like a categorical judgment, that even has a fixed place, has a proper-presentation such as "Socrates" as a quasi-subject, but what is presented is not posited. Now, we do not have a tautology or a contradiction, but we have a positive or negative functional-proposition with a definite term. It is categoroid, but not categorical. Therefore, it is not stated of Socrates that he exists, that he is some
191 thing, as it is stated of him that he was the teacher of Plato, the one who saved Greek culture from Sophism, and so on, as though both predications were on the same level. Were these predications equal, then "something" would be a predicate, and the predicate "something" is vacuous, a tautological predicate. Rather, there is not predication in the usual sense in the existential judging in general, but, as in the case of the particular judgment, a function–"Socrates is something"–receives a validity-value ideally, which is, so to speak, stamped in the empty place.

And we likewise see that what is crucial–what *Kant* gave precedence to in his multiple efforts to clarify existential propositions, namely, that existence is not a real, not an actual predicate–is not there. Existential statements do not in general have a predicate, any more than they have a subject. The much debated question as to whether there actually are subjectless propositions, and actually predicateless ones, is consequently evidently answered–and in the affirmative sense. On the other hand, *Kant* was again right when he said that existential judgments are not analytic (i.e., to be precise, tautological), but synthetic <judgments>–only, of course, he did not realize that not every synthetic judgment and, in general, every judgment is categorical, if it is propositionally simple.

You at once see what brilliant instinct *Brentano* demonstrated since he did not let simple judgments in his sense, i.e. judgments with affirmative positing, count as categorical, or since he actually first discovered the distinction between such simple judgments and categorical judgments, as well as already having discovered that for universal ones differences obtain between plain positings and positings built one upon the other.

The extremely remarkable thing about the judgments discussed–"Socrates is something"–in the sense of existential judgments is the fact that they occur as functional-judgments in such a way that their function contains a proper name without positing. In general, positings can occur in functions, for example, when we judge, "A triangle on the sphere has a sum of angles greater than two right <angles>." Only the argument cannot have any positing, but perhaps one can indeed appear in an occurring determination. The remarkable thing is that even proper-presentations can occur without positing in judgments and that judging is thus in general, possible or, meaning-theoretically speaking, that a 192
class of judgments is characterized by the fact that a coherent supply of meanings functioning as the substrate of the entire judgment-positing indeed contains various nominal presentations and other formations as direct components, but without any actual positing. And it <is> further remarkable that this supply and the overall positing have the specific relationship that the overall positing confers a multiplicity of positings in, so to speak, one fell swoop, as is the case for functional-judgments with their many arguments. We have precisely two different manners in which judgment-positing takes place. One is the fact that positing is synthetically bound to positing, namely, one is built upon the other. Various propositions can thereby be wound up in one another and connected with one another by means of insertion, but always in such a way that all together they combine in a unit of positing. A particular case of that is what is called categorical judgment. Another simple case is the causal judgment. The other possibility is the one we know from functional-judgments, the fact that func-

tions with full arguments are subject to an overall positing that all of a sudden, so to speak, aims into the arguments.

I must not linger any longer here. To close, I note in addition that I have only taken into consideration the simplest existential propositions to be stated in the forms of speech by "there is" and "there exists." However, the complex forms do not present any problems. So it is, when we introduce plural predications, or let the term A in "An A is something" become complex. If the range is sufficiently extended, then the fact that every arbitrary functional-judgment, however complex, can be equivalently transferred into an existential judgment, namely, every particular judgment into a positive existential judgment and every universal one into a negative one, is to be inferred from the theory of validity of judgments. It is also to be kept in mind in this respect that, as you can easily assure yourself, only the positive propositions are particular in our primitive forms. The negative ones are, however, always universal in nature. "An A is nothing" necessarily has the meaning "An A is
193 in general nothing," "No A is something," "There is no A."

<f) Dispute with Bolzano's Thesis: Existential Judgments as Categorical Judgments About Presentations>

The question is now whether my earlier analysis of existential propositions, of statements with "it is," "there is," is complete, whether the meaning of this discourse is exhausted in all cases, whether–apart from statements of inexistence, where completely definite judgments can indeed occur–it is therefore always only a matter of functional-judgments, be they particular or universal. After we have already been cautioned so often about how very much language loves, and how in our speaking we ourselves love, to keep directly equivalent thought transitions without differentiating expression (therefore, to retain the old ways of expressing, when the thought has shifted within the lines of direct equivalence), then from the outset we have a notion of the possibility that here too it is still many times a fact that, therefore, yet other judgments–whether closely connected to, or equivalent to, but structured differently, designated functional-judgments, or corresponding inexistential judgments–hide behind the familiar forms of speech.

It is in fact striking that, in pondering *Brentano*'s paradoxical theories, other interpreters have looked in such completely different directions than I have. So it is that the great logician B. *Bolzano* interpreted "A exists" as "The presentation A has objectivity." Related to that are similarly sounding and oftener concurring interpretations such as: To presentation A corresponds its object, corresponds something in reality, the presentation is a valid presentation, and so on. What should the "presentation A" signify here? I make a clear-cut distinction between the subject-presentation as experience and the nominal meaning, the nominal presentation in the meaning-theoretical sense. Such a distinction is not usually made (apart from *Bolzano*), and so the psychological and meaning-theoretical run together. People generally then further distinguish between presentations to which correspondingly objects exist or do not exist in reality, in truth (valid or invalid presentations), for example, "a lion"–"a centaur." *Bolzano* said: "objective" and "objectless" presentations. He then thought that existential statements only apparently predicate something of A and attribute the putative predicate "existence" to it. In truth, though, not A, but rather the presentation A would be the subject 194
and the predicate attributes "objectivity" to it. If I say "God exists," then I am not actually stating something about God, as if I said "God is good," but I am thinking the presentation "God" has objectivity. Similarly, "A real number exists" signifies "A real number has objectivity." Naturally objectivity is, as *Bolzano* keenly observed, not to be confused with real existence. In this manner, *Bolzano* is able to stand by his basic thesis that anchors such a wealth of most characteristic points in his theory of judgment-forms, namely, that every judgment is categorical.

At first, the interpretation is very shocking and appears to lead to an absurd infinite regress. Saying a presentation has objectivity is after all saying that an object corresponds to it. And what does this "corresponds" signify? Surely that there is an object of this presentation A that is in turn an existential proposition that would likewise have to be reduced. We would have to say on that account: There is an object of the presentation "object of presentation A," etc.

Bolzano would probably answer here that that infinite regress of reduction does not apply if one but grants that objectivity–and in the contradictory case objectlessness–is an essence-characteristic of the presentation concerned. Therefore, every presentation may have an essence such that objectivity or objectlessness may be attributable to it, precisely as truth and falsehood may be attributable to every judgment as an essence-characteristic. Since in each case it is a matter of an ultimate property that is not further definable. Any further inquiring and further reducing would be senseless. Now, were it surely correct that if a presentation A had this validity, the presentation "object of presentation A" would also have this same property of being valid, and so on *in infinitum.* But that would be a pattern that would be as unsurprising and also obvious as, for instance, in the case of judgments, the pattern that when U is true, the new judgment "It is true that U is" is true, and so on *in infinitum.*

As for the relationship to the particular or universal formulations of the proposition "An A exists"; or "A does not exist"; it could
195 then be argued that the meaning is indeed different on both sides, but otherwise plainly equivalent. If I can legitimately say that the presentation "God" has objectivity, the presentation "water-sprite" is objectless, then I can also say in an evident way, "God is something"; "A water-sprite is nothing". And if that is correct, I can also say what was said before: they are equivalent, yet different in meaning. One could also argue that the difference in meaning comes clearly to the fore when we think about the fact that people however often disagree about whether objects or states-of-affairs that occur in presentations <as> presented, in judgments as posited as true, could be addressed as realities or not. For example, realists and idealists disagree about whether phenomenal objects do or do not exist in reality. In the Middle Ages, nominalists and realists disagreed about whether the objects of general presentations existed in reality or were mere mental fictions, and so on. That is, however, defining the contrast between what is presented and what is real. We are contrasting real world–world of experience and empirical science–and world of the imagination. A painted landscape, we say, is not a real landscape. The image at the back of the mirrored wall is a mere image, does not exist in reality, but the actual body sending out light rays in front of the mirror <exists in reality>.

Therefore, a broad class of cases is marked out here in which the "A is" and "A is not" means "A is something real", "A is not something real", "is something non-real". Obviously the meaning is here different than in our functional forms. Nonetheless, the latter frequently serve to state reality. In terms of their purely functional meaning, forms such as "An A is something", "Anything is an A", "Nothing is an A" and so on do not at all contain the conceptual thought of reality. But, they occasionally acquire it and are also equivalent to such reality judgments. It is to be noted thereby that reality is always to be taken in the most general sense. We usually speak of reality in the sense of real reality (*realer Wirklichkeit*). There being reality (*Wirklichkeitsein*) is also being effective (*Wirkendsein*). But, on the other hand, in mathematics, we however also speak of existence and reality. If, though, we are clear about the fact that judgments about reality or non-reality actually have a different meaning-content than pure something-judgments, then their interpretation again by *Bolzano*'s formula seems very natural, almost a matter of course. A centaur is a fiction, something 196
non-real, a lion something real, Socrates something real–what is that to signify other than: The presentation "centaur" is something false, invalid, objectless; the presentations "Socrates", "a lion" <are> objective, correct, valid.

However, the matter is not so simple that one can side with *Bolzano* without further ado. I naturally do not deny that it is possible to judge about presentations. If by "judging about something", we understand categorical judging, then there is naturally genuine categorical judging about presentations, for example, when we judge, "The presentation 'a centaur' is a nominal presentation",– and likewise, "It is an objectless presentation". Here the presentation is posited exactly as Socrates is posited in the judgment about Socrates.

What appears questionable, though, is that existential judgments in the sense of reality- and non-reality judgments ought to be categorical judgments about presentations. The reason for interpreting them thus is obviously the quite conspicuous altering of perspective that is there when, for example, on one occasion, we judge categorically, "The big lion has broken out of the menagerie" and on another occasion, "A lion, that is something real, is not something fictional". *Bolzano* explains that saying that, in one case, one has actually judged about lions, in the other case, about the presentation "a lion". But, is it not also a difference of perspective whether, on one occasion, we judge, "'A lion' is a nominal presentation"–where we really completely confidently judge about the presentation "a lion" (and the expression "a lion", so to speak, brings quotation marks into our thinking) –and whether, on the other hand, we judge, "A lion exists"? Were *Bolzano* right that presentations are judged about in both cases, then no change of perspective would really occur when we go from one example to the other. It is clear that, when it comes to consciousness, we use the expression "a lion" in completely different ways there, and it already becomes apparent there that *Bolzano*'s interpretation cannot be right. (So, the predication of existence referring to an equivalent inexistential predication proceeding, so to speak, on the same plane would be incomprehensible.) Therefore, it is certain that when we make the existential statement "A exists" as a reality statement, we are not making it as a categorical statement about the presentation A.

197 Giving the matter closer thought, it occurs to me that *Bolzano* did not see some fundamental distinctions, which admittedly also were not for the most part seen after him either, and in any case their significance not recognized. It is at the very least due to this that people have not been able to come to grips with predications about realness and, parallel to that, with those about truth–by the same token, how the lack of clarity about the intrinsic structure of functional predications has not allowed the context of inexistential predications with the borderline cases of particular and universal predications to surface and consequently the meaning of particular and universal there-is-propositions to become comprehensible.

We can take up functional predications in like manner. There we encounter, I have pointedly emphasized this, the remarkable fact that nominal and propositional meanings having no inherent affirmative qualities occur completely differently than in the definite predications. When we state non-tautologically "Socrates is something", "A centaur is nothing", we do not judge categorically as when we say, “Socrates is a philosopher” or “A centaur is riding by.” The nominal subject-meanings are free of positing and not subjects in a serious sense.

So, when we say, “Socrates is something real,” he “really exists,” a judgment is not made about Socrates; the being of Socrates is not posited in the form of a positing-under. That becomes apparent in negation, in which the subject component obviously functions in completely the same way. Therefore, were we to state, “Socrates is nothing real,” we would have positioned Socrates to begin with as, for instance, in the judgment “Socrates is sick,” “He is no Spartan,” and so forth. Thus, he would really be intended as a reality from the beginning, and the statement that he is not anything real would be as good as if we said the real Socrates is non-real, which would be a contradiction. And, the positive judgment “Socrates is real” would likewise be a kind of tautology had we already posited Socrates as a reality. Now, one will not want to decide to identify non-tautological reality predications and functional-judgments here. (Apart from the fact that the question as to what corresponds to the functional-proposition “Socrates is something” as a definite proposition would cause unsolvable problems were we not to have other reality predications that are not functional predications.)

The question therefore is what the subject truly means by reality predication there in the existential predication, for example 198
“Socrates exists.” *Bolzano* says that what is posited as a judgment is not “Socrates,” but the “presentation Socrates.” That can, however, mean a variety of things. The subjective presentation as experience of presenting is naturally ruled out here. In *Bolzano*’s sense, “presentation in itself” is intended instead, and I have interpreted that as nominal meaning. But that is not the only concept of presentation, and precisely that concept has proved unusable. Presentation can also mean “the presented as such”–and this expression <is> taken in a specific sense. As long as we operate within the theory of forms, exclusively looking at the meaning-configurations, at their matter and qualitative nature, unconcerned about validity or invalidity, we have no reason to make use of these distinctions. But, it is conceivable that where there is talk of forms of statements in which the thought of validity or its equivalent

explicitly occurs, the distinction must be introduced.[1] It is part of the essence of the theory of forms that the series of original meaning configurations must be supplemented by a series of ones that draw in concepts that, as definite concepts, seem to violate the pureness of the configurations, yet must be drawn in as belonging to the essence of pure logic. That is the case with concepts such as "meaning", "proposition", "object", "state-of-affairs", and so on, but especially with the concept "realness"–"reality" and "non-reality", as well as <with> "truth" and "falsehood". These latter concepts are, however, those which are necessary to the distinctions indicated.

If we take any categorical judgment, then, I already said earlier, a distinction is to be made between subject-object and subject-meaning, likewise between predicate (property) and property-meaning in the judgment. We also distinguish between the judgment itself–as meaning–and the state-of-affairs judged. That was admittedly already said earlier, but closer reflection is required here.

These distinctions must be made whether the object does or
199 does not exist in reality, likewise, whether the state-of-affairs really obtains or does not obtain. We understand very well what is intended by that–since we constantly make use of that *in praxi*. It can be that we seriously make judgments about objects–therefore, in so doing, posit them as realities–, but it turns out that such objects subsequently do not exist. We consider states-of-affairs to be really obtaining, posit them as such, but our judgments turn out to be false. How then does this positing-as-real of objects that do not exist at all, or perhaps do not exist, the positing of states-of-affairs that possibly do not exist, happen? Well, very easily. We just judge, and while we are judging, we have the consciousness "S is p!" But, in the course of this, we distinguish between the meaning and the state-of-affairs meant, and on the nominal side between the nominal meaning and the object meant. How do we arrive at this distinction? Well, owing to the fact that there is something along the lines of identifying and differentiating judging. For

[1] Compare Appendix XIV, *The Nominalizing Transformation of Matter. "Existence" as a Predicate.* (Editor's note)

instance, we totally identifyingly judge "Socrates is the son of Sophroniskos"; "The victor of Jena is the same as the loser of Waterloo". And, what we then affirm of Socrates is obviously also carried over for us to the son of Sophroniskos. Every affirmation that is linked to the one subject is also linked to the other in our consistent judgment-consciousness. With Evidenz, we can *in abstracto* link the law of validity that what is identical can be substituted for what is identical in the judgment without changing its validity.

Figuring on each side of an identity judgment is, in a certain sense, the same thing and, in a certain<sense>, something different. In identity judgments, we judgingly say "the same thing"; and what is the same is object. On the other hand, a possible shift of one's gaze goes along with this, which says there that in this judgment itself something different is connected through the identity-moment–namely, the meanings. In this way, meaning and object separate for us. We say a judgment is made in the identity judgment about the object and about its identity with itself, and that has meaning due to the fact that two–perhaps very differently structured meanings–are linked by the meaning "identical". Judging about the identity of Socrates and the son of Sophroniskos is, however, not judging about these meanings, rather the latter takes place in new judging in which the meanings are objects, namely, via the medium of meanings of the second level. 200

If one has made this distinction, then three kinds of subjective judgment-consciousness are in general distinguished: subjective judgment-consciousness, judgment-meaning, and the state-of-affairs judged. And the same <holds> for nominal presentations: nominal consciousness, nominal meaning, nominal object. The nominal meanings occur in the same object as components of the identity judgments. This is the case for all judgments. For, properly viewed, identifying and differentiating play a universal role in the field of judgment. Especially in the case of the predicative basic form of propositionally simple judgments in the form of categorical predication, predicate-positing on subject-positing is not arbitrarily posited upon, but the predicate posited is precisely the predicate of the subject. The same thing meant in the subject-meaning is what is intended in the predicate-meaning. Indeed, when, for example, we have a property-meaning such as "red," "the table" is not primarily meant, but rather just "red," but, when we judge, "The table is red," the red is precisely red of the same thing that is intended and posited in the subject-meaning. The moment of redness is not grasped and posited in its own right, but rather as something about the table. It is, as it were, a partial identification underlying there. I go from the presentational table to the red-moment. I do not drop the positing of the whole thing, of the table, but what is presented overall coincides in a partial identification with the moment stressed, and thereby, in the meaning version, what is meant as "this table" becomes in-one-positing with what is meant and posited as "red."

Identification is again the underlying factor when we attribute several predicates to one subject. Several predication-rays coinciding in the subject position go out from the one thing posited as table, precisely from "this same table": "This table is red, the same one is made out of wood, and so forth." We can also specifically grasp in terms of meaning the identification indicated by the word "coincidence" as when we say, "This table is red, the same thing is made out of wood, and so forth," also when—transforming the form—we further say, "The same table, the same object, that is red there is also made of wood, and so forth." Those are evident transformations that are *a priori* possible on the basis of the meaning from which the first judgment proceeded.

In the judgment-context, the object as identical therefore stands out in contrast for us in different ways vis-à-vis the different
201 presentations or meanings. The same object confronts us as a unit vis-à-vis different meanings in which it is presentational, for example, vis-à-vis different predicate-meanings, which in the meaning relationship to the same subject indeed have the same, so to speak, focal point, but nevertheless are different, inasmuch as each introduces a different predicate-determination. Obviously, the differentiating, as negation, as cancelation of an identification figures as a counterpart of the positing of oneness, of the identifying in the broader sense—and that too can be useful to the contrasting.

Let us now take the case in which we do not simply judge, but judge affirmatively (approvingly) or negatively. In the face of a semblance of a judgment or passing through doubting and deliberating behavior, we adopt an affirming or negating position. Perhaps we had already simply judged–and in a simple categorical form "S is p!"–, but grown subsequently doubtful, or even, it initially only seemed that S is p, and we are now intending a definite position-taking. If we make a go of it, then we end up either with an affirmation, "Yes, S is p!" or with a negation "No, S is not p". Instead of repeating the complete expression, we also say something more indefinite and simpler (there the S and <the> p are really often very complex), "Yes, it is so", "No, it is not so"–or more simply "It is so", "It is not so". (The "yes" and "no" really go without saying when we repeat the judgment.) That will then even turn into expression of the affirmation and negation. Furthermore, it is also carried over from the categorical judgment to every judgment–"If M, then N: yes it is so". The way of expression may become still simpler. Instead of "It is so", more simply "It is" – negatively, "It is not". (I note in addition that when the affirmative word is brought in again, we do not say, "It is indeed so", but "It is really so", whereby the "really" precisely has the meaning of "indeed".)

Now predicative expressions can also be understood that arise on the basis of such experiences in predicative unfolding of the same by means of which characteristic forms of categorical judgments follow: "The fact that p is S, that is so" (or, "–that is really so") and "It is not–", "It is <not> right–". What are we making a statement about in so doing? What figures in the subject position? We are obviously not stating something about the meaning, not about the proposition, but about the state-of-affairs. When I judge "The fact that $2 \times 2 = 4$ is, that is so", I am not judging about the 202
proposition $2 \times 2 = 4$, but about the arithmetic state-of-affairs, about the equation < 2 + 2 = 4 > itself. On the other hand, we nevertheless again note that the way of judging is modified, namely, inasmuch as I nevertheless am not judging about the state-of-affairs in the normal sense, the fact that actual obtaining is attributable to it, roughly analogously to the way in which I judge that this table is red. If I judge categorically about S that it is p or not p, then that S is posited as existing. If, however, I judge "The fact that S is p, that is really so", "–that is not so", then I obviously have not already posited beforehand the fact that S is p on the subject side, since I am after all just deciding positively or negatively about, as

it were, imparting being to the predicate side. Actually, it is not merely the state-of-affairs that is judged about here, but the state-of-affairs as such presented in the meaning, the state-of-affairs as it is identity-point of the noema. The same goes for nominal meanings. Someone perchance speaks seriously about regular decahedrons. Our no is immediately directed at this. We say, "That is not," "–is not really." Again, the object involved there is not, however, the regular decahedron itself (as if we had made a statement about it–hence it would really be posited as being, while we really replied, "It is not"), but rather the presented decahedron as such.

Now you will, however, object: Is that not precisely *Bolzano*'s theory, just expressed in different words? No, and I again have to recall what was already explained: Namely, when we engage in nominal consciousness, then the noema, what was conscious in so doing, what was presented as such in so doing, is the meaning. In so doing, many a thing is of course left out of consideration. It does not matter whether we have clarity and distinctness or not, or <whether> we are more or less attentive, and so forth. According to the circumstances, what is presented is characterized very differently, but that does not change the identical content. It does not change the meaning, what is intended as such. But, then ambiguity enters in. Merely the object can be intended with what is presented, and the object can be <intended> in the how of its intendedness, and this difference becomes evident through comparison of the terms of identification. (While identifying, we were conscious of the same thing as the same thing, but the how, the meaning, is different on each side. The victor of Waterloo ≡ the loser of
203 Jena–attention can be directed toward the object, but it can also be directed toward the meaning. The meaning implies the object intended. The object as meaning-moment is not the object in the customary sense, but precisely the meaning-moment.)

<g) The Logically Different Proposition-Forms Converging Under the Heading "Existential Proposition">

I have presented several logically different proposition-forms that are all equivalent with one another (<or> may become equivalents after suitable extension of their originally narrower meanings) and that by virtue of this equivalence, which is immediately evident, converge under the heading "existential proposition."

1. A concept "existence" leads back to an original relation-concept, that of inexistence, the being-in-something, being-in-a-context.
2. Certain borderline cases of particular propositions have the peculiarity of only having one complete term (only one complete main component), and these are equivalents for existential propositions of other forms. "Something is A," "A is something," –negatively, "Nothing is A," "A is nothing." Through equivalent transformations corresponding to *Brentano*'s, the ordinary two-place particular propositions–"An A is B" or "Some A are B"–lead back to the aforementioned forms, "An A that is B is something," "Some B are A," "–are something."
3. The judgments about being right or not-being right, "That S is p, that is right," "–that is not right," can express the conviction that S is p is a correct one.

   If I come to doubt whether S is p, and then for whatever reason, I become convinced for any reason whatsoever, then through confirmatory coincidence, the consciousness of the conviction called into question–which is not judging, but a judgment-modification–changes into consciousness of being convinced about the same "content." The latter is confirmatory; the former retains the quality of being correct, right. Correlatively, the state-of-affairs called into question (which is posited here not merely as state-of-affairs, but as modified state-of-affairs in quotation marks) has the quality of "It is (really) so." So, we can say here, 213
   "It is right." In any case, its own correlative predicate arises. The "It is right" is also to be expressed by "It happens," "It obtains," "It is really so." Since it is a matter of the correlate of the affirming, we can also say: The state-of-affairs is worth affirming, being affirmed. The correlate of the yes precisely objectively suits it. The same is so for nominatives, "Socrates–that exists, he is real," "A god Jupiter does not exist."

We still have to deal with a particular case: Being right is not only being right in general as this occurs arbitrarily wherever deliberating, questioning changes into a corresponding decision, but it is a splendid case of where the questioning, deliberating intention is fulfilled in a presentive intuition, the questionable state-of-affairs (not the state-of-affairs that is simply posited–but the state-of-affairs posited in quotation marks) tallies with the state-of-affairs itself, with the intuitively given one. In the presentive intuition, we have the consciousness that the thing itself is physically there. And the state-of-affairs being questioned is identified with it and becomes evident. It is identified with the corresponding "true one". The questionable proposition-matter coincides with the truth given, the proposition given with the character of truth that not only in general has the judgment-quality, but the character of fullness bestowing guarantee or legitimacy.

The concept of truth that arises here is, as we see, polysemous, insofar as it follows the correlations.

1. We can designate different things as true. A proposition being deliberated, questioned, is true provided it can display its legitimacy with Evidenz. The state-of-affairs or proposition being questioned shows an ideal character in the consciousness of the fulfilling coincidence that one can call truth.
2. But, insofar as it has intuitive fullness, the fulfillment conferring judgment (the proposition) can also be designated as truth–therefore, the grasped state-of-affairs figuring here in its meaning and in its fullness. Then, by way of contrast, the proposition conscious in the mode of questionableness can be called "correct". It is directed toward the truth. Insofar as it is directed toward it, it is also, however, itself called true <proposition>. If we have judged beforehand, then insofar as in subsequent chal-
205 lenging and insightful reasoning, it coincides in the indicated ways with the intuitively given state-of-affairs in the meaning-form concerned, the unmodified judgment can also be called true. In the reasoning (which is not some coincidental fact, but grounded in the essence of the judgment and assigns an ideal predicate to it), it is proof of rightness, truth.
3. Finally, the state-of-affairs, which has its definite formulation in the proposition, can also be called true, truly obtaining, or <it can be stated>: The obtaining is <in truth> attributable to the

state-of-affairs. In the statement about obtaining as in-truth-obtaining, the state-of-affairs itself does not however figure in the subject position, but rather, as was already said earlier, the state-of-affairs in quotation marks, since the statement would otherwise really be tautological and the corresponding negative statement <self->contradictory.

What, moreover, these negative statements involve then corresponds to the fulfillment in which the consciousness of the question or the deliberation measures itself up to the intuition giving the state-of-affairs itself, as the parallel possibility the disappointment. The disappointment consists of the fact that a state-of-affairs is given in the intuition that conflicts with the judgment intention, and in this conflict, the intended or questionable state-of-affairs is legitimately cancelled out, rationally rejected.

Since language is essentially determined by the ordinary empirical intuition in which the states-of-affairs are those of natural reality (*Naturwirklichkeit*), of the sphere of effectuating (*Wirkens*), the words "true" and "real" (*wirklich*) become synonymous, and the meaning of the word "reality" (*Wirklichkeit*) then extends to all spheres of truth. We say, "That 2 × 2 = 4 is, the fact that an equilateral triangle is equiangular, that is true"–but also, it is really so, it obtains in reality, and so forth. Meaning extensions and shifts in meaning generally play a major role here. In ordinary language, the simple expression of the confirmatory assent, "It is so", which <is> in no way always evident–is–<correspondingly> manifests itself to experience or to some other sphere of Evidenz–becomes equivalent to the expression "It is true, it is in reality". And the same expressions are thereby used to show that judgments or propositions are right and that nominal presentations are right.

In addition, through a slight shift, the expressions, "It is"–and "It is true, real"–also change into the expression of inexistence, 206
insofar as intuitive being, as all-embracing spatiotemporal reality, makes up the domain of existence, in which there is what is specifically real, in which it is. And so, through broadened application–even where there is no talk of natural reality in the ordinary sense–the verbal forms named now and then change into forms of inexistence, become synonymous with them and therefore alter-

nate with discourse with “there is”: “There is a hexahedron, it is something real, it truly is.”

My reflections also explain how the judgment-moment that I called the judgment-quality can also be called character of “putative truth.” There before the judging person is: “S is p!,” “Gold is yellow.” We are then in the habit of saying that his or her judging is a considering-true. In it, the subject “gold” stands before his or her eyes as reality, the nominal consciousness in the categorical proposition is a considering-real, the positing an as-existing, as-existing-positing.

Naturally, someone who judges thus is not thinking of truth in the genuine sense of cognizing and proving to be true. But a law of validity says: Every judgment is either true or false in the sense of rightness; every merely putative state-of-affairs is true or false in the correlative sense of what is to be cognized or what is not to be cognized, and in the latter cases <of>what is conflicting with what is to be cognized. Provided that without any reason and cause, we then actually state our judgment in a more normal way and, either “cognizing” from direct intuition or on the strength of reasons—even if imperfectly—we genuinely still judge, we believe we can say that our judging or what is judged is something true and weighed over and over will fully stand the test.

(Also relevant to this are additional laws of validity that are quite evident, so that if our judgment “S is p” is true, the judgment that it is true is also true, and the other way around, likewise, the fact that if our judgment “S is p” is true, the judgment “It is so that S is p” is also true.) In short, the different judgment-forms operating with “being” and “truth”—when the matter is retained, naturally—are truly in agreement with one another. But considered purely conceptually, the pure “A is” that emerged for us in the
207 contrast between affirmation and rejection is in no way identical with “A is true,” where the concept of truth occurs, and where the relationship to agreeing with the facts is present. All the equivalent forms described (naturally excluding the functional forms) are categorical. In the case of inexistence-statements, relation-predicates are of a certain form belonging within logic. Especially, however, in the case of important statements about truth, rightness, obtaining, objective existing, we are indeed dealing with

special forms of categorical judgments, but with ones in which the predicates are specifically logical, i.e., those belonging in logic have a formal nature of their own. You thereby at once see that *Bolzano* came closest to the truth regarding the question of existential propositions (when we understand them as propositions about reality, truth, realness), especially also in that he observed that the subject of existential propositions is not that provided by the nominal meaning of the word, but rather that a meaning shift (that of the quotation marks) takes place.

I do not, however, want to say that all questions of impersonals are thereby solved and whether it must not be recognized there that one-place judgments are present here.

<Chapter 7

The Propositionally Complex Judgment-Forms>

<§ 41. Conjunction and Disjunction>

We now have to move on to a new area of the theory of forms, namely, in keeping with my plan, to propositionally complex judgment-forms. We can formulate the question here: How, in which forms, can a judgment come out of several propositions? Or, for the sake of simplicity, in which forms can a new proposition and a judgment be formed out of several propositions? We can distinguish between two cases here. Either both propositions are intrinsically related to one another in terms of their meaning-content, they have some component in common, or, they are propositions that are separate from one another.

Two propositions intrinsically and actually have a component
in common, stand in a relationship of community of meaning, 208
when not only a synonymous component occurs in each proposition, but when in one proposition, the component concerned refers to the corresponding component of the other <proposition> as something common, as "the same"– just as two propositional acts are really subjectively intrinsically bound by community of meaning only when the consciousness of the meaning-community with respect to what is in common is present.

Hereby, an important way of connecting propositions, the identity relation, is therefore indicated. For example, "A is α–the same thing is β" obviously says more than "A is α, A is β". In the latter case, we have separate propositions and no connection via the meaning-form "the same thing". In the former, the "same thing" is related to A, which is therefore marked out as foundational term of identity. It generally holds that in an identity relation, one proposition is the foundational one, the other the one founded. The one

E. Husserl, *Logic and General Theory of Science*,
Husserliana: Edmund Husserl – Collected Works 15,
https://doi.org/10.1007/978-3-030-14529-3_7

is independently detachable, the other is dependent, losing its independence through the identity connection with the foundational one. “The same thing is β” yields no meaning in the detaching when not preceded by an “A is α.” In our example, we observe another thing. If we say, “A is α,” “The same thing is β” alongside one another, then we sense an abbreviation. Both propositions ought however to be one, and if we state them in the unit of the judgment, then we realize that not merely two affirmations are linked to one another–namely, by virtue of identification–, but that one more is to be constituted, precisely the one we lacked in the expression. More completely, we shall for instance say, “A is α, and the same thing is β,” “A is α, and the same thing is also β.” Implied in the “and” and “also” is the fact that a unit of the judgment-consciousness actually runs connectingly through both judgments, that we can perhaps say by way of explanation that it is also intended that both hold, and that is something new in comparison with individual judgments.

We now observe that the same “and-connective,” the conjunction, is possible in the case of propositions based on one another by means of the identity relation. Two completely arbitrary categorical propositions “A is B,” “M is N,” enter into the unit “A is B, and M is N.” And thus then in general completely arbitrary propositions. It is now easy to see that, in general, every kind of connecting that unites arbitrary propositions, permits the identity relation
209 to be established between the components as well. As a result, I shall not take identity relations into account in any particular way now. It <is> further to be mentioned that we judge everywhere where we connect definite propositions, we can also connect functions, and the whole is a functional-judgment.

Conjunction is the first basic connection of arbitrary propositions. We recall a parallel way of connecting in the nominal domain, and there was an Or corresponding to the And there. That is so here as well. A second basic connection of propositions having arbitrary content is disjunction, “M or A.” While in the case of the And-connective, the union of the judgment “M and N” implies that “both hold,” and so each of the two propositions itself has the character of a judgment, the situation is different in the case of the disjunctive connective. We do not have a compound judgment here from which each proposition can be detached as a judgment. The meaning here is rather that of "one of the two holds", and it is naturally not judged there that M holds and that N holds.

However, with this talk of reflexivity, one must be mindful of the fact that, for instance, predications are not present in which we state validity as a predicate of L and of N. Those are merely equivalent paraphrases that should draw attention to inner judgment-moments or to qualities. Obviously, a certain layering of qualities one upon the other is present in the case of the conjunctive way of judgment—"M and N!", "God is just and evil people will be punished!" An overreaching union of the affirmation spans the affirmation-consciousness "God is just" and the second affirmation-consciousness, "evil people will be punished", namely, in such a manner that the overall unity of the spanning judgment quality aims at both matters, each of which, however, has its quality. The overall quality in a certain way shares in the two qualities. You see that the inner structure is actually completely analogous to that of a plural judgment, namely, with respect to the subjects conjoined.

I am skipping the corresponding remarks about the disjunctive judgments, and I only mention in addition <that> this plain Or-connective that runs parallel to the And-connective is not unambiguously indicated by our language, insofar as the often intended meaning of the manner of speaking "Either M is, or it is N" is that which, if the one holds, the other may not hold, and vice versa, which is in no way intended or logically validly decided in our
disjunction. 210

Finally, let it be pointed out that we do not have any categorical judgments in all these complex judgments since we can be assured of the fact that the reduction of the forms to "Both A and B hold", "Both propositions hold" is a paraphrase that changes the meaning and makes it more complex.

The connections are not limited to two components. Arbitrarily many components can be linked conjunctively and disjunctively, namely, either with the same number of places or associatively: (abc…) ((ab)c)…

<§ 42. The Hypothetical and Causal Connecting of Propositions into the Judgment Unit>

We now move on to a new basic form of the connection of propositions into the judgment unit, into hypothetical propositions. The verbal expression is the statement-form with “if-then.” “If M, then N,” “If A b, then C d.” But, the verbal expression is in no way unambiguous, and unraveling the ambiguities creates many a difficulty for logicians.

Finally, the causal connection, “Because M, then N,” “Because A is b, then C is d,” is to be mentioned as being closely related to the hypothetical connection. Causal judgments are compound judgments. One can detach both judgments “M!,” “N!.” Someone who judges “Because M is, then N is,” believes M is, believes N is. These judgments are in a certain way involved in the judgment-wholes. This is, however, not so in mere hypothetical judging. It is propositionally complex, but as judgment, simple.

The meaning of these judgment-forms, especially of the hypothetical ones, is very controversial. Clarity is of particular importance here, for only in that way can a secure foundation for an understanding of the inferences and laws of inference that have always made up a major part of traditional logic be acquired. Every inference is indeed a causal judgment.

<a) Definite Hypothetical Judgments and Hypothetical Functional-Judgments>

A major reason for confusions in the theory of hypothetical
211 judgments is the lack of a fundamentally essential distinction, which <is> moreover completely obvious to me, namely, of the distinction between the point of view of functional-judgments and definite judgments. A first class of cases is roughly represented by the example, “If God is just, then evil people will be punished.” Two definite propositions are connected with one another in a certain syntax, and the whole is itself again a definite proposition, namely, a free judgment, if we accept the affirmation-quality. The syntax is such that, besides its own content, each of these definite propositions receives yet another certain syntactical character. This is

again the relationship between syntagma and syntax, or syntactical stuff and syntactical form that we studied earlier. The syntactical form of the "antecedent" is thereby plainly different from that of the consequent. Moreover, the way of judging is quite remarkable. One can easily be tempted to view the judgment as being categorical, whether about propositions or about states-of-affairs, the one proposition being the necessary consequence of the other, the rightness of the one judgment being the consequence of that of the other, and so on. Naturally, in this way, people have easily had an opportunity to find confirmation for the belief that every judgment is categorical. Nevertheless, I cannot consider all such reductions to be correct. It is to be borne in mind that the positioning is undoubtedly not toward meanings, but toward what is meant, the intended "state-of-affairs" as such, after the fashion of truth judgments. And, if we judge, "If M is, then N is", then the "state-of-affairs" M figures as presupposition, and N is built upon that as positing-upon. But, the presupposition is not affirmative positing here, and the positing-upon is not either. It is a matter of plainly primitive forms not at all to be explained away that one must but make fully clear to oneself in their particularity. And, implied in the essence of these forms is that they establish such a unity between presupposition and positing a consequence, or what was presupposed and what was posited as consequence as such, that equivalent relation-judgments, such as, "The presupposition M has the positing of consequence N" can be formed, and so on.

One must then differentiate well between genuine hypothetical judgments, in which definite propositions come to be connected, from the corresponding functional-judgments, thus, for example, general judgments of this kind, "It universally holds that if M is, N is." They are the exact parallels of the corresponding functional-judgments in the area of conjunctive and disjunctive thinking, 212
therefore, "It universally holds that M and N is, M or N is." It is naturally to be borne in mind here that both the hypothetical and the conjunctive and disjunctive thoughts are below the layer of universality-positing, i.e. belong in the function. For example, "It universally holds that, if a > b, <then> b < a–i.e., here we have the two functions a > b, <then> b < a" with the arguments a and b, that mean arbitrary quantities. The two functions are combined by the

form “if–then” into a single hypothetical function, and then this signifies that this hypothetical-functional combination holds universally for arbitrary pairs of quantities a b. We very often omit the expression of universality and simply say, “If $a > b$, then $b < a$”; and associated with that is confusing these functional-propositions with definite hypothetical propositions.

Certain verbal, often identically expressed, universal functional-judgments in which the hypothetical thought–that of the conclusion–is missing must not be identified with these hypothetical functional-judgments. The statement-form, “It universally holds that if M is, then N is” can signify: Every case of the validity of M is also a case of the validity of N. For example, “If the atmospheric pressure drops, the barometer drops.” The arguments are not clearly expressed here. They naturally lie in the intensity of the atmospheric pressure and the level of the barometer, in what can differ on each side, and whose difference is something undetermined here. This is then, merely a being-together and being-given-with, regular accompaniment (where the one is, the other is too) stated in general terms, a “whenever the one, there is the other.” The hypothetical thought in a concise sense indirectly stated therewith that a thesis necessarily follows from a presupposition is missing there.

Naturally, particular forms often loosely expressed in language also occur. So the “if-then” does not always mean strict universality, that strict “always–if”, but also “often and most of the time.”

In terms of their thought-content, therefore, not all these judgments belong with genuine hypothetical judgments. On the other hand, it is nonetheless good to discuss them both together because they are connected, namely, for reasons of logical validity. That in fact especially holds for unconditionally universal judgments, on the one hand, and necessity-judgments on the other. I stress,
213 “unconditionally universal” because universality can truly merely mean empirical generality, i.e. mere factual generality, as when we speak of all roses of our garden, or make a statement about regular accompaniment of the kind, “Generally, a carnation is planted next to a rose”–the same goes for, “*Everything* in the world.” In all such judgments, overt or unstated individual existence is posited. However, universality is unconditional if it is “purely conceptual”

as in pure mathematics, for example, "In every case, in which it holds that a triangle is equilateral, it also holds that the triangle is equiangular." In itself, as was said, the difference is irrelevant here, but to be stressed, namely, because strict universality and necessity overflow indiscriminately into one another in our thinking owing to the one's equivalence with the other. The triangle theorem mentioned is obviously fully equivalent to the proposition, "It holds universally that if a triangle is equilateral, it <is> equiangular" (understood in the sense of the necessary conclusion).

I, in addition, now briefly touch upon the fact that "M and N" become additionally propositionally complex, can even again be conjunctions or disjunctions.

If we now compare definite hypothetical judgments and the corresponding functional-judgments, then they stand in meaningful, and not merely external, relationships to one another, precisely by virtue of certain intrinsic relationships of necessity and regular universality.

1. Let us take as example the trivial inference–but just grasped and understood as hypothetical judgment–"If all humans are mortal and Socrates is a human being, then the necessary conclusion is that Socrates is mortal" or, "If $2 > 3$, then necessarily $3 < 2$." Here, we have definite propositions and the thought that at once unites them as judgments is that of necessity.
2. Let us contrast the corresponding functional-judgment. We can then take the one that no longer contains the hypothetical thought in the function at all: We would then have: Every argument system that satisfies the propositional functions "All A are B" and "G is A" also satisfies "G is B", where G is any object-presentation, A and B any property-presentation.

A remarkable relationship belonging in the logic of validity obtains here that is only mentioned here to direct your attention 214
to the unique nature of the structure of the judgment-forms now in question. Namely, if the definite hypothetical inference-judgment is correct, then the functional-judgment is also correct, and vice versa; i.e., in this case necessity and universal validity are equivalent. The necessity pertaining to the particular case, to the inference about humans, Socrates, and mortality points to general

validity for arbitrary terms of this kind. And if this obtains, then the particular case naturally does as well.

Now that is by no means a situation acceptable for every definite hypothetical judgment. You see that already from the fact that if we strike out one of the so-called premises of the inferring hypothetical judgment and then form the hypothetical judgment, "If all humans are mortal, then Socrates is mortal," we may on no account say that if the judgment is true, then the law "If all A are B, then every G is B" must hold, which would obviously be false. And yet, there is still some truth there. Namely, we had struck out a premise, and correspondingly a part of the putative law is missing. That points to a certain possible supplementation, to certain extended relationships that join in and co-determine the meaning of every hypothetical proposition that is not inferential. Connected with that is the fact that, in the case of every hypothetical proposition, we must speak of terms in similar ways, as we earlier spoke of arguments in the case of functional-propositions. We therefore speak of terms or bearers of conditionality to which the arguments in the pertinent universal proposition correspond.

<b) The Terms of Conditionality in Hypothetical and in Causal Judgments>

Many kinds of terms can occur in hypothetical propositions, but not all are terms of conditionality. It is to be noted that when we actually infer, i.e. actually carry out the inferential judgment, the terms are marked out for our consciousness and must be since the terms are to signify something in the meaning. If we had inferred, "If all humans on Earth are mortal, and Socrates is a human on Earth, and so forth," then, for example, "Earth" is not a term of
215 conditionality, but the whole–"humans on Earth"–is a term, and as that, figuring as a whole before our consciousness. With the abridgement of the inferential hypothetical judgment "If all humans are mortal, Socrates is mortal," "human," "mortal," "Socrates" are still marked out as terms. Wherever we have clear judgment-consciousness, we judge in such a way as to relate the positing of necessity to these terms with rays, although by no means in a uniform manner. This indeed also becomes apparent in

the fact that antecedent and consequent are overall very different in nature, precisely by virtue of the diversity of this positing, and that the consequent has the character of conditional necessity, that of necessary conclusion.

It is all the more so in all that in the case of inferences, of the corresponding causal judgments. The terms are marked out there in a very clear-cut way, and it is moreover to be borne in mind here, as also in the case of hypothetical judgments, that where the same terms occur in antecedents and consequents, they are consciously in the judgments as the same, therefore, linked by identity consciousness (even though explicit formulation by the word "the same", and perhaps also by the meaning-moment, is lacking; however, the judgment-meaning would then not be perfectly complete). In causal judgments, every propositional component (as was already observed earlier) has its affirmative character, but this is specifically modified as premise and conclusion, and each of the conclusions has the character of "validity as necessary consequence."

It is now to be observed that hypothetical and causal judgments are not brought to full clarity in all cases, and pertaining to that is above all the actual performing of the correlating necessity-consciousness with the actual enunciation of the terms. Corresponding to them are then also differences of meaning. If we ask what is intended in non-enunciating hypothetical judgments, then concluding from the presupposition is certainly intended, but in an indefinite way, which enters into the matter as a distinctive kind of indefiniteness. We would then have to explain the meaning circumscriptively something like this: With respect to certain, but not more precisely determined, terms, M follows from N. Belonging with the meaning (if it is totally complex), we therefore then have a particularity-thought, which in the other case, in that of clear judging, is missing. We have here before us the definite terms as bearers of conditionality.

Let us pause in addition to consider another case. From the 216
outset, it is clear, and would also be evident as a law of validity, that depending on more precise determination, which ought to be that of the different possible groupings of terms that bear conditionality, the entire meaning of hypothetical or causal judgments must

be essentially different. In cases in which multiple options of this kind are possible, therefore, in particular, where the propositions have a more complex structure and contain many kinds of terms, there can be a completely valid conclusion in one interpretation, a fallacy in another. This is also a source of bewildering logical paradoxes that can only be solved by resorting to the theory of terms.

I cannot go into the complexities of the basic forms of propositional composition dealt with up until now and mention only that ordinary disjunctive judgments are obviously combinations of hypothetical ones and our disjunctive ones. Finally, I have to say in addition that traditional logic deals with hypothetical and disjunctive judgments under the heading of "relation" while overlooking the primitive conjunctions and disjunctions. *Kant* places "categorically", "hypothetically", "disjunctively" alongside one another, as though the one is a classification implemented from a definite viewpoint.

<Chapter 8

The Cardinal Differences Within the Field of Meaning>

<§ 43. The Modal Distinctions>[1]

Up until now I have aimed at a systematic combination of simple and compound judgment-configurations, whose differences were purely determined by matter, namely, in such a way that any drawing in of concepts through contrasting matter and 217
quality and the different modifications that the quality can undergo within each genus (such as judgment, presumption) was still to be avoided. Only once have I of necessity violated this principle, namely, in order not to have to separate existential judgments as judgments of being from their equivalents in the domain of particular and universal judgments. Likewise, I have also barred all complications and modifications that arise when concepts are drawn in that have arisen in reflecting upon the matter or upon the whole configurations and such, therefore, concepts such as "proposition", "presentation", and so all purely logical concepts. The first transformations, however, are especially of great importance and to some extent hard to clarify. And they are those that can be called modal distinctions in the concise sense. Historical logic presents a heading "modality", and it is handled so unclearly that it includes the forms interesting us here. *Kant* arbitrarily grouped judgments into four groups from the standpoint of quality, quantity, relation, and modality, something which again gives rise to a false idea of equivalent treatment that is in no way present.

---

[1] Subsequent note by *Husserl*, "From the start, the basis for classifying modality is to be discussed before beginning the theory of forms: Forms that do not draw in 'qualities', that are not reflexive and <ones> that are". Editor's note.

E. Husserl, *Logic and General Theory of Science*,
Husserliana: Edmund Husserl – Collected Works 15,
https://doi.org/10.1007/978-3-030-14529-3_8

In introducing modal differences, people commonly say that the judgment-mode, the mode of connection between S and p in the categorical judgment (which really counts as basic type) could be different depending on the mode of validity intended. Either "S is certainly p, it is actually the case that S is p", or "S is possibly p", or "S is necessarily p". And, the three *Kant*ian headings, assertoric, problematical, apodictic judgments, arise accordingly.

First of all, with respect to these judgment- or statement-forms, any mixing up of psychological and meaning-logical distinctions is again now to be avoided. It may therefore not be said that it is a matter here of expressing *our* certainty, *our* considering-possible, necessary, as if that should be predicated about. Naturally, one can predicate thus, but that is of no concern to us here, and in discourse, reality or certainty, necessity, and such are normally not 218 intended.

<a) The Possibility-Statement in the Sense of Seeming and in the Sense of Not-Being-Excluded>

If we remain within the pure theory of meaning, then for the time being we come to the actually completely justified forms I dealt with earlier: "It is so, that S is p, it is actually so"–or "That S is p, that is, that is certain, that is true". "Certain" and "true" do not signify exactly the same thing here, but it indeed appears that the "certain" and the pure Is mean the same thing. It is the objective correlate of certainty brought to predication. Certainty is, however, nothing other than character of judgment, that is, naturally of judgment in the meaning-theoretical sense. When it comes to validity, this "certain" is equivalent to "It is true".

The precisely analogous statements for the sphere of possibility-statements[1] and probability-statements would be "It is possible

---

[1] Subsequent note by *Husserl*, "'Impossibility' must be precisely discussed: 'compatibility', 'incompatibility'; <a> law that <when> an a and <a> b <are> united, a whole of the form (αβγ) is possible; a law that such a thing is not possible, that there is no such whole in unconditional generality. There is no entity that this entity unites, etc." Editor's note.

that S is p,” “It is probable that S is p,” and for questions, “It is questionable–,” “It is doubtful–.” The same thing holds of all those judgment-forms, as was explained for the reality judgments with that-propositions in the subject position, which naturally do not confer any being-value upon the merely thought “states-of-affairs.”

Very important now is an equivocation in possibility-statements. (1) In one case, we have the meaning just now indicated, “Something is to be said for that,” which plays its role in the theory of probability. The “It is possible” is the correlate of a seeming, just as “It is probable” is the <correlate> of a presumption. (2) <In the other case, we have> the totally different meaning that in certain spheres makes the possibility-judgment totally equivalent to particular judgments. In mathematics, one says, “It is possible.” For example, “It is possible for a triangle to be isosceles,” as much as “There is such a triangle.” Here possibility and being do not coincide conceptually, but in equivalence. How does this equivalence come about?

On its own, the meaning of “It is possible” is frequently “It is
not excluded.” Here “not excluded” would naturally signify 219
excluded by the axioms, which alone are what excludes in mathematics. Likewise, “It is impossible” <would signify> it is excluded (by the axioms). The difference between the two earlier meanings of “It is possible,” between the one associated with seeming and the present one that signifies “not being excluded,” consists precisely of the fact that in one case the meaning is positive, “Something speaks in favor of that,” in the other, negative, “Nothing speaks against that.” The two are not, however, equivalent. It plainly holds of mathematics that every proposition that <is> not ruled out by the axioms is true (while in natural science something analogous would hold if, with respect to the basic laws of nature, we knew that all being in nature strictly obeys laws, and if we had at our disposal a complete set of laws of which we could say that they were *all* basic laws).

<b) Necessity as Apodicticity and Relative Necessity. Its Relationship to Laws>

Closely related to these judgments about possibility and impossibility are <judgments> of necessity and conformity to laws and their opposites. To be distinguished as well is the case of the statement, "That M is, is a law" and the statement, "That M is, is a necessity," although the modes of expression become very mixed up. Every necessity points to conformity to a law. Necessity and conformity to laws are correlates, and this gives a concept of necessity a fixed sense. The "It must be so" expresses that what was stated may count as a conclusion, as a particular case of a law, for example, if we say of the quartic equation under consideration that it would have to have four roots, namely, bearing in mind the law that every algebraic equation of nth degree has n roots. The opposite of necessity would there be a more definite concept of possibility, namely, related to the same matter. In this sense, "It is possible that the points of a triangle indicated (point of intersection of the altitudes, the centroids) coincide" was tantamount to saying: It is not a law of triangles that these points always coincide. And so, conveying the meaning more loosely, "It is possible" and "It is necessary, that an A is B" are often tantamount to "It is universally so" and "It is not universally so" (which, since the negation
220 is merely directed against the universality, does not imply that it is so in the particular case).

In relation to all that is then, however, a concise concept of necessity, that of apodicticity, which is a genuine modal concept, but equivalent to the concept of validity on the basis of a law. If, for example, we really and perspicuously carry out an inference, then the relationship of the conclusion to its terms is there before our eyes, but not only as existing in the universal nature of judgment, but in addition characterized in specific ways, precisely as the necessity-meaning. We are thereby not thinking of a law, but a law pertains to this. One cannot reasonably carry out an inference without recognizing that inferring in that way is universally valid and sanctioned by laws, that substituting terms by indefinites therefore yields an unconditionally valid law-judgment.

In meaning, corresponding to this consciousness of necessity is a specific coloration of the Is that expressed predicatively results in the reflexive judgment, "That M is, that is a necessity", "That S is p–this being of a state-of-affairs–is a necessary being". We also, however, say equivocally, it is a necessity that a + 1 = 1 + a is, that two straight lines intersect in one point, and so on. Those are, however, laws. The equivocation lies in the fact that we call the laws as source of necessity itself a necessity. Actually, we would only have to say in the particular case, "That these two straight lines must intersect at a point is a necessity". We in fact see that without thinking about the law, in the consciousness of the apodicticity alone. At the same time, we observe that every individual case of an axiom yields an apodictic necessity and that necessities of arbitrary universal forms occur, as also laws either have a hypothetical content or do not.

All individuations of laws are necessities in themselves, absolute necessities. Along with this are relative necessities, namely, a proposition also figures for us in specific ways as apodictically characterized if it is proven to be a "necessary consequence", therefore is a component part of an inferential necessity. Necessity-consciousness is related here, albeit nevertheless something different in comparison with the case in which a proposition is merely an individual case of a law. Relatively necessary judgments are only characterized as necessary in relationship to their premises.

So, each necessity in certain ways indeed has a relationship to 221
laws, but remaining is the clear-cut distinction according to which the one has a correlate in a law whose individual case is that indicated as necessary, while in the other case, the necessity that is relative to the conclusion, and consequently brings with it a relationship to premises of a certain determined kind whereby the law of necessity lies in the relevant law of inference according to which this conclusion from the premises could be characterized as being in conformity with laws. The validity of the predicate "necessity" is still equivalent to the validity of the pertinent law.

<§ 44. The Idea of Law. An Apodictic Necessity Corresponds Only to Pure Laws in the Individual Case>

Where do things then stand with the Idea of law that has played a specific role here? Can we not characterize it more precisely? It is clear that a law is a universal proposition. Conversely, not every universal proposition is a law. The word "law" is furthermore in many cases used in the sense of a universal requirement, that is, of a community-requirement, as proceeding from the bearer of the community-will, as proceeding from an authority, which has to place requirements on the community to which those living in <the> community feel bound to submit. Those would be laws in the political sense, church-religious laws, also mores, and so forth. That does not concern us here, since we are in the realm of judgments and not in the realm of practical requirements. If we remain in the realm of judgments, then a proposition's universality can be empirical or pure (unconditional) universality. In both cases, a "law" requires that the universality be infinite, not confined to any thesis of a single individual. ("All roses in this garden"–"all things in the world" and so on.) Laid down then is the concept of pure law purely rooted in the concepts occurring in it.

Someone will perhaps say: Law is a universal proposition that is stated in the consciousness of unconditional universality of validity. This subjective expression is, however, not completely clear. Can we state the same universal proposition–for example "All humans are mortal"; "All trees in the forest are lindens"–in a dual consciousness, on one occasion, in conditional, and then in uncon-
222 ditional universality, and naturally do so retaining the content, namely, of the identical function? That naturally does not work. On one occasion, we can speak loosely of universality, actually merely believing it holds for the most part in that way, and on another occasion of real universality, but then the first judgment is just only a particular judgment. Clarity with respect to the law-concept requires that we naturally already have definite, complete meanings before us and that we therefore do not operate with loose terms.

The distinction <between> unconditional and not-unconditional universality–which is often called "empirical universality"–must therefore refer to something else. We can thereby rather set aside talk of consciousness which, as we saw, so easily leads astray, since everything that interests us must be demonstrable in the realm of meaning. The universality, that precisely characterizes a judgment as general, universal, can be restricted or unrestricted. It is restricted when we speak of all Europeans, i.e. all the people of Europe, provided the qualifier "living in Europe" occurs, something which conceptual universality no longer allows to appear as pure. Such restricting does not admit the presentation that we associate with the word "law". So if we were to have understood the talk of "all humans" to mean actual humans living in historical time or those on Earth, then a pertinent universal proposition with "all humans" could no longer count as a law. Full universality in the sense of law requires that it absolutely read, "*All* humans", disregarding any limitation to any individual existence whatsoever. No nominal presentation with an individual-nucleus may therefore occur in a law-judgment, no proper name, and likewise no "this" that refers to individual existence, and in turn no particular term of the kind that would refer to individual being may occur in the law.

In short, pure laws = universal judgments–but the universality must be pure, unconditional, not limited by any overt or hidden individual-positing and particular positing.

We therefore see that consideration of the difference of nuclei has a role in determining the law-concept. Only pure general-nuclei are admissible. All individual-nuclei are excluded. In this 223
purest sense, all mathematical-universal, as well as arithmetic, geometrical, kinematical, theorems are laws, and so forth, as also are the universal propositions of the theory of meaning, in addition also the laws belonging to the essence of nature in general, not though, the particular laws of nature, which are therefore not laws in the pure sense. They still implicitly contain reference to the actual world, which is only one and the same in all judgments of the natural sciences, so that here too we speak of laws.

An (absolutely) pure apodictic necessity corresponds to every pure law of this kind in the individual case. A law does not correspond to every statement containing a reference to individual existence, and consequently not everyone can <be> made with the consciousness of necessity, or have the validity-character of necessity either, thus, for example <the statement> that the weather is beautiful today, likewise <the statement> that all bodies have gravity, and so on. Being able to be characterized as necessary and valid is therefore a definite property of a proposition. For example, "These two apples and those two apples together make four apples"; or "This thing there is extended, has causal properties"– are necessary, obviously hold purely by virtue of laws. If a judgment is made in the consciousness of purely apodictic necessity, the restricting terms, as opposed to the bearers of necessity, must (as <is> easy to see) in certain ways be used as the restricting terms in a particular consciousness, in the consciousness of contingency or actuality, namely, of the contingent restriction of necessity.

It is further to be observed that even purely mathematical propositions that do not include anything about individual existence can also be made in the consciousness of necessity, precisely by virtue of the relationship to a higher consciousness of laws, for example, the <relationship> of number 2 + number 3 = <the relationship> of number 3 + number 3. Naturally, that holds for finite numbers generally and in conformity with laws. And so, we are likewise in general able to have woven the universal into the particular, to see or to suppose it to be necessary, not merely something valid, but something valid as an individual case of the law, but without our predicatively stating or thinking that this proposition is anything particular, etc.

It is moreover to be said in addition that singular mathematical judgments have the property of not being judgments about facts,
224 and yet they are not laws. But they are nevertheless equivalent to laws. More universally: There are pure concept-judgments that exclusively contain pure concepts, pure general-nuclei and that are of course not in themselves of the nature of laws, but yet can always be rephrased in law-sanctioned universality, in full equivalence. If the sound c, taken as idea, is lower than sound d, then it is naturally to be said with unconditional universality that every real or possible individual sound that corresponds to the Idea c, is lower than the very same individual sound of the quality-Idea d, and so in general.

Therefore, the distinction between pure laws and pure concept-judgments (law-sanctioned states-of-affairs and states-of-essences) is not as essential as the difference between the two kinds of judgments and factual judgments. We need the law-concept in order to mark out a kind of universalities–<the> purely conceptual universalities, especially in contrast to the necessary particulars. But, it is otherwise important to bear in mind that every essence-judgment, every purely conceptual one, has the value of a law, though not always the form of a law.

<§ 45. a) The Distinction Between Judgments in Pure Concept-Judgments and Factual Judgments (A Priori and Empirical Judgments)>

The cardinal distinction running through the entire realm of meaning is that which is essentially co-determined by the separation of nuclei into individual nuclei and general nuclei, although not exclusively determined is, namely, the separation of judgments into pure concept-judgments and factual judgments or, as we can also say, <into> a priori and empirical judgments:

1. By an empirical judgment, a factual judgment, I understand a judgment in which individual existence is posited, in whatever way, whether definitely or indefinitely. Belonging there are, therefore, ordinary existential judgments, but not, for instance, those of pure mathematics, in which pure numbers are posited. Every categorical judgment of the empirical sphere, such as "Today the sky is blue", and so forth, also belongs here.
2. On the opposite side are the pure concept-judgments that simply do not posit existence in any overt or implicit form, therefore, all the purely general judgments and their equivalents.
Bear well in mind that we are not identifying the concepts "gen- 225
eral" and "universal judgment" here. Every functional-judgment is universal that has the form of what is universal in general, whereas a judgment about ideal objects is general. Corresponding to every purely conceptual nucleus, as we know, is a nominal proper-presentation, for example, corresponding to the nucleus of the adjective "red" is the noun "red", where the

object named there, the "essence red", is the Idea. Every judgment, then, that judges about such objects, general objects, is a general judgment. And such a judgment can be universal and particular, singular and plural. It can have each of our logical forms, something of which we can certainly assure ourselves in sciences such as the mathematical sciences.

Every general judgment can be changed in such a way that it no longer judges about general objects, but rather about individual ones, but in unconditional universality (therefore, not as existential judgment). Since this change is grounded in the universal essence of such judgments, it is not then very important if we also call these judgments changed in this way general, therefore, do not rigorously keep general and pure concept-judgments apart, although nevertheless differences must again be seen.

I called the difference between existential judgments and pure concept-judgments a cardinal difference. One immediately sees that accordingly all sciences separate in a clear-cut way into ones in which only pure concept-judgments occur, in the others existential judgments as well. The former would investigate what holds "*a priori*", what is grounded in the essence of pure concepts and holds for particulars in general in unconditional universality, precisely exclusively insofar as they are particulars of the pure concepts concerned. The others would investigate matters of fact, quite simply, what holds for matters of fact. If it has already been ascertained that they are particulars of certain pure concepts, then pure concept-judgments or pure laws apply to them. However, investigation into matters of fact does not begin with such things, but first ascertains what holds for what is given in experience, which alone can give matters of fact, i.e. under which concepts it is to be grasped and how it is to be further determined in accord with the essence-insights of pure concept-judgments.

I described the difference between pure concept-judgments and factual judgments also as a difference between a priori and a posteriori <judgments>, wherewith a fundamental, really the most fundamental, meaning of this ambiguous distinction is specified.

The difference is actually indicated in this way with respect to substantiation, etc. *Kant* also talks of a priori and a posteriori concepts, just as the tradition prior to him had. Corresponding to them in the realm of my meaning analyses is naturally the conceptual distinction between essence-concepts, pure concepts and individual concepts with which we have become acquainted as the difference between pure nuclei. (Every nominal presentation has its stuff and its form, and eventually, we precisely come to differences between presentations in terms of their ultimate stuff, therefore, <to> differences such as those between the presentation-content "Socrates"–or the concept, in the sense of the nucleus–and the content "green," "similar," and so on. Both can be linked in complex presentations. However, as soon as an individual-content occurs, the concept is impure.) Therefore, it is the difference between concepts that is in turn the basis for the distinction between a priori and a posteriori judgments. 226

A very important–indeed fundamental for all of philosophy–special case of a priori judgments is denoted by the heading "analytic judgments." And the very controversial distinction between analytic and synthetic judgments, which *Kant* related to the distinction between a priori and a posteriori judgments, plays a major role. Before examining this more closely, it is to be borne in mind that the universal distinctions pertaining to the sphere of pure concept- and law-judgments recur in the sphere of analytic judgments.

Therefore, we have to distinguish analytic laws, or analytic pure concept-propositions from analytic necessities. And again, in the sphere of existential propositions, we have to distinguish those having the character of analytic necessity and those not having it. Thus, any denial of an explicit contradiction is an analytic necessity and any explicit contradiction itself an analytic impossibility. For example: That the theater is full and not full is an analytic impossibility, and if we state that that is not true, then that is a necessity. The corresponding universal law "It is not true that any object a is and at the same time is not" is then an analytic law.

227 <b) Concept-Truths–Factual Truths. Analytic–Synthetic Truths>

The outcome of the reflections of my last lecture[1] was as follows: Truths break down A) into pure concept-truths, truths that do not include any kind of positing of individual existence; B) <into> ones that do include some such thing, namely, factual truths, existential truths. Or epistemologically expressed: Truths break down into a priori truths and a posteriori truths (or <into> non-empirical truths and empirical truths).

The pure concept-truths are either pure laws or to be converted equivalently into pure laws. There are various gradations here. Pure laws can relate to singular particulars that are themselves of a purely conceptual nature (namely, ideal objectivities), or they can relate in unconditional universality to individual objectivities. The latter holds indirectly for all pure laws, insofar as ideal particulars, such as, for instance, the Idea “two”, the Idea “sound c”, and so on have a pure extension of empirical particulars.

<Regarding> A) Looking now at pure concept-truths, i.e. a priori ones. These break down into those that a) exclusively contain formal categories, and those, that b) contain other pure concepts.

a) Pure categorial concept-truths as a whole. <These> essentially belong together and form a single system of scientific disciplines, all of which I deal with under the broadest heading of “formal logic” or “analytics” (= *mathesis universalis* in *Leibniz*' sense).
b) If we now look at the pure concept-truths that do not exclusively contain that remarkable group of pure concepts that I called formal categories, then they again break down into two groups: first of all, those that remain truths when we replace their material concepts by pure categories; second, those for which that is not the case. For example, “Two colors the same
228 as a third are the same as one another” is only a special case of “Two objects the same as one and the same third one are the same as one another”. Or, “If one spatial distance is greater than a second one, then the latter is smaller than the former”.

[1]The text is published as Appendix XV. (Editor's note).

That is naturally a special case of a universal principle regarding magnitudes that holds for any kind of magnitude and not merely for spatial distances. "Magnitude" is however a categorial concept. All a priori individuations of pure laws of meaning belong here, all a priori individuations of arithmetic laws, and so forth.

Every special case of a law is a necessity. Therefore, these pure concept-statements are analytic necessities. Therefore, that characterizes the one group of a priori truths in completely definite ways. The others are the a priori truths that are not analytic necessities–such as, for example, geometrical and kinematic axioms. We call these synthetic or synthetic a priori concept-truths.

<Regarding B> As concerns factual truths, they too can be categorized into two groups: mere individuations of a priori truths and those that are not. a) Those which are empirical individuations of analytic concept-truths, i.e. arise through substitution of a term positing existence for a pure concept; b) those which are not, but are empirical individuations of synthetic (synthetic a priori) concept-truths; and c) those which are neither of the two, are pure factual truths.

Let us now contrast analytic and synthetic truths as categorization.

I. By "analytic truths" in the broadest sense, I at one point understand, analytic concept-truths, therefore all pure categorial truths, therefore, the entire pure mathesis, pure logic, then however, also their a priori and empirical individuations, therefore, the analytic necessities. In the case of empirical individuation, we have a mixture of what is a priori and what is empirical, insofar as the carrying over to the case of an individual existence–for example, <the carrying over> of a logical law to empirical propositions–precisely brings in empirical existence, but thereby at the same time necessity, necessity of the being-so, in this case *hic et nunc*. We therefore then have analytic-necessary existential truths.

II. In like manner, by synthetic truths, on one occasion, I understand synthetic concept-truths, such as, for example, the axi- 229
oms of the pure theory of time, of pure geometry, and so forth,

then their empirical individuations, therefore, the synthetic existential truths, which are empirical owing to the introduction of existence, but otherwise express the fact that what is existing is necessarily this way or that, namely, according to a synthetic necessity (therefore corresponding to a pure non-categorial law). Therefore, the carrying over of any geometrical law to a given instance of nature also belongs here.

If one understands the heading "analytic" and "synthetic" so broadly, then one must nevertheless constantly keep in mind the categorizations falling under it, for they are the most important. To bear in mind in addition is that if someone says that all analytic truths are *eo ipso a priori*, but not all synthetic truths *a posteriori*, then that is to be taken *cum granis salis*. An analytic existential proposition holds by a priori necessity. It states something that holds *hic et nunc*, because in general it holds by virtue of analytic laws. Provided that it is not then to be stated about the particular function of such a proposition that what is existing actually is, but that it is necessarily like this and that, talk of the a priori validity of the proposition is justified. For example, "This thing is extended" – obviously because thingness is an extended existence, and the fact that an extended existence is extended is a particular case of the logical law of the tautology that what is an a being b is also an a. Otherwise, the entire proposition as it is there is, however, a posteriori, provided it is speaking of this thing. And, if in stating it we had succumbed to a hallucination, it would be false. The fact that the subject-positing is valid can naturally not guarantee the analytic law. Only experience and empirical substantiation can show that.

Now, "mere" factual truths that are altogether "contingent", where precisely no component of necessity is <given>, separate themselves <off> from such analytic existential truths. These separations affect the objective nature of truths. They turn into separations of judgments when we substitute the supposed truths for the actual truths, and then therefore distinguish judgments that are supposed as concept-truths of an analytic or synthetic kind, are supposed as existential truths and as analytic necessities with respect to their being-so-composition. Then, also judgments in the meaning-theoretical sense acquire difference and classified separation, with a minor alteration, which I am not going to go into more closely. 230

<§ 46. The Laws of Apophantic Logic and Those of Formal Ontology>

A new distinction that I make within the analytic sphere, whereby we can limit ourselves to analytic concept-truths, is now of the greatest importance from the standpoint of formal logic, but then in addition also from the standpoint of higher philosophical interests.

The separation is determined by that of the formal categories themselves into the two groups correlatively corresponding to each other: meaning categories and object categories. On the one side are concepts such as "meaning", "judgment", "proposition", "nominal presentation", in short, all the concepts we acquired in the theory of forms of meanings. On the other side are concepts such as "object", "property", "relation", "unit", "multiplicity", "cardinal number", "whole and part", "magnitude", and so forth. From the perspective of the theory of meaning, we start with the Idea of judgment and pursue the different judgment-forms. The laws we obtain in so doing are laws for meanings, as the concepts that originarily and immediately arising here are concepts of meanings. Two kinds of laws are then possible.

<a) Purely-Grammatical Laws and Laws of the Apophantic Theory of Validity>

To begin with, the laws with which we have been concerned *in extenso* up until now: the laws of the theory of forms of meanings. Knowledge of them is fundamental for both logicians and grammarians, and in the latter respect, I also call them purely grammatical laws. In accordance with our terminology originating in the nature of the matter, these laws are also to be called analytic laws. If neither *Kant*, nor one of his successors, thought of such a priori truths under the heading "analytic laws", that is because they allowed themselves to be led by very narrow considerations and

did not arrive at the cases of natural unity of the Idea of the "analytic". Above all, however, it is because they did not in general see, and for this reason never dealt with, the distinctive nature of the
231 meaning-laws of the indicated groups.

The theory of validity directly attaching to the theory of forms of meanings–which I also discriminatingly call the apophantic theory of validity and combine with the theory of forms under the heading of apophantics–forms a second area of analytics in the broader sense. If at one point we examine the different judgment-forms in the theory of forms, we can then ask whether—viewed as forms for universal judgments–they unconditionally permit the formulation of universal truths or falsehoods. In fact, there are several kinds of such law-truths. If, for example, we take the pure judgment-configuration "S is p and is not p", then for this purpose, we can state the law that, for any S and p that would be substituted in it, every judgment of this form is false. Every contradictory categorical judgment is false. Likewise, every tautological judgment, every judgment of the form "Sp is p" is correct, if the judgment "Sp exists" is correct. Again, we can state that if two judgments of the form "All A are b" and "All b are C" are valid, the corresponding judgment "All A are C" is implied as a necessary consequence. The entire analytic theory of inference belongs in the domain of apophantic logic. Every analytic law of inference states that in two judgments of pure forms, for instance U_1, U_2, with regard to definite terms, the conclusion of pure form U_3 is implied.

It is, however, to be noted here that in all laws of this sphere, the defining concepts are indeed formal categories, but categories that definitely belong to the class of meaning categories. For example, in the law of inference "If all A are B and all B are C, then All A are C", the concept defining the extension of the variables is the concept of the concept itself. That means that the hypothetical law is valid for any three concepts A, B, C. If we infer in accordance with the formal law, "If from proposition M, proposition N follows, and from proposition N, proposition P follows, then from proposition M, proposition P follows", then we are inferring in accordance with a law where the terms are propositions, i.e. for any propositions M, N, P, this mode of inference is valid. Once again, "proposition" is a category of meaning.

<b) The Formal-Ontological Laws as Equivalent Rephrasing of the Apophantic Laws of Validity and as Laws for Objects That Arise Through Nominalization of Dependent Proposition-Forms> 232

In comparison, there is, however, a wealth of laws that do not make pronouncements for meanings in general as regards their categories, but make pronouncements about objects in general, properties and relations in general, about sets in general, about what holds for sets in general with regard to their containing one another or being mutually exclusive, or what holds for numbers in general with regard to their different relationships grounding in the essence of numbers. Likewise, <there are> propositions about the relations between whole and part, about sequences and ordinals, and so forth. That is the field of formal ontology. A great percentage of such propositions simply arise from the fact that we equivalently rephrase apophantic laws of validity into laws for objects, for properties, states-of-affairs, and so forth, because one can do that *a priori*, as is immediately obvious–instead of speaking about arbitrary concepts as meanings and about what holds for propositions of one form or another with respect to their truth and falsehood. If they are satisfied by such and such concepts, we can speak of arbitrary properties and states-of-affairs in relation to obtaining and not obtaining. Instead of saying, for example, that every proposition of the form "S is p and not p" is false, one can say, "For no object does it hold that it may have some property and not have the same property." Likewise, the ontological proposition that every object has some predicate (or that an object without a predicate is a *Widersinn*) can be looked at as the equivalent rephrasing of the meaning-theoretical proposition: If the proposition-form "X is something" is satisfied for some X, then there is a concept-meaning a that satisfies the proposition-form "the same X is, is a."

In addition to that comes, however, a multiplicity of new laws of the formal ontological sphere that pertain to all the concepts that originate in judgment-forms through the nominalization of moments. The form of the indefinite plural can be nominalized in such a manner that the concept of multiplicity results, whereby

multiplicity turns into the object (<and> multiplicities fall under this concept as objects). Likewise, the form of "two" that, for
233 instance, occurs in the definite plural form "Two A are B" can be nominalized into the number "two", where "two" turns into the object. The objects that we call numbers of the number series in general arise in this way. And then belonging to these new concepts (which, as one sees, have their essential relationship to meaning-forms) are pure concept-propositions that judge about sets in general, about numbers in general, and so forth. Likewise, nominalization of the form of the identity and relations judgments "identical", "equal", "unequal" yields the hypostatizations equality, identity, difference, inequality. The relation-form of the "in something" yields inexistence (being contained), the concepts "whole" and "part", and so forth. The concepts "sequence" and "ordinal", the concepts "quantity", "magnitude" arise in like manner. And, the formal mathematical concepts in general arise in this way. And, in formal mathematics, judgments are then made about cardinal numbers, ordinal numbers, and other pure numbers in general in formal universality, likewise about sets, orders, and order types in general, about combinations and permutations, about quantities in general, and so forth.

The theory of forms is now no longer directly at the basis of the judgments, but one is operating within a realm of concepts that ultimately lead us back to certain meaning categories, but arise out of these through a certain transformation, through a certain nominalization. And, in addition to that come indirect concept-formations having a relationship to meaning-forms, such as, for instance, the concept of quantity in the original sense, as that of a whole of equal parts.

That may suffice to show that with formal mathematics we do not actually enter into an essentially new domain, but are here dealing with a field of pure concept-truths whose conceptual matter is inseparably linked to the original matter of the logic of meaning. The philosophical import of this realization can of course not be assessed here. The great interest attaching to discovering and soundly determining the fundamental demarcation in the overall realm of knowledge cannot in general be discussed here. Such a fundamental demarcation, which ranks as the lowermost and perhaps the most important, is that which separates the domain of analytica priori knowledge from synthetic-a posteriori knowledge and isolates the entire domain of a formal a priori ontology in a clear-cut way under the former heading. 234

<c) Disagreement with Kant>

Kant, who first saw the difference between analytic and synthetic a priori and yet did not understand its full scope and essential demarcation, not in vain called it a classic distinction for transcendental philosophy. What he lacked was the genuine concept of the analytical, which is determined by the conceptual sphere that I called that of the formal categories. What he further lacked was the understanding of the essence of formal logic, of traditional logic to begin with. If we look at what it offered to purely logical truths and, therefore, just as *Kant* wished, separate the logical a priori from everything empirical-methodological, especially from everything that is a matter of the theory of the art of logic, then we soon realize that what is purely logical according to the tradition exclusively belongs in the apophantic sphere, and that therefore formal logic to a certain extent sought to be apophantics, without being able to define its goals and its natural limits scientifically. Apart from *Leibniz* and a few of those influenced by him, no one had suspected–and especially not *Kant*–that pure arithmetic and all the disciplines essentially related to it, such as pure combinatorics, pure theory of magnitudes, and so on, intrinsically belong together with the old formal logic. Unmistakable for *Kant*'s readers is furthermore the fact that for him the concept of the analytical extends as far as his concept of pure logic, i.e. to be precise, of apophantic logic. One does not at first notice that his definition of analytic judgments oddly limits itself to categorical judgments (as though *Kant* himself would not have coordinated categorical, hypothetical, disjunctive judgments under the heading "relation") and implies that every categorical judgment is analytic whose predicate concept is overtly or covertly contained in the subject concept, therefore, for example, "A body is extended". The denial of such a statement produces a "contradiction", and *Kant* also said in the *Prolegomena* that the principle of all analytic

judgments is the law of contradiction. Exactly this, however, leads us to an understanding of the implicit identification of analytic judgments and all formal logical judgments, more precisely, of analytic laws with formal logical laws and of analytic necessities with formal logical necessities. It is in particular a tenet of the apophan-
235 tic theory of validity that by virtue of its basic principles–to which the law of contradiction itself belongs–, the denial of the law of contradiction is deducible from the denial of every purely analytic concept-proposition (analytic law) and that, consequently, the denial of any analytic truth, i.e. the assertion of an analytic falsehood, is traceable back to an analytic contradiction. One can accordingly also say that what is characteristic of analytic truths is that a contradiction may be derived from them, or that they evidently "imply" a contradiction, whereby of course the validity itself is and must be analytic. Connected with that therefore is the fact that for *Kant* analytic judgments and formal logical judgments (in the dual sense of law and necessity) coincide. And, it is consequently understandable why he did not call arithmetic truths analytic, but synthetic. Actually, no arithmetic proposition can be understood apophantically-logically. At least that has never been demonstrated up until now. Only the fact that every theorem follows from the arithmetic axioms can be deduced formal-logically. The synthetic (in the *Kant*ian sense) is then, however, implied in the axioms, and from there is carried over to the content of the theorems themselves. However right *Kant*'s observation is then in itself (when one takes into consideration the separation between what is merely apophantic and what is ontological in the broader sense), by means of it, *Kant* blocked the way to an incomparably more important realization, namely, to the fundamental separation that dissevers everything belonging to the realm of formal category from the sphere of the non-formal a priori. The consequence of this for *Kant*'s critique of reason is a completely inadmissible equation and equal treatment of the arithmetic disciplines with the remaining purely mathematical disciplines, with geometry and chronology, and the combined severing of both from what he called pure natural science. Nevertheless, I must not dwell longer on this point.

<Chapter 9 236

Inferences and Proofs as Judgment-Units>

<§ 47. Clarification of the Essence of the Logical Theory of Inferences from the Highest Vantage Points>

<a) The Modality of an Inferential Judgment Must Have the Value of Being Genuinely Apodictic>

I shall now move on to a series of comments that will be useful in clarifying for you the essence of the logical theory of inferences from the highest vantage points and at the same time open up to you the prospect of understanding a remarkable, and in a certain sense the highest, area of analytics, namely, the formal theory of theories or theory of manifolds.

Let us go back again to hypothetical judgments. I shall first recall what is familiar. Hypothetical judgments are necessary judgments, judgments which state a necessary conclusion. Corresponding to necessity is a law whereby corresponding to the terms of necessity are those of conformity to laws. Law in the strictest sense is a purely conceptual law, an a priori law. However, the concept of law is commonly used more universally, to such an extent that in the factual sphere, we speak of law in the case of every unconditionally-universal, and even only roughly-universal, rule of existence and events within nature, for example, of strictly physical, but also of meteorological, laws etc. Hence, hypothetical statements sometimes indicate strict, purely conceptual necessity, sometimes necessity related to the natural sphere, but unconditional within nature, but often also to vague necessity, i.e., we attribute necessity when only a particular case of a rough conformity to rules is present, and as a result significant probability.

E. Husserl, *Logic and General Theory of Science*,
Husserliana: Edmund Husserl – Collected Works 15,
https://doi.org/10.1007/978-3-030-14529-3_9

I now again stress the case of purely conceptual necessity, of a priori necessity. Then, the necessity is apodictic. It is then to be said <that> wherever a hypothetical judgment ought to be called an inferential judgment, or wherever a causal judgment with
237 "because–then" is to be called an inference in the concise sense, the condition to be satisfied is that the modal nature have the value of genuine apodicticity. This therefore implies that corresponding to every inference in the genuine sense is a law of inference as an a priori law.

<b) Direct and Indirect Hypothetical Relations>

I now distinguish between direct and indirect hypothetical relations, and then also direct and indirect inferences. A direct relation is for instance present–to illustrate the matter immediately using an example–when we infer in accordance with the form "if (or because) a > b and b > c, then a > c"; an indirect relation, however, were we to infer, "a > b and b > c, therefore c < a". And, as one readily sees, both inferential relations are thereby completely equivalent. The second is however indirect, because we see that actually lying in the premises –and precisely directly–is not c < a, but a > c, and that we only come to the other concluding proposition through a further turning around of this proposition. In this manner, we generally see in certain cases that the concluding proposition proceeds directly from the premises, that the conclusion is a direct conclusion from the arguments, the premises. However, as soon as we replace one of the propositions–or <one of the> judgments–with an equivalent one, this this directness is lost, though the inference may otherwise remain obviously correct. Every indirect necessity-judgment refers back to a direct one in the following way. If we have the direct inference, "If M holds, N holds"; and from N follows, for instance, N_1 (with the subscript 1), then we can always infer, "If M holds, then (considering that N_1 follows from N) N_1 holds"–or, "If M holds, N_1 holds, considering that N holds".

$$\left.\begin{array}{l} M \in N \\ \\ N \in N_1 \end{array}\right\} M \in N_1 \qquad\qquad \left.\begin{array}{r} M \in N_1, \text{ considering that } N \in N_1 \\ M \in N \end{array}\right\}$$

In our example, from $a > b$

$c < a$ follows, considering that

$b > c$

$a > c$ (this follows directly from our premises and that in turn has $c < a$ as consequence).

It is also to be noted that wherever a conclusion follows from more than one premise–let us say from two–, we can steer over 238
into the thought of indirect inference, namely, thus: If N € P follows from M, then, when M and N together provide consistent grounds for P-ness, we can always establish in thought an indirect necessity-relation between M and P and between N and P and then form the inferential thoughts, "If M holds, then in consideration of the fact that N also holds, that both hold, P holds (or: given that N also holds)"; "If N holds then, in consideration of the fact that M also holds, P holds (or: given that N holds)".

<c) Imperfect and Perfect Hypothetical or Causal Apodictic Judgments>

From there, we shall then be led to those hypothetical or causal apodictic judgments that, as opposed to perfect, I call imperfect. To begin with, subjectively-psychologically speaking, it can be that the one premise is a well-known truth, and what is new and interesting for us is that a relation of necessity obtains between the other one and the conclusion, in which that established truth is of course involved. Complete necessity binds both premises with the conclusion. However, we drop the one. We still retain necessity, but, in any case, an indirect necessity. Here, though, the following very often also enters in. When we infer, "All humans are mortal, therefore, Caius is also mortal"; with the dropping of the premise no longer of interest to us that Caius is a human, we can no longer replace the terms by free variables in the abridged inference. The

law corresponding to necessity would indeed moreover read, "All A are B, therefore, C is B"; which would indeed be false. Therefore, with the suppression of a premise that is in itself true, and the suspension of a necessity-relation at the same time partial and indirect, the immanent relationship to laws is as a rule lost. I then define an inferential relation, whether it be direct or indirect, as a perfect one when it is of such a nature that, when we replace the terms of necessity by variables, we obtain a corresponding valid law of inference, when therefore the necessity is perfect in nature, when it is a particular case of perfect conformity to laws. When
239 that is not the case, then I call the inferential relation, or what <is> the same thing, the necessity-relation, an imperfect one. So, the famous inference with "human-mortal-Socrates" is obviously perfect. Likewise, the inference "a > b > c – a > c" is, but also that directly attaching indirect inference, "c < a". With this example, one also immediately understands the law that a perfect necessity-relation remains perfect when a proposition is replaced by a directly equivalent one.

What is though more pertinent to us here is the law that every direct inference is perfect, although as can be seen from the last example, the reverse does not hold. It is further to be seen that every imperfect inference arises through the destruction of a direct inference, namely, in accordance with the law that by striking out a thetically true premise from a necessity-relation, a necessity-relation remains. A true premise can always be struck out, and that usually gives an imperfect necessity-relation.

Many important things are still going on here that I cannot discuss more closely, such as the distinction between simple and composite inferences, and that in a different sense. I only draw attention to the fact that by inferences in the concise logical sense we are accustomed to understanding only direct inferences and thereby those not containing any extra-essential premises. Among them, I understand those premises that can be eliminated without disturbing directness, and hence perfection. It might also be good to incorporate this property into the concept of perfect inferences, perfect and irreducible. We say (as a direct causal judgment) of a direct valid inference that the direct arguments for the consequence posited in the concluding proposition lie in the premises.

The arguments are obviously the premise-judgments, or -truths themselves, provided that we take the word judgment and truth in my meaning-theoretical sense. I further say of every perfect causal context that a perfect argument for the consequence is given in the set of premises. Every imperfect argument therefore points to a perfect argument or to a perfect set of arguments.

Since the word "inference" is still used in a sense that reaches
beyond the sphere of a priori necessity, the inferences under dis- 240
cussion are to be called a priori, or also "deductive"; the latter in the sense of traditional, though not clarified, talk of deduction.

<d) Eliminative Inferences>

A further distinction that can already be cited as being of great interest here in the sphere of universality in which we are now operating is that of perfect inferences in eliminative inferences and other inferences. In such an inference–as is to be seen in connection with laws–arguments and consequences must have something in common with one another, i.e. a perfect argument not having any terms in common with the consequence would be a *Widersinn*. Furthermore, when several arguments work together in a perfect argument, in other words, when the perfect inferences have several premises, the latter will generally have different terms, generally, however, also some terms in common with one another. When, then, the inference is of such a form that such common terms of the premises drop out in the conclusion,[1] then the inference is called an eliminative inference. For example, "If all (no) A (are/is) B and all (no) B (are/is) C, then all A (no A) are <(are/is)> C." Here the concept B is, mathematically speaking, eliminated.

When traditional logic distinguishes between mere conclusions and inferences in the genuine sense or syllogisms, that is admittedly certainly not a sufficiently general, well-founded distinction, but one can say that all "syllogisms" are eliminative inferences, mere conclusions however not. Belonging to the latter are, for example, inferences from "All A are B" to "No A is not B", or from "All A are

[1]The terms that drop out are called middle terms.

B" to "Some B are A", and so on. I would add to this, for example, all inversions like a > b <in> b < a, and countless other forms.

<e) Purely Formal and Material–Analytic and Synthetic Inferences>

Of greatest importance is now a further distinction that is of course completely obvious to us, that of the separation of infer-
241 ences into purely formal and material (or analytic and synthetic) –and thus correspondingly the difference between analytic and synthetic necessity-relations, analytic arguments and synthetic arguments, or analytic and synthetic consequences. The difference is simply given with the fact that the law that gives necessity meaning and character can be precisely either a formal, analytic one, or synthetic. In the analytic sphere, apophantic inferences then especially stand out. They are what the inferences that traditional formal logic have dealt with exclusively since *Aristotle* and have also not been dealt with in sufficiently theoretical completeness there. One can also characterize "apophantic inferences" in this way: An inference is apophantic when the law-bound sphere of variability of its terms of necessity lies in the purely apophantic categories, therefore, when one can mathematize them in such a way that an apophantic analytic law arises, a law whose functions are the apophantic proposition-configurations.

The examples I have repeatedly used are of this kind, for example, "All A are B–therefore: No A is <not> b", "Some A are b–therefore, There are A that are b", or "If no A is B and no B is C, then no A is C either". We can then especially say: "Syllogisms" are to be defined as apophantic eliminative inferences. Arithmetic inferences, for example a > b > c – a > c (here > means greater than), are not apophantic, but analytic in our broader sense. The "greater than" cannot be replaced here by the general apophantic category of "relation". Sound-color-inferences etc. provide examples of non-analytic (non-categorial) inferences, "a louder than b"; "a brighter (higher) than b".

It was a great mistake on the part of the old logic that it did not realize that its set of inferences dealt with anything but all inferences, that with its apophantic inferences it believed it was dealing with all inferences in general. Even *Sigwart* lapsed into error and fallacy here.

What is deceptive here is the fact that we can transform every non-apophantic inference, therefore, also every a priori inference of the synthetic sphere, into an apophantic inference, for example, 242
when in the field of sounds we infer, "Since c is higher than d and d higher than e, then c is also higher than e"; that is certainly and evidently a direct inference. The law of inference is synthetic a priori here. We can, however, make an analytic inference out of this, namely, if we infer, "Since c is higher than d and d higher than e–and since a law universally holds that if, of three sounds, the first is higher than the second, and the second higher than the third, the first is also higher than the third–, then c is higher than e". In this case, we have transformed the synthetic laws of inference themselves into a premise and added onto the given premises, and then one immediately sees that the inference is analytic, namely, that it holds for sounds and their pitch relations, *because* it holds for arbitrary objects and arbitrary relations, provided only that they comply with the prescribed sentence-forms.

We moreover immediately see that formulating laws of inference and introducing them as premises is also possible within the analytic sphere and that, so proceeding, we can trace every inference back to a single operation, to the subsumption of the particular under the universal. Instead of simply concluding, "There are unbounded conic sections–therefore, not all conic sections are open", we can say: "Since the law for arbitrary terms A B holds that, if there are A's that are not B, not all A are B either, then the same thing can be stated of conic sections as A's and for the property of boundedness". That is an apparently meaningless observation. We shall however see that it is of importance for the understanding of the deductive sciences.

Of further general distinctions belonging to the theory of inferences in general, to begin with, I in addition mention the distinction between inferences in terms of the kind of terms of necessity. It is to be borne in mind there that not just isolated terms in the premises–be they terms with their own nuclei or with adjectival ones or with relation-nuclei–can be bearers of necessity, but also the whole propositions themselves.

<f) Propositional Inferences>

A class of inferences in the apophantic sphere is precisely characterized by the fact that the whole sentences out of which the unit of the inference is built are terms, whereas how we organize the potential complex propositional wholes into sentences is, however, not a matter of indifference. So every hypothetical inference of the
243 forms bearing the names *modus ponens* and *modus tollens* is a propositional inference. It reads, "If it is true that if A holds, B also holds, and if B does not hold, then A does not hold either. <If it is true that if A holds, B also holds,> and A holds, then B also holds." Here A and B are signs for propositions. (Naturally, in actual inferring, we need not just draw in the concept of proposition and that of validity.) We infer, for example, "If in the present figure, angle α and β were equal, then two right angles would have to occur in triangle ABC. That is, however, not the case. Therefore, α and β are not equal." Here α, β, and so on are not the terms, but the whole proposition. Likewise, if we infer in accordance with the form "If A holds, then B holds, then C holds–therefore: If A holds, then C holds." In other cases, however, as in what are called categorical syllogisms, concepts–understood in their syntactical category–are variables.

<§ 48. The Concept of Logical Equivalence>

It is also now time to discuss the oft-mentioned concept of logical equivalence in its systematic place. In the broader sense, two propositions M and N are called equivalent when they are related to one another in such a way that if M <is>, N is, and inversely. However, one also speaks of equivalence in the case of concepts. Namely, it universally holds that if something is a, it is b, and if something is b, then also inversely a is, then the concepts are called equivalent or concepts having the same extension. For the theory of equivalence, it is particularly important to be mindful of the ambiguity and indeterminacy of the hypothetical statement-form. The mere statement "If M € N" does not tell us which are the terms of necessity and whether the relation is perfect or imperfect. That implies essential indeterminacy, because the terms and law-

relationship upon which necessity is based is decisive for its full meaning. It would therefore be utterly wrong to relate the two interrelations to different terms and laws. Where it is a matter of equivalence, the basic requirement is–though it may remain implicit–, that we operate on the basis of the same perfect necessity-relation, so that therefore the meaning is: B follows from A and A from B necessarily directly perfectly, or indirectly, and when the latter <is the case>, then in relation to the same arguments to be complemented. For example, two geometrical propositions are equivalent if they follow from one another. So, when 244
they do so indirectly, it is however understood that the implicit auxiliary arguments, which add every proposition to the complete argument of the other, lie in the axioms. Conversely, "following from one another by means of the axioms" is therefore geometrical equivalence. And that is the situation in every mathematical discipline, for each one has its unity in a closed axiom system. In this manner, one can say of two theorems quite remote from each other in content–such as the theorem of the sum of the angles of a triangle and the *Pythagorean* Theorem–that they are equivalent to one another in *Euclidean* geometry. Naturally, the analytic-synthetic distinction also carries over to the concept of equivalence. A narrow concept is that of apophantic logical equivalence. Logically equal in the concise or formal logical sense are, for example, propositions such as "All men are mortal" and "There are no immortal men", but not propositions like "a more intense than b", "b less intense in comparison with a."

I have repeatedly had reason to talk about the dangers involved in confusing equivalence and identity of meaning. This especially concerns direct equivalence, because where–as in objective research–interest is directed toward the things and not toward meanings, what is directly equivalent is in fact equivalent. In every inference, in every comprehensive theoretic system, direct equivalence can (in accordance with an unconditionally valid law) be used for direct equivalence without further ado, without violating the truth. Since direct equivalence then also signifies direct obviousness of the transition from one to the other, it is simply called the same. Whether we say, "Something is A" or "There is an A", whether we say "Some A are b" or "There are A, that are B",

that is "in essence" the same. Or "in essence", one says "it is the same factual situation". One can actually also seriously find arguments for this talk, provided that the presentation-foundation is the same on both sides, and the same term content occurs in judgments on both sides, only expressed in different logical forms. That holds literally and strictly especially of apophantic-analytic equivalence. The same nucleus-content that is freely variable in its nucleus-category is there on both sides, and only the apophantic
245 form has changed in the sense of a purely apophantic law.

<§ 49. The Concept of Proof>

I am now going to discuss briefly the concept of proof. Inferences can link up with one another. From one or several premises, we say A, B follows. If we then connect B with the new premises B′. Then, for instance, C, from C and C′ then D, and so forth can follow in an ideal manner. The whole construct of inference-combinations ends, for instance, with a final concluding proposition Z. Every proof looks something like this. Psychologically speaking, we wish to prove Z. The truth of Z is not directly given. We do not see it directly. We then seek a way of proving it. We look around in the sphere of knowledge we have already secured to see whether we find truths in it that as premises yield Z as concluding proposition. And often needed there is such an indirect route, such a linking of direct inferences. For, direct inferences are directly perspicacious. We see directly that the conclusion lies in the argument. Z therefore becomes indirectly perspicacious to us, either when it can be directly perspicaciously inferred from premises that are themselves directly perspicacious, or when it can itself be deduced from premises that are directly deducible, and so forth.

We do not call every arbitrary combination of inferences a proof, but rather such combinations of inferences that are constructed out of direct and rigorously perfect ones. Superfluous premises will not be tolerated in a proof. What is not necessary for the argumentation is a hindrance and must remain struck out. So, one will not, so to speak, go round in circles and want to advance from equivalent to equivalent, introducing new terms and then eliminating them again, without anything essentially new resulting. In this way, whole portions of the proof can be superfluous, which would be a flaw in the proof. In terms of meaning-form, proofs are no doubt always proofs by elimination, and the new thing they teach, or that we want to learn from them, consists in the fact that groups of terms that were indirectly linked in the arguments of the proof come directly into combination through the elimination of multiple intermediary ones in the concluding proposition. The whole proof–as many partial inferences as it contains, with the judgments constructing it–is obviously a judgment-unit. It can be looked upon as a judgment. 246

Obviously derivable from every proof is then a consistent causal judgment whose causal antecedent consists in the conjunction of all of the arguments of the proof, i.e., all of the premises that are not concluding propositions, while the causal consequent of the final concluding proposition is Z. A B C…

$$\left.\begin{array}{l} A \in A' \\ \quad A'B \in B' \end{array}\right\} AB \in B'$$

$$B'C \in C' \quad ABC \in C'$$

Finally, "Because A B C … Y is, Z is." In a concrete proof, this causal judgment is perfect and irreducible. None of the antecedents is thereby expendable. Furthermore, this causal judgment cannot be of the nature of a direct inference, because that would mean that the whole proof is superfluous. We call this causal judgment, which is therefore of the nature of an indirect inference, a contracted proof. It assembles all arguments together and directly attaches the conclusion on to them. We can also say that in a correct proof, all the premises not deduced in it, in their conjunctive unity, form a perfect and irreducible argument for the proven proposition as conclusion. It remains to be noted that when the proof is also in its way a kind of fixed unit that does not permit the adding or discarding of arbitrary proof-arguments, for this reason, the same concluding proposition is very frequently provable in many ways, and that in many proofs the same premises can in part occur as arguments. Every perfect whole argument is something, so to speak, compact, but that does not rule out the fact that different arguments of this kind have partial arguments in common.

<§ 50. Deductive Theory as a Web of Proofs. Analytic and Synthetic Deductive Theories. The Systematic Construction of Analytics Is a Matter for Mathematicians>

It is now to be pointed out that the fundamental distinction between the analytic and the synthetic carries over to proofs, and when we take one more, final, step further to deductive theories and theoretical disciplines as well.

From the point of view of the logic of meaning, a deductive theory is no more than a web of proofs by means of which an
247 essentially related group of truths leads back to one and the same store of basic truths as the consistently perfect irreducible basis out of which they are all provable and by this means are (as people also say) explained. The theory's system is a necessary system. What is explained theoretically, what is deduced, is derived as a necessary conclusion. And depending on <whether> only pure concept-truths (noetically speaking, a priori truths) are found among the proof-arguments or also factual truths, the whole theory is a priori or a posteriori. That does not, however, change anything about the fact that deductive theory is a genuine necessity-system, every step of which is subject to a priori laws, and the fact that what is theoretically substantiated is substantiated as an a priori consequence, namely, as following *a priori* from the arguments.

Of course, to characterize the genuine concepts of argument and consequence, some supplement is still needed here. A theory, a theoretical discipline, has its chief grounds upon which it rests. The order of a deductive argumentation is not always a genuine one that traces back to genuine arguments. What characterizes genuine arguments or genuine basic principles in a priori disciplines?[1]

All mathematical-physical disciplines are of the nature of deductive theories and also have their relatively independent focus. Thus, all of mathematical mechanics since *Galileo* is a single deductive-theoretical system that traces all mechanical events of

---

[1] Compare Appendix XVI, *Indirectness Belongs to the Essence of Theorems, Directness of Insight to the Essence of Basic Principles*> (Editor's note)

physical nature deductively back to what are called the basic laws of mechanics where, arithmetic and, in general, mathematical analysis, geometry, pure theory of time, kinematics, of course serve as auxiliary disciplines. These are for their part themselves also purely deductive disciplines, that is, a priori, provided the store of basic laws in them is a priori in nature, while the major premises of mechanics are at least partially a posteriori in nature.

The difference admittedly appears somewhat obscured by the fact that the basic laws of physics are stated in the form of pure concept-propositions. It only becomes apparent upon reflection that these propositions thetically imply the existential positing of nature. One can also say that the difference becomes apparent 248
through noetic reflection, insofar as basic laws, which really are of the nature of pure concept-propositions, must be able to come to givenness in directly evident insights, while the opposite holds for factual laws. But this difference does not fall within the scope of the logic of meaning itself, which we really go beyond wherever we resort to knowledge and the characteristics of knowledge, of consciousness.

If we carry the analytic and synthetic distinction–apart from the a priori and a posteriori–over to deductive theories and disciplines, then these break down into analytic (or categorial) and non-categorial or synthetic. The non-categorial ones <break down> into synthetic a priori ones and into a posteriori ones. All the analytic disciplines (disciplines bringing all the laws to ultimate theoretical unity) that belong to the formal categories, form, as I have already repeatedly mentioned, a homogeneous unit, precisely analytics in a broader sense.

Were it in general my task to construct systematically the theories of analysis, then it would now be the time to begin this construction and to implement it, because I have worked out and I have elucidated all systematic basic concepts that are destined to play their role in it. However, the construction of these theories and theoretical disciplines is in my opinion not a matter for philosophers, but a matter for mathematicians. Just as from time immemorial, arithmetic, and since the Renaissance, algebra and analysis, as against philosophy, have developed as independent sciences, so they must also remain independent. And, if in the last

century, mathematicians also adopted the deductive theories of traditional syllogistic logic and gradually developed a mathesis of propositional, conceptual, and relational meanings, namely, in the spirit of the solely proper mathematical method, they have only laid hold of a field that was their very own possession. In my opinion, all polemicizing against mathematizing logic testifies to a lack of understanding. Constructing deductive theories is definitely the business of mathematicians. They are, as it were, the solely competent engineers of deductive structures. The task of philosophers
249 lies in another direction, above all in the complementary reflection on the essence and meaning of the governing basic concepts and basic laws and not in the least in weighing the intrinsic relationships of such disciplines to all other disciplines.

I am therefore skipping all of traditional syllogistic logic, which is precisely a piece of pure mathematics, namely, of the pure mathematics of propositions and of predicates of possible subjects in general, and I am skipping the all the rest of theoretical analytics in order to remain within the sphere of my own philosophical interests. Proceeding upward, I wish to give the idea of the theory of science, which I had placed at the head of these lectures, the greatest possible breadth and depth and seek to press forward from the philosophically preparatory considerations to the intrinsic, genuine domains of philosophy itself, that is, of philosophy as critique of knowledge.

<Chapter 10 249

The Field of the Theory of Probability>

<§ 51. The Difficulty in Grounding the Theory of Probability>

For the discussion of the important field of the theory of probability, I only still wish to identify very succinctly the place where it originates.

Up until now, I have exclusively given priority to the sphere of judgments as affirmative judgments, i.e., meaning-theoretically speaking, the sphere of supposed truths. The modalities of necessity also belong in this sphere, because necessity is a characteristic of certain truths. We know, however, from earlier lectures that there are still other judgment-qualities. Without being more specific, I spoke at the time of possibility, probability, questionableness. I also mentioned that these qualities occur in the sphere of judgments as a result of the fact that possibility or probability come to be expressed and then introduce a distinctive kind of meaning-content into this sphere. Subjectively speaking, we make an affirmative judgment about possibility or probability when we say, "It is possible that S is p", "It is probable, questionable, that S is 250
p". The concepts "possible", "questionable", "probable" are thereby not specifically apophantic categories, if we understand apophansis as judgment, as supposed truth. They just arise from their own sources. Supposed truth in certain ways stands apart in a parallel manner: supposed possibility, probability, questionableness. Insofar, however, as what is supposed thus is expressed and occurs in thought-formulation in the affirmative judgment, judgment-forms arise that contain the new categorial moments. And afterward theories–precisely theories about possibilities and probabilities–apply to such judgments or to such categories. However, <the> situation is again not the same as, for instance, for values, pleasing, displeasing, and so forth, which also occur in the apo-

E. Husserl, *Logic and General Theory of Science*,
Husserliana: Edmund Husserl – Collected Works 15,
https://doi.org/10.1007/978-3-030-14529-3_10

phantic sphere, as one might guess, as a result of the fact that values, pleasantness, unpleasantness are expressed as "It is pleasing that," "It is unpleasant that," and so forth. There is, however, an intrinsic connection between possibility, probability, questionableness, doubtfulness, and the plain certainty whose correlate is "quite simply being." It is not without reason that in logic it is not a matter of values, but of probabilities, and so forth. Plain believing in the mode of being certain is essentially and generically related to considering-possible, presuming, questioning, while feelings make up another genre.

The philosophical and factual difficulties connected with grounding the theory of probability that are unsolved up until now are all due to the fact that, on the one hand, people were not able to effect the separation between what is psychological and meaning-theoretical, and, on the other hand, that people did not understand how to sunder the concept of possibility that is fundamental for the theory of probability from other concepts of possibility of a logical nature. This must be pointed out here to begin with and my earlier remarks on modalities recalled.

<§ 52. Possibility as Presumability. Precedence and Gradation in the Area of Presuming. Indifference as Absolute Questionableness>

There is an analytic possibility, the formal logical one, accord-
251 ing to which every concept and every proposition is possible that does not conflict with any analytic law. There is a possibility in the broader sense, a synthetic a priori possibility, according to which what does not conflict with any synthetic a priori law in general is likewise possible. In the latter case, for noetic reasons, possibility is equivalent to intuitive presentability. What can be made intuitively explicit or, if it is to be something individual, present itself intuitively, cannot involve any incompatibility of essence, therefore, not conflict with any essence-law.

Parallel to that, there is also a natural scientific possibility and impossibility, for example, a case of perpetual motion. Namely, what is indeed intuitively presentable, but contradicts the basic laws of natural science secured by countless experiences and verifications is natural scientific nonsense. Beyond that, it is moreover clear that talk of possibility in the natural sciences has something inherently vague about it, and indeed when in popular language everything conflicting with any narrow, rough empirical rule is declared impossible.

All these possibility concepts have nothing essential to do with the possibility concept of the theory of probability. It need only be said that for what is possible in the latter sense, we also frequently say: "what is presumable" cannot be anything impossible in the other sense–and above all not in any a priori sense. What is possible as what is presumable is the exact correlate of presuming, and just as judging can be correct and incorrect, so presuming can, and consequently in the meaning of correctness, there is also something presumable that not only counts as presumable, but truly is presumable.

To begin with, let us look at the presumable that <is> the validity-correlate of plain–not of relating and preferring–presuming. It is the correlate of the plain consciousness, "Something argues in favor of that." Empty possibilities are possibilities in whose favor nothing argues. They are mere figments of the imagination or imaginable things. A well-founded "possibility," or something presumable with a reason for it, is something entirely different. In all presuming, something figures to the consciousness as presumable, i.e., just as something argues in favor of this or that. And this argument can be correct. Seeing the barometer rise, we for instance presume that there will be good weather. A rise in the column of mercury perceived argues in favor of that, testifies in favor of that. But, of course, it is a presumption and not a belief. <It> is not expressed as a judgment, as an affirmative positing. In 252
our eyes it is not yet truth, but only "probability."

Probability–originally and understood outside of any relationship–is the same as the possibility occurring here, and again the same presumability. And, for its validity, every probability in this sense requires a reason for being probable.

Now, however, presumptions come into relationships. You immediately understand what is meant if I point to the familiar way of speaking, "This argues in favor of it, that other thing also argues in favor of it, and still more is in its favor". I am speaking from a subjective perspective of stronger and weaker presuming. We say that the presumption grows stronger with the number of reasons for presumability. However, we also say that the more there is that argues in favor of something, namely, in favor of the fact that something exists, the greater is its probability; the more that argues against it, the more the probability diminishes. This is an objective way of speaking.

Anyone who has learned to distinguish between what is a matter of consciousness and what is a matter of meaning does not then find it hard to understand that presuming-acts stand opposite to presuming-meanings, furthermore, that a unique originary relation of precedence and gradation belongs to the presuming-meanings, and accordingly also objectively valid laws of gradation.

In these relationships of gradation what, considered in isolation, was already called probability, is then referred to as mere possibility, and possibilities are weighed against other possibilities, that is, also possibilities <of the kind> that A is B, and possibilities <of the kind> that A is not B, with respect to the reasons for them or the weight that they carry. Therefore, a distinction between positivity and negativity occurs in the area of presuming just as in the domain of affirmation, only that in the latter there is no talk of a precedence and gradation that plays such a great role in the area of presuming. To be noted thereby is that a negative presumption that argues against "A is B" is equivalent to the positive presumption that argues in favor of "A is not B".

Furthermore, possibilities designated as equal "add together" with respect to the weight they carry. Possibilities referring to the same matter behave in relation to their signs just as positive and
253 negative quantities do–at first only in the adding and aside from the question of dividing. Two antithetical possibilities can keep the same weight. There is indifference. That is the absolute questionableness occurring here. Absolute questionableness is thereby in no way to be confused with empty indifference, namely, < the indifference> "Neither does anything argue (naturally, for my consciousness, <for myself,> the indifference I presume) in favor of A, nor anything against it". No indifference obtains means that with

respect to the preponderance of arguments of a positive nature, "it is more probable that A is B than that it is not <B>"; or simply, it is probable. People particularly simply say it is probable where the sum of negative reasons seems "very negligible" in comparison to the combined force of the positive ones. The concepts "strong", "weak", "very negligible", "very strong", and so on occur here with the same vagueness as elsewhere in the other areas of gradation. That does not, however, rule out that, here as elsewhere, under certain circumstances, the vagueness would not be kept within limits or indeed transformed into exactness.

<§ 53. Seventeenth Century Mathematics' Discovery: In Certain Spheres Probabilities Can Be Transformed into Exact Quantities>

And there, seventeenth century mathematicians precisely made the discovery that such comprehensive classes of probability judgments can systematically be made exact, in other words, that in certain spheres, one can transform presumptions or probabilities into exact quantities. In this way, the mathematical theory of probabilities, the theory of exact (quantitative) probabilities arose. It is thereby all a matter of establishing a field of equal possibilities in objectively valid ways–of possibilities in whose favor that very field precisely argues in exactly provable ways, which carry the same weight, positive and negative–and then of referring every complete cycle of weighing probability back to this basic field. Since many logical considerations and potentially very complex deductions are otherwise to be engaged in when working out probability problems, the basic requirement that always gives unity of meaning to all the considerations is always that all concepts of possibility and probability used within the same problem 254
and set of problems have to refer back to one and the same basic field of equal possibilities that, as it were, acts as the basic norm for all relative assessments. All the notorious, but not always elucidated, fallacies of probabilities and of the theory of probability are themselves due to the fact that either the fundamental field of equal possibility was not exactly and unambiguously defined, or that in the course of examining them, in spite of rigorous defini-

tions, the field was changed unawares. In the weaving of probability inferences built one upon the other, different fields do indeed as a rule occur, but always in such a way that one field must be the basic field and all other fields have been singled out from it in exact, legitimate ways.

What separates the field off is thereby always an exclusive disjunctive judgment that says: There are n and only n possibilities that are equal to one another. One of them is certain truth, but which is undetermined. If, then, for instance, it holds for p of these possibilities that they bring with them the actual being of a state-of-affairs, for instance, of M, and it <holds> of n – p that they bring with them the converse, then p/n is the mathematical probability for the occurrence of the event called M.

I cannot go more deeply into this. The immense significance of the theory of probability is apparent without further ado if you think about the role presuming plays in our thinking. On the other hand, the wealth of original concepts is no longer very great here, simply because the entire field of the other apophantic concepts also come into consideration for the theory of probability. What it introduces that is new is really only some qualities and associated modal characteristics.

<SECTION III 255

THE GENERAL IDEA OF THE THEORY OF SCIENCE>

<Chapter 11 255

The Pure Theory of Manifolds as Science of the Possible Forms of Deductive Theories in General>

<§ 54. The Science-Theoretical and Science-Practical Function of Analytics>

I now turn back to the general idea of the theory of science that was to be the governing idea for the entire content of my lectures. I now leave behind all particular considerations. We are sufficiently prepared to proceed upward in systematic steps to inquire into the sequence and interrelationship of the disciplines that lend the Idea of the theory of science the greatest possible breadth and meaning.

My reflections naturally connect up with my idea of analytics, which I am now situating in relationship to the remaining sciences, especially the theoretical sciences in the specific sense of the word. In the broadest sense, theoretical sciences are those that stand in contrast to the normative and practical disciplines, in which therefore the governing interest is purely factual and epistemological, not normative or practical. In the specific sense, theoretical sciences are, however, those that aim at explanatory theories. They are, on the one hand, purely nomological sciences that aim theoretically, i.e. deductively, to trace an infinite supply of laws back to 256
a finite foundation of basic laws that united present the perfect basis for all other laws as necessary consequences. On the other hand, belonging here are the concrete sciences that set themselves the goal of not only describing concrete individuations of a field, let us say of some natural sphere, but of explaining them as necessarily being so through subsumption under the laws of a corresponding nomological discipline.

E. Husserl, *Logic and General Theory of Science*,
Husserliana: Edmund Husserl – Collected Works 15,
https://doi.org/10.1007/978-3-030-14529-3_11

In the natural sphere, the goal of objectively valid, exact description automatically leads back to the goal of explanation. But, only owing to an unjustified partiality in favor of nature is it comprehensible that people are inclined to identify science and theoretical-deductive explanatory science. In actual fact, not all sciences are theoretical in the specific sense indicated. This does not hold, for example, for phenomenology and for what is called theory or critique of reason. All nomological-explanatory sciences obviously have a very close relationship to analytics, a relationship that is of less significance for all the remaining sciences, but also can nowhere be completely lacking in them.

At the beginning of these lectures, I already pointed to the most universal science-theoretical nature of analytics which, since we have in the meantime become better acquainted with it, must for many deep reasons be intelligible to us. To begin with, the all-embracing nature of analytics is clear as a result of the fact that, with its theoretical validity, it obviously extends precisely as far as people judge, affirm, presume, ask, and as such things take place conceptually (I could just as well say, explicitly). Therefore, insofar as statements occur in it, every science falls within the territory of analytics. Its laws, first as laws of meaning, as laws of affirmative or presuming meanings, serve as principles for each and every definite meaning present in any scientific statement. And, in this respect, a special position is occupied by the basic laws of apophantic logic, among which, through a suitable extension of the concept of apophansis, we can also from the start rank the basic laws of probability-statements.

In the explanatory sciences, deducing furthermore plays a leading role. If we leave out of consideration the admittedly very extensive cognitive activities in the empirical sciences that must
257 have already transpired for an empirical statement (and especially a statement of a law) to have been able to be formulated, then in all deductive theories and disciplines, thus, for example, in the whole of mathematical physics, knowing is continual deducing. Each step falls under a law of hypothetical necessity, and if we conceive of the synthetic laws of this kind as relocated under the axioms, then each step becomes analytic.

Now, we of course need not do this formulating, and in actually inferring, we need not emphasize over and over in pure universality the law of inference that it cannot lack if it is genuine inferring, a capturing of genuine necessity. In this respect, therefore, the development of analytics appears to be of theoretical interest only. It is indeed inherently a matter of the greatest interest that in all explanatory thinking, the meaning relationships are governed by a priori, analytic laws, laws having their own a priori, analytic system.

However, to be pointed out is that analytics, the formal mathesis, assumes a tremendous practical role for the scientific enterprise in all theoretically explanatory sciences. And, in further consequence, people need to be made aware how that scientific-practical role had to prompt the development of a new and highest kind of formal mathesis there, a science of possible forms of theories and theoretical disciplines in general. Let us reflect together upon the following: A theoretical discipline, for instance, a discipline of mathematical physics, or even simpler, a discipline like pure geometry (but also an analytic discipline like pure arithmetic) breaks down into a system of proofs and does so in a system of primitive inferences. If we conceive of the law of inference of every primitive inference as formulated, then this would initially play a very meaningful role in knowledge in shielding us from fallacies that famously occur in verbal, and often so vague, thinking because we are compelled to formulate the hidden premises, etc.

In the transition to the law, we have to replace the definite concepts of the discipline with indefinite variables. Out of the individual propositions that we encounter in the axioms as truths then come truth-forms. If from the outset, we then remember to proceed in such a way that we substitute logical letters-terms for all 258
definite concepts in the logical discipline and always the same ones for the same ones, what do we gain in so doing? Well, just as every proposition has its proposition-form, every inference its inference-form, so every proof has its proof-form, and finally the whole theoretical discipline its discipline-form. Like the simplest judgment-form or judgment-configuration, such a form is a function. It is a mathematically-universal presentation subsuming under itself not only the given discipline, but infinitely many

possible disciplines, namely, disciplines of exactly the same form. If the discipline was really one proceeding validly, if all inferences were valid formally, i.e., necessities obeying laws–therefore, also all proofs and systems of proofs–, then universal validity is attributable to the discipline also. That is to say that it stands in precisely the same way to possible particular disciplines of this form as a valid law of inference does to all possible inferences falling under the same one. If we go from an inference to its law, then the latter, for instance, states, "From two premises of such and such a form, a concluding proposition of such and such correlate form generally validly follows". Likewise, if we go from a given proof to its form, it is maintained that, "One can universally validly prove propositions of such and such a form in a system of proposition-forms and inference-forms to be described in such and such a way", and again likewise in a whole discipline.

If, at the outset, one already has at one's disposal all basic laws of analytics, then without starting from given proofs or theories and mathematizing them, one can obviously conceive of valid inference-forms combined in free combination into proof-forms, and proof-forms into theory-forms, in valid ways. And, if one does this in systematic ways, then one obtains multiple theory-forms of a valid kind and can then say of them that every such form can subsume under itself infinitely many possible particular cases of potentially actual theories, provided that occasion for theorization only presents itself in factual domains that would lead to similar theories. Of what use can this general consideration then be, of what use such an a priori construction of valid theory-forms and discipline-forms? Well, of quite tremendous use, provided only
259 that the circumstances are favorable.

<§ 55. The Nature of Deductive Theory, Taking Euclidean Geometry as an Example>

We first think of a deductive discipline formulated in that form that mathematicians always strive to give it, albeit more in compliance with some obscure compulsion than logical insight. It is here a matter of a tendency that first became one of the predominant ones in the development of mathematics in the last century. *Au*

fond, it is ancient. Having brought it into the world is the everlasting merit of *Euclidean* geometry. You all know from school about the specific makeup of this *Euclidean* geometry. It proceeds in a purely *a priori* manner, i.e., it thoroughly formulates laws pertaining to the pure essence of space or to the configurations to be delineated *a priori* in it. Thus, it is definitely a matter of pure concept-truths. Now, the most obvious procedure would, however, be for one to begin by securing a few pure concepts such as, for instance, "straight line" and "angle", "plane", "solid", "congruence", just as many as one needs in order to establish first, a priori, directly perspicacious axioms. From then on, one advances by recognizing that other truths are included in the first truths as necessary consequences. Then, one draws in, for instance, new concepts and new directly evident concept-propositions and advances again on to new, no longer obvious, ones by drawing inferences out of those already established. Now, *Euclid* did not abide by this procedure. It undoubtedly also preceded him, or was undoubtedly observed by the first pure geometries of ancient times. For it is inescapably the first. But since it had already proved abundantly successful, it precisely made possible another, so to speak, more refined procedure. *Euclid* did not make further pronouncements about this, but it is apparent in the manner he presented the famous *Elements*. A set of so-called axioms is placed at the top. In *Euclid*'s system, axioms are propositions that are stipulated, but not proven. Here stipulation implies that it is easy to convince oneself that these propositions directly state given pure concept-truths, that they are direct *Evidenzen*. All of <the> concepts occurring in those axioms are listed and defined. And the definitions claim to secure concepts 260
that are also directly obvious as purely conceptual givens. And now we come to the main point. While, then, these–let us say thirteen (or how many?)–axioms are placed on the top, and conclusions drawn from them, and from the theorems deduced ever new conclusions <are drawn>, and so again and again, this happens in such a way that in the whole broader system of the *Elements*, with all the many theorems after theorems, no new axiom is formulated. Accordingly, this geometry claims to have all direct law-truths in its axioms–that is, independent from one another–pertaining to the essence of space from which all other geometrical laws, all in

general pertinent to the essence of space, can be obtained through pure deduction.

It was only very late, only around the end of the eighteenth century, that it became apparent that *Euclid* had not actually been fully true to this aim. He had indeed not formulated any further geometrical axioms, but he had passed over many an axiom in silence. He had made use of many obvious steps of inference whose principle is specifically geometrical–and one that was missing in the axioms formulated. This is true in every place in which a *Euclidean* proof cannot dispense with an illustration, where therefore the consistency is not purely formal and is to be formulated.

Let us then imagine *Euclid's* having been improved in this respect (as has in the meantime long since happened, whereby the order of exposition was also improved at the same time), but while retaining his great idea. That is of course not as easy as it appears, because naïve inferring and proving are relatively easy and suited to our nature. However, here we are unaccustomed to doing what is required, subsequently analyzing the proof process, breaking down each inference into really primitive elementary inferences with painstaking care. No abridged inference is admissible, no main premise may remain concealed, and one is to see to it, time and time again, that every premise used is either one the axioms enumerated or one of the theorems already derived from them. If that is not the case, then if the premise is in general correct, it must be an axiom. One must then further constantly compare the axioms adding on with one another and see whether or not one or another of them can be derived as a theorem from the remaining
261 ones. And finally, in the case of inferences, one must also continuously go back to the inference-forms and thereby distinguish between analytic and specifically geometrical ones. A specifically geometrical law of inference–such as, for example, the law that two geometrical figures that are congruent with a third one are also congruent with one another–is just not analytic. It expresses something specifically pertaining to the essence of space and must consequently find its place among the axioms. Those are therefore very laborious zigzagging investigations.

<§ 56. Taking Definite Deductive Disciplines Back to Their Discipline-Forms. Equiform Disciplines. Inclusion and Exclusion of Discipline-Forms. Operating with Imaginaries>

If then everything has turned out for the best, then we have a system unfolding itself in pure deduction into infinity, in which all further propositions follow purely analytically–and fully explicitly–from a finite number of independent basic propositions as a perfect, irreducible basis. Precisely for that reason, it becomes possible to formalize the whole discipline with all its objectively secured propositions, proofs, theories, i.e. to go back to its pure discipline-form, while indefinite-general variables are simply substituted for the factual concepts.

Now, one can however easily assure oneself that for the form of such a deductive discipline purified in this way, it is exclusively a question of the form of the axiom system. Namely, if we have two scientific fields–be they even worlds apart from each other–to which axiom systems of the same form pertain, then theoretical disciplines of the same matching form must also correspond to both. To each theorem in a science, a theorem of the same form must correspond in the other, and the other way around. In other words, they thoroughly coincide in terms of form, in terms of basic principles, just as in terms of theorems, in terms of proofs, of theories. If we exchange the designating terms for the corresponding concepts here and there, then each proposition of a field changes into the corresponding one of the other field.

Now, it is, however, also evident that it would be foolish to construct two such disciplines separately from one another. And, it is 262
at the same time clear that when we switch from an axiomatically reformed discipline to its purely formal generalization of its axiom system, we thereby have a law-form of the discipline that generally validly deals with not only the concrete <discipline>, but all disciplines of the same form once and for all.

It actually frequently happens that disciplines are of the same of form, thus, for example, all the disciplines that are called arithmetics, while upon closer inspection, the "numbers" with which they operate–such as cardinal numbers, ordinal numbers, large numbers, and so forth–are essentially different conceptually.

However, remarkable relationships thereby emerge. Generally speaking, it can happen, and actually does happen, that several deductively theorizable domains have a set of axioms in common that are formally the same, but in one <domain>certain axioms occur, in the other domain others that do not have any formal equivalents in the parallel domain. Then, the discipline-forms with their proof-forms, theorem-forms, and so forth will not completely coincide on both sides, but the one discipline-form <will>, for instance, be completely contained in the other. All formal consequences that result through the addition of new axiom-forms and < from> the associated relation- and conjunction-forms do not then have any interpretable meaning in the parallel domain. So it can be that axioms in a theoretical domain are subject to formal restrictions and have a meaning in terms of them, those in the parallel domain possess an unrestricted meaning. Such are, for example, the relationships between the theory of cardinal numbers and that of ordinal numbers. If we correlate ordinal numbers to the array of elements in a two-sided infinite series, then from one hypothetical zero point on the series of ordinal numbers, there is +1, +2, +3,... and the other side −1, −2, and so forth. The relationships of each such series, say, of the series of positive numbers, correspond precisely to the relationships of the series of cardinal numbers. The latter, however, is not two-sided and not to be supplemented by a second series. In the arithmetic of cardinal numbers, the axioms have such a restricted meaning that they bar any subtraction of the kind 3–4 as nonsensical. There is nothing negative there. In the arithmetic of ordinal numbers, however, all subtraction has a good meaning, whether the minuend is greater or smaller than the subtrahend. Similarly, in the series of cardinal numbers and the arithmetic of cardinal numbers, a genuine frac-
263 tion has no meaning, and an irrational number no meaning, a quantity $\sqrt{-1}$, and so forth. But all that has a meaning when we form magnitudes referring to a multidimensional continuous domain of quantities on a line.

Now, however, the remarkable thing happened that, with a few precautions, all calculations in the arithmetic of cardinal numbers could in practice be performed as if the laws for operations were unrestrictedly valid and meaningful, and so in general that within a restricted domain of deduction, one could disregard the restrictions and so proceed as if the axiom system were a broader one. People have gone to infinitely many pains to justify the puzzling operating with imaginaries of every kind, especially since it has proved to be highly fruitful in a technical respect. If one has ascended to the knowledge of the essence of the theory-forms, then the present problem is to be formulated as follows: Two valid discipline-forms may stand in such a relation <to one another> that the axiom system of one is a formal restriction of that of the other. All deductive consequences of the narrower axiom system are then obviously contained in the deductive consequences of the broader one. Now, however, all theorems deducible in the broader system must (1) break down into one exclusively containing concepts of the narrower one, which <for their part> are valid in terms of the narrower axioms, therefore not imaginary, (2) into theorems containing such imaginaries. Therefore, if we compare, for example, the arithmetic of cardinal numbers and the arithmetic of ordinal numbers–and correspondingly compare their discipline-forms, then we have, on the one hand, formulas, and in general theorems, which contain the negative that is imaginary in the sense of the narrower axioms of the arithmetic of cardinal numbers, and on the other hand, theorems that do not contain what is imaginary, therefore no difference $a - b$, where $a < b$.

If, in order to connect on to a traditional way of talking, we call what is non-imaginary the real (*das Reale*), then we therefore have to speak of theorems that have real meaning in every respect. In the broader system, such theorems can now be obtained, either by using imaginary axioms–, the proof is then imaginary, goes through the imaginary throughout, but the imaginaries, so to speak, self-destruct in the proof process and ultimately drop out of the concluding proposition–, or the proof process is only real (*reell*) from the start and evermore. It does not overstep the real domain (*reelle Gebiet*). There the problem is then under what circumstances can 264
one be completely sure that the theorem derived is valid independently of the proof process, i.e., valid independently of whether it goes through the imaginary or not. Or, more clearly, when can one be sure that to every proof being conducted via the imaginary, a proof exclusively conducted via the real (*reelle*) domain with the

same final success as the real (*reellen*) theorem must be comparable? Obviously, a proof of cardinal number propositions going through negative numbers will indeed be completely reliable when we know *a priori* that corresponding to every such proof must be a proof that draws on purely real (*real*)-valid means, therefore, exclusively on the meaningful axioms for cardinal numbers.

Those inexperienced in mathematics will ask why people spend so much time thinking about this. Proofs are precisely conducted rationally. In the sphere of cardinal numbers, though, it is rational to use only propositions that have a meaning for cardinal numbers. However, a cardinal number–an answer to the question "How much?"–can, never read 3–4 or ¾, and so forth. Such number formations have no meaning there. Why, therefore, introduce meaningless things into the operations and deductions? Certainly, though, the answer lies in the fact that what is outwardly nonrational, and also more complicated in the theoretical execution, can be incomparably easier and more convenient. With the extension of the discipline-form within which one is actually conducting the proofs, the power of mathematical thinking increases in a completely extraordinary way.

I cannot go more closely into the systematic solution here. I only want to mention that the justification of the free formal operating is obvious when the axioms break down into disjoint groups that, for instance, refer to operations that do not intertwine with one another. It is otherwise when the axioms intertwine with one another and sometimes have narrower, sometimes broader, validity. Arising there as a result is a formal equivalent of the right to operate freely lying in the by no means obvious property of all mathematical disciplines of being complete (*definit*). Passing through the imaginary is permitted in every complete domain of deduction. A domain is, though, called complete when every proposition arbitrarily put together with axiomatic concepts, and only out of them,–therefore, every proposition having grammatical meaning in this domain–is either true on the basis of the axiom system of this domain or false on the basis of it. In other words, the truth and falsehood of every grammatically producible proposi-
265 tion that exclusively speaks the language of this domain is decided from the outset by the axioms–either necessarily following from

them and then true, or not <following from them> and then contradicting them, then false. One, however, soon satisfies oneself that not every formally thrown together system of axiom-forms allows one to define a discipline-form that is complete in this respect.

<§ 57. The Construction of Complete Discipline-Forms Starting with the Euclidean Manifold>

The circumstance then that, instead of in the specific fields of knowledge, one can operate rather in the spheres of pure forms, and if they are complete can freely operate, then leads to systematically broadening the complete discipline-forms there and to pursuing all possibilities of the construction of complete discipline-forms. Correlatively corresponding to each such discipline-form is an object-domain conceived in indeterminate universality that is fully indeterminate in terms of content and only receives more precise determination through those correlations of categorial concepts that are present in the axiom-forms. Here mathematicians speak of a manifold, for example of three dimensional, four dimensional, n-dimensional *Euclidean* manifolds, or of non-*Euclidean* manifolds, etc. What therefore is a *Euclidean* manifold? The term refers to *Euclidean* geometry, to the geometry of space.

Space is the form comprising all spatial configurations, each of which is a continuous configuration out of ultimate spatial elements, points. Every configuration that ought to be able to belong with space is subject to fixed laws, the main features of which were discovered by the Ancients, and vice versa. These laws permit one to generate constructively other configurations, indeed all others, out of given configurations. The accompanying a priori theory is *Euclidean* geometry, understood not exactly as the system of geometry that *Euclid* presented in his *Elements* in its limited scope, but as every theory that unfolds the law-governed essence of space in proper ways. One then says space is a *Euclidean* manifold. What
does that mean? 266

If, however, we move from geometry to its theory-form, only assuming that the previously described distillation process has been carried out on this geometry, then the following generalization results. We define an indeterminate multiplicity of elements, or what is the same thing, a manifold, using the following property: Any two elements, called points, determine a configuration called a straight line. Any two straight lines have one and only one point in common, and so forth. In other words, we define a multiplicity of something in indeterminate generality by means of axiom-forms that are precisely formalizations of *Euclidean* axioms. The manifold is now no longer the manifold of points in the ordinary sense. There is no longer any talk of space at all. Before formalization, it was stated "Two points determine a line." That is a geometrical truth that has a clear meaning. Everyone knows from geometrical intuition what point and line mean. And precisely the same thing holds for all other basic concepts and axioms of geometry. After formalization, the words "point," "straight line," "angle," "intersect," and so forth are completely empty signs that only have the purely formal meaning that the axiom-form prescribes for them. It is then said, <a> certain something called point, a certain something called straight line, is to stand *ex definitione* in a certain relationship <to one another> in the manifold to be defined, described by the words: Between any two points there is a straight line that "runs through them." In the place of space, we have a concept defined by purely categorial concepts which is purely set forth, namely, defined, as a formal possibility. And the definition of a manifold as *Euclidean* does not state anything about existence any more than the definition of a golden mountain does about <a> mountain made of gold. Indeed, even more than that. It is not only not a matter of existence here, but also not of real, actual essentialness. Whether there is not only something individual-real (*Reales*), but also something actual-essential that corresponds to the form defined remains completely open. Only a form is defined, but whether axioms have lasting existence as truths in any actual, real, or ideal, sphere corresponding to the form indicated remains open.

The *Euclidean* manifold defined is just exactly to space as the number 2 is to any concrete two, to any concrete group of two
267 things or two ideas, and so forth, or as the form "categorical judgment" is to an actual categorical judgment, and so on.

The *Euclidean* manifold also has a "being", in certain ways, namely, provided it has been correctly defined, provided that the way the axiom-forms are ordered together does not contain any formal contradiction, any violations of analytics. This being is, therefore, only the being of formal, analytic concord. So, when we then say of a specific categorical judgment, for example, "The sky is blue", that it is a concordant categorical judgment, or for instance of a specific inference that it is a valid eliminative inference, then we have then to say of space that it is a valid *Euclidean* manifold.

On the basis of the definition of this manifold, on the basis therefore of its axiom-forms put together by definition, we can draw deductive conclusions–still remaining in the form–, construct proof-forms, and it is then *a priori* certain that something in the theory of space corresponds to everything obtained in that way. We need not then in general practice geometry any longer. We can just practice the theory of the *Euclidean* manifold and obtain everything that we would ever be able to obtain in actual thinking about space. If we use the terms of the theory of space in the formal sense, then we need not change the terms in moving from the theory of manifolds to geometry. On one occasion, the words "point", "straight line" etc. mean concept-forms of specific formal-axiomatic properties and, on another occasion, actual concepts belonging to the essence of space.

If, though, we then remain with the form, we soon find opportunities to vary the system of the axiom-forms defining the manifold in different directions. Instead of three dimensions, we choose four, or even n, i.e. arbitrarily many, dimensions, and then elaborate a theory of manifolds for all manifolds of n dimensions, which are all still called *Euclidean* because, independently of the number of dimensions, the axiom-forms do not undergo any essential change. We can, however, also try other variations. We derive one axiom-form after another for ourselves and watch how we can modify them without the harmony of the entire system suffering from it. For each such modification, we follow up the changes that the original theory-forms undergo in deductive unfolding. We then 268
realize that, for example, the n-dimensional *Euclidean* manifolds can be understood as borderline cases of a law-governed infinity

of other manifolds, whereupon it is characterized as being manifold of curvature 0, in contrast to the infinitely many manifolds of changing positive or negative curvature. And all these infinite manifolds are characterized by common properties, for example, by the fact that in them every configuration can "shift" within the manifold without "straining" and "distortion", where "shifting" ("straining", "distortion") is obviously a purely formal concept, a formal generalization of what we know in space as movement. In short, by proceeding in this way, we are no longer practicing a concrete deductive science like geometry, formulating actual axioms and deductively deriving truths in a concrete factual field, but we are practicing a kind of scientific-less game with the forms of possible theories, with whole discipline-forms. In a glance, we survey infinities of possible discipline-forms, or infinities of manifolds defined by axiom systems and explore the laws governing in the relationships and variations of manifolds. And since our procedure is purely formal, since we do not use a single concept not arising out of the analytic sphere, then this whole theory of manifolds, this highest level of mathematics, is nothing other than a supreme consummation of analytics.

That it is not here a matter of a mere game, but of a sphere of insights worthy of the highest theoretical interest, I need not say. So, the practical significance is inexhaustible, because the analytic relationships are such that, the further our theoretical insight reaches, the more perfectly we master the theory-forms in accordance with their law-governed variations, the more perfectly we can also perform deductive work in a concrete theory. In the concrete spheres, we are as a rule constrained. The formal limits that the requirement imposes upon us not to pronounce any factual *Widersinn*, therefore, not to use any *widersinnigen*, imaginary concepts, hinder us in deductively theorizing work. And there is pre-
269 cisely the marvelous thing that the transition into the clear infinities (but also law-governed infinities) of pure forms and variations of forms frees us from such conditions and at the same time explains to us why passing through the imaginary, what is meaningless, must lead, not to meaningless, but <to> true results.

<§ 58. The Universal Scope of Possible Formalization. Notwithstanding, the Theory of Manifolds Is Itself Subject to Analytic Laws>

It is so remarkable that the meaning of their own logical discoveries has become so unclear to mathematicians themselves that we hardly ever have an opportunity to hear a proper interpretation of them from them–indeed, in the place of that, false and wrong theories. It is the case, here as elsewhere in science, that realizing creative achievements and providing proper information about the meaning of those achievements are two different things. What is creative is precisely accomplished in naïve activity, in an instinctive sphere. For the philosophy of analytics, absolute clarity is, however, indispensable. How much absurdity have mathematicians and natural scientists as great as *Gauß, Riemann, Helmholtz* fallen into through their misinterpretations! There, one generally speaks of n-dimensional spaces, of non-*Euclidean* spaces (*Lobatchevsky*), and in so doing, takes this talk–which however should only have formal meaning–seriously, as if the possibility of varying the analytic form of geometry in different ways–or the possibility of varying the axioms' forms-system–had the least significance for the possibility of a space constituted differently from "our" < space>, or even for the possibility of conceiving of a different kind of straight line–for instance, even a straight line that is curved. That is all *Unsinn*. It is just as little to be deduced from the possibility of arbitrarily changing–for instance, from affirmative to negative–the form of a proposition (let us say, that "red" is not "green") within certain limits that it would be conceivable and thinkable that "red" would one day be "green", as the possibility that a curved space or of straight lines that intersect one another a second time at some very great distance, the distance of the fixed stars, and so on, follows from the possibility of continuously transforming the *Euclidean* manifold into a non-*Euclidean* one.

The lack of clarity in the arithmetic sphere is no less great. The 270 difference between manifolds <within> the theory of manifolds–which naturally requires the lower concepts and laws of analytics in each step–and those manifolds concrete in their way that belong to the store of analytics itself is also disregarded there. Or, expressed correlatively: People disregard the difference between the theory types of the universal theory of theories and the actual theories belonging to the store of analytics itself, whose independent realm cannot disappear through any formalization. Considering the course of our work and our careful demonstrations, we ourselves naturally never fall prey to such a temptation.

In contrast, many mathematicians–and precisely those doing creative work in the domain of the theory of manifolds–however believe that it includes the entire *mathesis universalis*. The following is to blame for the error. If they were to proceed and speak in a logically correct manner, mathematicians would actually have to distinguish among the arithmetic of cardinal numbers, the arithmetic of ordinal numbers, and the different kinds of arithmetics of large numbers, and they must distinguish between all these actual arithmetics and the arithmetic theory of manifolds or, as I would prefer to say, the discipline of arithmetic-forms, of the forms of possible arithmetics in general, for as usual, here too, arithmetic itself, for example, the arithmetic of cardinal numbers itself, is indeed to be distinguished from the form of this arithmetic as discipline-form, and so it is everywhere. And accordingly, also to be distinguished is the systematic exploration of the possible variations and relationships of these forms–maintaining a general type that in general defines the formal genus-type arithmetic–from the systematic exploration of truths that are valid for cardinal numbers, ordinal numbers, and other categorial configurations, that <in turn> fall under a formal type called number. How this type is to be defined out of purely formal considerations–and the formal Idea "arithmetic" likewise–is itself a matter of the pure theory of forms of deductive theories, just as the formal concept "geometry" is defined in formal generality, or if one wishes to avoid the word, the science of formal topological spaces. But as was said, just as, on the one hand, the topological theory of manifolds is to be distinguished from the geometry of space, so, on the
271 other hand, the arithmetic theory of manifolds is to be distinguished from the arithmetics themselves.

However, *de facto*, mathematicians never speak of the arithmetic of cardinal numbers and the arithmetic of ordinal numbers, and so forth. They normally do not at all see that different arithmetics are to be distinguished here, and that is because in their theoretical constructing, they constantly operate within the arithmetic manifold and that if they take any special interest in cardinal numbers—or in ordinal numbers or in applied arithmetic for numbers on a line segment, and so on—they overlook a step in reasoning, namely, just that of the application, of subsumption. What they call arithmetic and algebra is from the start a hypothetical theory of manifolds. Positive and negative numbers, their rational, irrational, imaginary numbers, are all only number-forms, are formal configurations of this theory of manifolds. All possible calculations in all the analytic domains named are then performed through subsumption under these forms without further ado. However, note well, in every actual arithmetic, axioms such as $a + 1 = 1 + a$, or $a + (b + 1) = (a + b) + 1$, are precisely actual axioms, in the theory of manifolds, though, mere axiom-forms, therefore, not truths, but mere components of the definition of manifold and antecedents for the hypothetical, formal judgments that the theory of manifolds makes. It says over and over that when something in some domain of deduction—an axiom system of such forms $a + 1 = 1 + a$, and so forth–is valid, then theorems of the corresponding forms formulated by it in the theory of manifolds arise in it.

In the cat<egorial> sphere of cardinal numbers such axioms then hold—although within certain restrictions—, therefore, everything holds that formal arithmetic, the discipline of the number forms and forms of possible number propositions, hypothetically and precisely formally deduces, also for the domain of cardinal numbers, namely, not hypothetically, but thetically.

All actual theories, therefore, also the analytic theories, can precisely be formalized in the sense of the theory of theories. Even syllogistic logic is not to be made an exception here. In its case, formalization even leads to a theory-form that can be understood as a special case of the formal genus-type "arithmetic". All the well-known algebraic propositions $ab = ba$, the laws of association, distribution, hold, and the brilliant *Boole* saw that two closed domains 272
of ordinary syllogistic logic can be dealt with as if it were an arithmetic, only that the number series reduces to the numbers 0 and 1. All arithmetic laws are valid there if we but add the fact that $1 + 1$ does not $= 2$, but $= 1$. If one knows nothing about the theory of

theories that seems to be perfect nonsense, and it then seems absolutely strange when one hears of machines, of a kind of piano, that analogously permits one to resolve complex webs of syllogisms by means of a mechanical game, the way one can randomly mechanically carry out additions, multiplications, divisions, and so forth with calculating machines. However, the matter is no more miraculous there than here, and in syllogistic logic, the 0 and <the> 1 are signs that have very little to do with the arithmetic 0 and <the arithmetic> 1. They are only selected, just as the signs +, ×, = are, in order to allow certain formal analogies to come to the fore and in further consequence to make perceptible the fact that the theory-form agrees to the extent indicated with that of an arithmetic. How little, on the other hand, thinking within the hypothetical reflection of theory-forms can exhaust everything, how little the formal theory of manifolds can confirm all of analytics, and along with that also the special domains of analytics that we call arithmetics, follows from the fact that the theory of manifolds nevertheless also proceeds deductively, that it is for its part based on pure analytic categories and that each step it takes is subject to some analytic laws that are precisely laws and not merely forms. If it draws an inference, then some laws of inference belong to the inference. If in solving problems, it weighs the possibilities combinatorially, then it uses propositions for cardinal number and combination, and so forth.

<§ 59. The Tasks of an Ideal *Mathesis Universalis*. The Theory of Manifolds as the Consummation of All Purely Categorial Knowledge>

An ideal *mathesis universalis* would have the following task (beyond the theory of forms of meanings about which I spoke earlier). To begin with, the major task would be that of obtaining a comprehensive set of axioms formally independent of one another
273 that state all the direct law-truths pertaining to all the formal categories, therefore, as much to the apophantic categories as to the formal ontological categories. The direct axioms of cardinal numbers, of ordinal numbers, large numbers, and so forth would therefore also occur there. Since the categories themselves interact with

one another in complex ways, some entering into the store of the others, the complexities would have to be kept track of systematically, and the axioms <would have> to be laid down step by step. This task is however not wholly independent of the other one, which is now to be discussed.

Mere axioms are not theories. It is therefore vital to construct the deductive consequences that are decided in the axioms systematically. Here, one is again to proceed in systematic ways, i.e., not <dealing with> what is dependent before what it is dependent upon, therefore, certainly not syllogistic logic after arithmetic. There exist here however peculiar dependencies arising out of formal logics' peculiar self-referentiality. Logic indeed lays down laws to which all particular logical relationships must conform if they are to be able to be valid. While logic is doing that, though, it itself states judgments, derives judgments from judgments, deduces, proves. Therefore, what it lays down–since it is indeed to be true, logically correct–must itself conform to those laws that it lays down. That cannot, however, be a vicious circle. It would be so if logic were not permitted to take any step before having already been developed as a complete logic, something which is obviously not necessary.

It must accordingly build itself up in the following way. It must begin with completely direct axioms. Each step of indirect thinking that it takes must be directly perspicuous. It is only valid if its law is valid. This law must occur among the axioms, therefore, be directly warranted in and of itself. If one already had all the axioms given beforehand, then one would not need to say anything further. But, in the initial presentation of the problems of logical theory, one does not yet have all the axioms and so the laborious work ensues of reducing each step in the inference to its law and seeing whether that law is already formulated as an axiom,<is>, however, potentially to be transferred to the axioms.

The main interest for philosophers, naturally lies in the system
<of> categories and of axioms themselves and in understanding 274
the procedure. In the formal categories and the axioms implicitly lies all the conformity to laws that the analytic form *a priori* prescribes for all possible thought-meanings and correlatively for all pure thought-objectivities.

One can then, however, not only make statements about objectivities thought in isolation in formal generality, but <one can> think such objectivities as components of formally defined manifolds. Manifolds are thought, infinitely open, sets of objectivities that are defined by axiom-forms, i.e., by forms of laws that are thought as valid for objects, or for their otherwise undetermined and formally thought combinations and relationships. Particularly most meaningful there are the specifically mathematical manifolds, the complete ones. In them, the objectivity-groups are formally fixed by complete axiom-forms out of whose defining essence theory-forms are then to be derived as their pertinent deductive consequence-characteristics. All possible actual deductive disciplines, even those of analytics itself, then fall under the theory of manifolds, wherewith the ultimate consummation of all purely categorial knowledge is obtained.

<Chapter 12 274

Broadening the Idea of the Theory of Science Beyond Analytics>

<§ 60. Logic as Formal Theory of Meaning and as Formal Ontology Is the First Manifestation of the Idea of the Theory of Science. The Most Universal Concept of Theory of Science>[1]

If we again take up the Idea of the theory of science guiding us from the beginning, then, as the formal theory of meaning and ontology described, logic is the first manifestation of this Idea. Knowing (*Wissen*) in the sense of science (*Wissenschaft*) is thinking or thought-state-of-mind that refers back to thinking. 275
Corresponding to thinking is something thought, and so corresponding to every science is a system of judgments in my meaning-theoretical sense, a system of postulated truths and probabilities, and these refer to objects and states-of-affairs. The science of meanings in general, of truths, possibilities, probabilities in general, of objects in general in absolutely pure, formal universality, yields a system of absolute truths to which every science is obviously bound, and which are prior in terms of validity to every science in general–as already to every judgment in general.

I now ask whether we cannot broaden the Idea of the theory of science, whether we cannot characterize new fields of investigation, and perhaps whole disciplines that–even if in a different respect–are prior to all particular sciences and consequently do not presuppose any of them in terms of their content.

I add an explanatory remark here. We can initially call generally science-theoretical the exploration of everything concerning the Idea, the essence of a science as such in unconditional general-

[1] The original version of §§60–62 is published as Appendix XVII. (Editor's note).

E. Husserl, *Logic and General Theory of Science*,
Husserliana: Edmund Husserl – Collected Works 15,
https://doi.org/10.1007/978-3-030-14529-3_12

ity. That initially refers to knowledge which concerns what is constitutive of the Idea of a science, i.e. concerns essence-moments that necessarily construct the Idea of science. The Idea of meaning, more precisely of judgment, is such a constructive moment, and so the whole of apophantics is theory of science in a magnificent sense–and likewise the whole of formal ontology, since the Idea of object in general, along with all related Ideas like property, relation, and so on, is constitutive of the Idea of science. Science is indeed science of any objectivity that is called its domain.

We can then, however, surely stretch the concept of science-theoretical to such an extent that <it> includes everything that, according to its essence, no preexisting science presupposes, or that belongs within the sphere of any particular science and is otherwise of such a nature that every particular science could in principle make free use of it, without for that reason forfeiting its own domain, whether it actually finds reason for that then or not. In this sense, the formal theory of theories is already science-
276 theoretical. It is not part of the constitutive essence of every science to be theoretical science in the precise sense I have described, therefore, to derive theoretical deductive systems out of basic laws as a priori or empirically obtained foundations, and thereby to have to engage in theoretical explanation. That does not however change anything about the fact that the theory of theories is science-theoretical, because the Idea of deduction, of proof, of theory, or the Idea of manifold is by essence prior to any science, and it is thereby a matter of higher essence-configurations that further form constitutive concepts of science in pure universality.

We shall from now on hold fast to this most universal concept of "science-theoretical" as theory of science and reflect together about what may emerge *a priori* as science-theoretical fields of investigation. It is clear that from now on, we are changing outlook, that we must forgo abstraction from the nature of the nuclei. Analytics is the scientific field of pure form. We now have to take into account what is called the matter of knowledge.

<§ 61. The Science-Theoretical Nature of the Synthetic-Formal Ontology of Nature. Kant's "Pure Natural Science">

Let us take up the Idea of mathematics. According to my analyses, there was a difference between analytic-formal and non-analytic mathematics. Everything that pertains to analytic categories in mathematical disciplines figures then on one side, therefore, the arithmetic of cardinal numbers, the arithmetic of ordinal numbers, and so on, as well as the formal theory of manifolds, not, however, geometry as theory of space, pure chronology as theory of time, and so forth.

Are these disciplines then not science-theoretical? They are obviously not so in the most universal sense, because not every science deals with spatiotemporal objectivities. That is certainly not true, for example, of the *mathesis universalis*, or of analytics itself with all its disciplines. On the other hand, one will nevertheless be able to point to the fact that, within the vast complexes of sciences of nature referred to as non-analytic (therefore synthetic), the mathematical disciplines play a role similar to the analytic 277
mathematical disciplines, therefore, that within the bounds of nature and natural science, they are science-theoretical in nature.

If we survey the host of natural sciences, then each one of them theoretically explores some domain of the totality of nature. In theory, though, each one of them can make just as free use of all the synthetic-mathematical disciplines as of the formal-mathematical ones. The reason is obvious. Just as the objectivities falling under the heading of "nature" are precisely objectivities–subjects of properties, relation terms of relations, components of conjunctions, parts of wholes–, in short, have an analytic-ontological form on account of which they come under analytics, so it as nature has a form. *Kant* spoke of *natura formaliter spectata*. Everything worthy of the name thing has a bodily figure that takes its place in space, that has the formal properties of a three-dimensional *Euclidean* manifold, and consequently requires the familiar geometry as unfolding of its essence. Furthermore, any concrete being exists in time and moves in space, and is consequently subject–something which concerns the temporal-figures and movement-figures–to chronology and the theory of motion (kinematics). Furthermore, everything has its real properties and is a temporal unit amid change in these real properties. In this respect, we call it a substantial unit. And substance again stands in a close relationship to causality. The thing changes preserving its real identity in the changing, but it changes in such a manner that its changes are dependent on those of other things, namely, in accordance with fixed causal laws. In other words, every change has its causes. Each is necessary. Each points to laws of change that unify and govern the functional dependencies of the changes in different concrete things. Nothing is isolated. Everything is a component of a causal-coherent nature. Similar characteristics belong to the essence of thing and nature.

Specific natural sciences explore specific things, specific real properties, real changes, specific real laws as causal laws of the changes. Every <natural science> accepts things experienced simply as things and only asks what holds of them more specifically. In
278 all its ways of being experienced, in all its statements about things, each is, however, bound to what the Idea of thing and nature prescribed *a priori*, or in *Kant*ian terms, to the form of nature. The knowledge proper to the form, to the spatiotemporal form, as well as, *formaliter*, to the store of real spatiotemporal plenitude owing to which the thing is a natural reality, all that is prior to the specific natural sciences, is a common base of scientific knowledge of which each specific natural science can make free use and which does not belong in any of them. That also holds for the most universal, what are called abstract, natural sciences. They concern specifically given nature, though with the greatest universality. They already presuppose that nature exists as a system of things that is subject to the form of nature as an a priori, as an *eidos* that is prior to all specific nature to be given empirically.

All that is beyond doubt, and so in the realm of natural knowledge analytic-formal ontology comes before synthetic-formal ontology, namely, the ontology of nature. It would break down into several separate disciplines, <into> the disciplines that explore the a priori of space and time, therefore geometry, chronology, and kinematics, on the other hand, <into> those disciplines that would correspond to *Kant*ian "pure" natural science, therefore, explore the a priori of spatiotemporal reality (matter)–except for pure space- and time-form. One then observes that the analytic categories, the categories of objectivity in general in the natural sphere would come before the categories of nature. We could say: the categories in the original *Aristotelian* sense, where we would, however, have to exclude all "psychical nature", to which *Aristotle* surely thought his categories also referred. We could again say the *Kant*ian categories, since in his theory of categories, *Kant* was essentially guided by nature in the usual sense of physical nature.

As we clearly see beforehand, the Idea of nature–now still understood in the concise sense of physical nature–that gives unity to the many sciences we call natural sciences gives them, not contingent, but necessary, unity. And again, we see that the heading "nature" subsumes ways of being in such a way that everything otherwise called being is cut off from it, as if by chasms, that therefore the most radical demarcations define natural being. That is 279
exactly why the a priori that is in general proper to nature has such a magnificent meaning, why the talk of "nature in general" as form and the parallelization with form in the analytic sense. And, again associated with that is the magnificent and, as I could say, science-theoretical, nature of the disciplines proper to the essence of nature as form, in contrast to the natural sciences in the usual sense, the empirical natural sciences.

<§ 62. The Task of Defining the Highest Regions of Being Starting from the Idea of What Exists Individually in General, as Well as of Unfolding the Regional Basic Concepts in the Regional Categories. The Concept of Region>

A major task emerges from this point on. If we start with the universal idea of that which exists, more precisely, of what exists individually in general, one can then inquire into the radically distinct, highest classes of being of all, or as I am also accustomed to saying, the regions of being. Nature is such a region. We cannot generalize further here without completely losing our grip on actual matter and arriving at the formless analytic generality "what exists individually in general". In the formal sphere, a series of variations of the original category "what exists individually" is made available to us: property, relation, conjunction, and so forth. If we keep to the nature region, we then obtain (surveying this highest form of reality) the natural categories "physical property", physical relation", and so forth. And since <it> is now nature's job to sketch the a priori truths proper to the highest natural categories, those truths that make up the a priori of nature and are "science-theoretical" for specific natural sciences, then something similar must after all hold for other regions of being. Therefore, the task arises of a general theory of the regions of being and their categories and the a priori truths proper to each region. First of all, a radical "theory of categories" would be necessary, but better, I say, a systematic exploration of radical regions of being and of the categories specifically pertaining to each region.

280 Initially, in further consequence, there would be the task of the systematic grounding of the a priori disciplines that pertain to each region of being and that, *a priori* and for the most radical reasons, prescribe to all individual objectivities of the region concerned their form and, as a consequence, are science-theoretical in an extended sense for all specific sciences of each of these regions to be conducted empirically. Instead of science-theoretical disciplines, we can also speak of ontologies, as we have already done for the nature region.

Now, yet another thing for the explanation of the conception of the region concept, for example, <of the region> nature. <One sees or, so to speak, feels that> "nature" is not some fortuitous name, like the name "oxygen" or "fir tree". There is certainly knowledge that especially applies to the species fir tree or oxygen just as to every species of objectivity. But, those are empirical names, and from experience we draw different special kinds of objects of this kind and determine their characteristics through experience. But all the experiential kinds of things, and not merely those, <but> unlimited varieties of possible new kinds are included in the Idea thing. This concept obviously occupies a position completely different from that of the definite concepts I call species-concepts of things. And the way in which it concerns all things is obviously different in nature from, for instance, the empirical generic concept mineral that subsumes the experiential classes and species of minerals. This contrast makes it understandable that we say: The concept thing, natural object is so unique that it founds an ontology that *formaliter* prescribes a meaning and formal laws for all sciences of empirically existing things. However, it would of course be wrong to want to draw an a priori ontology out of the concept mineral. This concept merely has empirical meaning, and everything that holds of minerals in general, how they split into genera and species, which universal characteristics are attributable each one (genus and species), all of that, only experience can teach and is not to be drawn out of pure concepts.

The same holds for the concept human being or animal, ζϖον, and again for the concept soul, as for the concept "physical thing". Certainly, the concept human being is a concept of something that has a body, and the body is a physical thing, a spatiotemporal real- 281
ity. However, the body as such is more than <a> thing. It is an ensouled thing (feeling, freely moving), and the participial "ensouled" does not signify anything to be paralleled with predicates like "heavy", "transparent".

Every factual physical-concrete predicate corresponds to an Idea prescribed by the physical thing-concept. But ensoulment is found in, so to speak, a completely different dimension. Taken as a physical thing, the body, with all known, or yet to be known, properties to be experientially ascertained–be there infinitely many of them–does not contain anything psychical. Investigating physical nature and natural causality, we never come across anything psychical. And nevertheless, what is psychical belongs experientially to the thing that is called human body there. And what is psychical belongs to it as something ensouling, but a soul also belongs to it. The body is not ensouled by an isolated thing qualified as psychical, say an isolated feeling, but by a soul, to which a unit of "experiences" of that sort belong. Psychical experience requires a stream of "consciousness" and belonging to the essence of such a stream are many kinds of general characteristics. Belonging to this are rules that ground the talk of states of mind, natural tendencies, and so on.

<§ 63. The Ontology of Spiritual Being as a Priori Science of Spirit and Consciousness>

And again, similarly to within mere physical nature, what is regionally eidetic and empirical part ways there. The fact that human beings are what they are, what they experience, what makes up the specific content of their flowing overall consciousness, what kind of specific states of mind, natural tendencies they have, how natural tendencies as a rule develop in people, and so on, only experience can teach all that. On the other hand, however, standing above everything is the Idea of this basically essentially unique psycho-physical unit that I am calling "ensouled being", with the associated Ideas of "flow of consciousness", "soul", "character" and so on. And consequently, here again we have an a priori framework, an a priori form and accordingly an a priori science, the ontology of spiritual being.

It is owing to the naturalistic tendencies of our time that <it> is
282 not willing to do justice to this a priori, or indeed to the a priori in general, and that there is a general inclination to deal with everything psychical as something physical of a higher stratum and consequently to alienate it from its essence.

Belonging to the realm of the spirit that we encounter in the a posteriori as ensoulment of physical things, of bodies is, I said, what we call "consciousness", wherein the life of the spirit as spirit consists—a unit of the flow of consciousness, into which all particular experiencing fits without being detachable from this unit. And, in this sphere occur the remarkable differences between act and what is conscious <in> it, indeed, considered more precisely, the differences between act as consciousness, consciousness-meaning and conscious objectivity as such. In the a posteriori science of the spirit–which views this spirit, and consequently also the actual flow of consciousness belonging to it, as something bound to the body and something psychophysically dependent on the body, consequently indirectly bestows upon it existence in objective time, and even in space–consciousness is just explored as something existing in terms of its spatiotemporal coming and going, in terms of its existential patterns and potential laws. The functional dependencies of the body and the bodily occurrences are especially a vast field of investigation, that of psycho-physics in the narrower sense. Insofar, however, as consciousness and spiritual subject can be the object of a priori investigation, we would have a clear-cut distinction between empirical psychology on one side and a priori science of the spirit and consciousness on the other <side>. Obviously, considered theoretically, the position of the latter in relation to the former would be analogous to that of the a priori essence-science of nature in relation to each empirical natural science.

<§ 64. The Ontology of the Communal Spirit as a Priori Essence-Analysis of the Collective Spiritual Life and of Its Objective Correlates>

The individual mind is a component of a spiritual realm. We shall become aware of the fact that reality is not a mere unit of nature in which isolated bodies occur as bearers of souls. As full bodily-spiritual units, animals and humans engage in two-way communication, form societies, and "society" is a word for many kinds of different and higher units, which are units through agree- 283
ment. However, these units have characteristically different types and forms. Friendship, family, association, state, people, and such come to mind.

Human community is defined by these, as it were, objective immaterial powers: rule-governed speech, customs, rules of law, <of the> Church, and so forth. They make up forms of community life and, on the other hand, they are something objectively contrasting with the individuals of the community, objectivities that can be considered in their own right and made into objects of investigation.

All other intersubjective spiritual objectivities which, along with those discussed, we are accustomed to refer to under the heading culture, "creations" of the life of the spirit–of humanity, not individual human beings–come into consideration in addition. Thus, literature, more specifically Greek, German, French literature, and so forth, are distinguished in terms of different periods and are differentiated in accordance with their factual content as belles-lettres, scientific literature, technical writings, and so forth, in addition, the creations of art, of scientific or pre-scientific technology. All these creations are obviously not independent of individual people and their minds. Individual minds participate in their development, and conscious creations of individuals enter into them to a considerable extent. So literature or art is a self-developing stream of creations of individuals. Each of these creations–or rather <each of these> created things–is not, however, a being in its own right, and above all they do not have physical being. Rather, they are spiritual units of a peculiar kind that lead their own manner of intersubjective existence. And yet, it is once again an existence that continuously retains its relationship to the individual minds within the context of the societal unit, further exercises its influences upon the new creations of the remaining minds and potentially exercises additional influences on all of the further creations of the unitary life of communities and their further developing objective forms.

All that is an immense field of empirical investigation, the field of what are called the humanities and social sciences, which do not view the individual human being and his or her inner life in connection with his or her body and nature, but put him or her into the great context of the life of the community and view all his or her thoughts, value judgments and achievements within the relation-
284 ships of the spiritual life of the community, namely, from the perspective of community thinking, valuating, creating with all its remarkable objective correlates. One sees thereby that just as the body is a component of physical nature, the mind is a component in the unit of the communal spirit as a higher-level mental unit.

Belonging to individual human beings and the individual mind would be an a priori Idea that implies a whole wealth of a priori Ideas, namely, with regard to all the multiple forms of consciousness and <the> forms of their interweaving, or <the> forms of their correlates, as well as with regard to the pure Ideas of ego, personality, character traits etc. The individual mind is never experientially isolated there. Individual reflection abstracts from the societal relationships and the relationships to societal objectivities. But this abstraction is only empirical. *A priori*, we can surely not say that a subject is inconceivable without relation to other subjects. If we do conceive of it, however, as it really is experientially, woven into the societal context and as co-bearer of the societal consciousness and of its cultural correlates, then we realize that new Ideas arise here, foremost, the Idea of the communal spirit with all related Ideas. So, opening up here is a field of a priori considerations, an ontology of the communal spirit, an a priori essence-theory that is prior to all empirical humanities and social sciences, precedes here, the way the a priori of nature does the natural sciences. This essence-theory builds itself up as a higher storey above the essence-theory of the individual mind, is therefore not independent of the former–just as the communal spirit is indeed a higher-order objectivity founded in the objectivities that we call individual minds.

In all these respects, we are still in the early stages with vague notions. The ontologies are still in need of development. Only in the natural sphere do we have fully developed ontologies at our disposal: with respect to space, time, and motion. However, even pure natural science is still completely insufficiently delimited and developed. All the more, the essence-theory of the individual mind and of the communal spirit! The Idea of an a priori analysis of the collective spiritual life and its objective correlates is so far from the thoughts of sociologists and historians especially that probably
just the contention that there could be anything of the kind and 285
that <it> might be the necessary epistemological basis of all genuine social science would undoubtedly be declared by the sociologists to be mysticism or scholasticism.

Up until now, I have considered the scientific fields of nature, of the human mind (with respect to its characteristics, its act-life, and its relationship to nature) and finally of the human and animal communal spirit–therefore <the> natural sciences, psychophysics, and what are called the humanities and social sciences, among them both the historical sciences and the general social and cultural sciences.

In all these sciences, it is a matter of existential truths. Ostensibly, we do not overstep the bounds of this sphere if we now direct our gaze to value-truths in the broadest sense. Valuating is an act of the individual consciousness and plays its well-known great role in the community-consciousness, for example, in the talk of goods in the economic sense. Things or accomplishments that actually as a rule have customarily come to be valued in a community are as a rule coveted and strived after under recurring circumstances, have value, and inasmuch as agreeing actually and in empirical scope such values are exchanged for one another and measured against one another, one speaks of market values and such.

With all that, we are not overstepping the bounds of the empirical sciences discussed any more than when we speak about what is judgingly supposed in the intellectual sphere, let us say, for instance, in the natural sphere which had factual validity for a people or era, therefore, when we, for example, describe the cosmology of the ancient Ionians or Hindu mythology and such.

It is completely different when we assess judgments (as supposed truths), supposed values and activities of human beings and of humanity in their different historical periods and when we inquire into rightness and wrongness, validity or lack of validity from the standpoint of reason.

In judging with rational insight, in the empirical-intellectual sphere of human community life during different stages of its development, we assess the store of what is objectively valid that mankind has embraced and turned to good account. And, with respect to that, the concept of development immediately acquires
246 a definite meaning as intellectual development. We practice the history of science in this way, that is, not merely in the sense that we describe the course of human beings' theoretical certainties and by penetrating into it understand the defining communicative motivations, but in the noetic sense that we investigate the development of scientific truths, scientifically valid theories, and so forth, whereby we measure supposed validity up against insightfully embraced validity, objective truth.[1]

<§ 65. Formal and Material Axiology and Practice>

<a) Mere Value-Takings (*Wertnehmungen*) as Grounded Acts. Esthetic Values and Ethical Values>

Up until now, I have remained within the realm of extra-valuating existence or, from the standpoint of consciousness, within that of experience and experiential thinking. People do not, however, merely experience. They also engage in other acts, position-takings. Living is also valuating, valuating as beautiful and ugly, as good and bad. Valuating can be explicit or inexplicit, therefore, connected or not connected with thinking. In valuating, something however figures as a value, as something beautiful or ugly, as some-

[1] Compare Appendix XVIII: *The Problem of Reason.* (Editor's note).

thing good and bad, namely, provided that the regarding gaze is directed straight through the valuating to the value as such. We normally speak of perception and experience only where we are directly conscious of a spatiotemporal existence, in perception especially conscious as directly self-given. We have here a precisely parallel, originary presentive act that we could, for example, call value-taking, the act of direct value-givenness. (And more generally, in parallel with experience in the ordinary sense, we could speak of existential experience, of value-experience). Also here, direct value-taking and value-positing can be more or less clear or vague, distinct or confused and, as in the case of plain experience, also here, plain-direct value-positings are to be made without any admixture of thinking.

We observe that these plain value-takings are always founded 287
acts, i.e. necessarily underlying them are other acts and ultimately existential experiences, <acts> making existence presentational, namely, either unmodified or modified ones. You readily understand what that means with the help of the correlative value-objectivities. In valuating, something figures as value, or a value figures. A value is, for example, a work of fine art. At first, this figures as a thing, as the picture hanging on the wall, a thing with a frame and canvas, and so forth. But in seeing the picture, we are not turned to the thing hanging on the wall. Appearing to us in the sensory consciousness of the picture is the painted or drawn object, in the latter case, perhaps a small figure sketched with gray strokes. This is not the thing depicted. Rather, the subject of the picture, for example, a suprahuman heroic figure, is presented to us by means of this picture-object. We therefore have different sensory phenomena here and we can be oriented toward these experientially, just as the orientation toward the subject is an experiencing, potentially a modified, consciousness, namely, if what is portrayed is not taken for real. With all this, there is still, however, no talk of valuating. The valuating is, however, built upon the experiencing imagery-consciousness, and in it figures something being of value. The value-bearer is what is conscious in the mere imagery-consciousness, not simply the object that is portrayed there and that, if it is, is a mere natural thing, a mere existence. It is indeed not, as it would be for esthetic valuating, indifferent how the object

is depicted and whether it is depicted exactly by this picture or not. Instead, it is precisely essentially a matter of that. And, whatever conceptions may then still be aroused and associated with it that make what is appearing pictorially as such first and foremost into an esthetic object, this much is certain, that the esthetic object is the portrayed object, as long as it is pictorially depicted in this way and further conceived of in this way. Therefore, the esthetic object is founded in certain, and as a rule, quite complex ways, just as correlatively esthetic valuating is a founded consciousness that has another consciousness, and ultimately below it experiencing–or quasi-experiencing–consciousness, as substratum.

The same is so when we valuate a landscape as beautiful. We naturally do that again for the sake of its way of appearing as the one that is just now showing itself in this way in this light, from this
288 standpoint, where again various moods, more or less vague presentations, and so forth will also be aroused and contribute to the esthetic conception. It is the same in the case of valuating as good, be it valuating as ethically good or as practically good or useful.

We thereby observe an essential difference. To begin with, it proceeds from our reflection that the value grasped in valuating is something of value. Valuating is, we see, a founded act that, taken fully concretely, has a lower layer (of experience and potentially further acts already built upon that) and then on top of that, as a specifically new phenomenon, the valuating position-taking. Hence, something figures there that is not yet value, but just figures there as being of value by virtue of this position-taking. Associated with that is the fact that the word "value" has two meanings. What has value means–as far as it has value, is of value–quite simply a value. We can distinctly distinguish: the thing, the value predicate, and the concrete value as the thing of value understood in terms of its value-quality.

Now, the cases part ways according as the potential doxic existential positing, <the> thing-positing, by virtue of which the thing figures before us as reality, belongs essentially to the founding substratum of the valuating, or only belongs to it extra-essentially. So it is completely irrelevant for the beauty-value of a portrait whether, for our doxic supposing, the person portrayed exists and ever existed, or is a mere fiction. This is proper to all beauty-

valuating in the concise sense, whether esthetic or not, and as a result we can directly define the concept of "beauty-value". On the other hand, it is characteristic of good things that they are existential values, i.e. that existential positing is proper to the essential foundation of valuating position-taking. Therefore, in one case, something presented as such, a thing, provided it is depicted in such and such a way, conceived of in such and such a way, is merely valuated, and its being is itself not valuated. In the other case, precisely the being is valuated, the being of this thing in such and such a state. Naturally, there are then also mixtures, inasmuch as one and the other can be valuated at the same time. So it is when a handsome person is valuated as handsome and as something good at the same time, the latter namely in the joy over the existence of this handsome thing. Joy in general is a term for existential valuating. Of course, nothing of this changes when we valuate in our imagination, as indeed also when we judge in our imagination. A 289
kilogram of radium that I imagine appears to me as being of immense value, namely, as something good by virtue of the immense sources of energy lying in it. But then its existence is posited in a modified way in imagining, and <that> essentially consolidates with the good-valuating. We then say hypothetically, "A kilo of radium would have great value as something good". Something good exists when the good thing has existence. Now, one will perhaps object that something beautiful, for instance a work of art, only exists when the thing exists. However, we should not jumble up different concepts of things. The existing work of art, in the sense of the canvas thing hanging on the wall and so forth is not the beautiful thing, but a thing that by its nature arouses image-consciousness in us in which we are first conscious of something beautiful, the esthetically valuated thing and values. By virtue of this relationship, the canvas thing painted with colors in such and such a way is of utilitarian value, commercial value. It is something good. It is, however, not the thing having esthetic value in the genuine sense. It is not the esthetic object.

We at once recognize here an essential relationship between valuating beauty and valuating good, namely, the fact that, inasmuch as it is beautiful, every beautiful thing quickly turns into a good thing when the valuating is directed to its existence. Every existing beautiful thing is something good, namely, <with> regard to its beauty.

Just as the experiencing expectation is directed toward the non-being of the experience, so–in the present sphere of valuating–desiring, wishing are directed toward the non-being of the valuation. Every good-value, if it does not really exist, is a possible desire-value. And when what is desired occurs, then it figures as good.

<b) Values as Founded Objects Constituting Themselves in Value-Taking. The Psychological-Naturalistic Misinterpretations>

We considered all that to start with in the sphere of mere valuating, of what is not yet mixed with thinking. However, just as in the empirical sphere, consciousness can find its expression, thinking form, in the valuating sphere, and so a value, an esthetic value or a good-value, not only figures in plain and potentially fully intuitive form, but it is then also a thought-object. It turns into the
290 object of the statement and of the logical determination.

And all that signifies that values are objects. There are not only existential, natural objects, not merely things, but also values. And, even if values refer back to things that have value, so precisely a fundamentally essentially new predicate occurs here, the predicate “beautiful” or “good”, which belongs, so to speak, in a totally new dimension and is not to be mentioned in the same breath as thing-predicates such as “heavy” or “hard” or “colored”, and as psychological predicates such as “angry”, “merry” “perceptive” and so on–in short <as predicates> of the kind all existential predicates in general <are>.

Without doubt, a new domain of objectivities opens up here, and it is the one modern psychologism has sinned against most of all. For reality in the usual sense, the sphere of the existence of things, had already long ago won acceptance scientifically as being nature, and the psychological misinterpretations did not hamper the natural sciences, but only interfered with the correct interpretation of their essence and that of the associated ontologies. In contrast, psychologism–and the more universal naturalism allied

<with> it–hindered the development of the axiological disciplines and completely blinded people to axiological ontology.

In our time, there is admittedly much talk of value theories, but they are all wholly given up to psychologizing the objectivities that I call values, therefore senselessly mixing up value and valuating. The predicates beauty and goodness break down into feelings and desires, as little as the predicates of concrete existence, the predicates "being concrete", "being spatial", "temporal", "being hard", "heavy" and so forth break down into sensory presentations, into perceptions, memories, expectations and such. Just as things are units of sensory experiences and not themselves sensory experiences, and just as things are what they are whether they are experienced or not, so a value is a unit constituting itself in valuating experience and not itself valuating. And a value is what it is whether it is grasped in a valuating manner or not. Being a value, being good or beautiful, does not mean that anybody considers a thing to be of value, or that there is a widespread tendency in a community to value something, to love it, to be fond of it, or accordingly to desire it–thus, in accordance with the sole authority 291
that can decide there, in accordance with the authority of consciousness itself and of the meaning inherent in it.

This indeed emerges completely clearly when we try to carry statements about values over to valuating acts or states of mind. The concepts for the value-objectivities lose their meaning for valuating acts and likewise for the corresponding states of mind. So it is that belonging, for example, to values as such are relations of gradation. As concerns consciousness, besides individual valuating as such there is also relational valuating, that is, a higher-valuating and a lower-valuating intertwined in the unit of this relating valuating. In this unit, we encounter "a more beautiful than b", "a better than b", and so forth. But being-more-beautiful is a predicate of the value and not a predicate of the valuating, just as in the relating consciousness of number comparison being-more is not a property of consciousness, but something conscious in this consciousness. Desiring may <be> "more ardent" or less "ardent", but if it is thereby really characterized by intensity, higher intensity really pertains to the more ardent desiring, lower to the less ardent. On the other hand, however, we meaningfully differentiate between greater and slighter value and <between> more or less intense values. The ardor of the desiring does not increase the value itself. Chilliness does not make it a non-value.

It is also evident that the attitude adopted in valuating is itself an immanent directing toward the value, namely, toward the thing in terms of its value-quality, while the attitude we adopt in making the valuating into an object, or in positing the object in relation to the fact of the valuating is a totally different attitude. The more impartially one studies the phenomena and delves into them and their intentionality, the clearer it becomes–and finally completely evident–that valuating is consciousness precisely in the sense perceiving or presenting is and that it stands in the same relationship to heeding, believing, thinking as the latter. And that is then further manifest in the fact that values and value-relations are judged and disputed whether the values are really lasting, whether the value-relations, the higher values and hierarchies of values are really as they are supposed and affirmed to be or not. Therefore, just as in the sphere of experiencing consciousness, we also find here the difference between what is supposed and what is real, what is non-rational and rational, what is lawful and what is unlawful, what is valid and not valid. So here the distinction transitions into the thinking building itself up on valuating: On direct value-truths are built indirect ones, and the latter must show
292 themselves to be valid or invalid. In addition, also here, as in the experiential and thought spheres, is the distinction between positivity and negativity. In contrast to consciousness of existence stands consciousness of non-existence, in which <the> tendency to existence-positing conflicts with certain existence, with certain existence-positing, and cancels it out. Likewise, standing in contrast to positive valuating is negative valuating.

While psychology and the social sciences and humanities have to do with the facticity of human valuations–and with their value suppositions as such–, a new science, a science of reason is necessary that makes the reason-problems of valuating and value into an object of investigation.

What of course makes the situation appear especially difficult here, and initially clouds one's clear vision for apprehending a distinct object- and reason-sphere, is the fact that we are accustomed and inclined to accord thinking a special affinity with experiencing, consequently to count them as one. Thinking, judging are built on the basis of presentations. What is to turn into the object-about-which must be presented–and presented means, I believe, perceived, remembered, or otherwise experienced in the broadest sense. However, thinking is also built on value-taking and valuating presenting. Objectifying positing is something which can pass through value phenomena as well as through perceptual phenomena and then turn into the substratum of thinking.

So, we therefore have value concepts, or value thing-concepts, and belonging to them are in fact most concepts of everyday life. Pure natural concepts are actually more or less artificial and have their dwelling place in pure value-free existential science, <in> physics especially. In the humanities and social sciences, the distinction is not at all as clear. Interpretations especially continually coalesce with existential statements in history, so that, when the question of the scientificity is raised, it must be twofold and sharply distinguished: that of the scientificity of the historical facts and that of the scientificity of statements of values. We therefore now see that cleaner separations would be required under the heading 293
"humanities and social sciences", precisely the separation into existential investigations and value investigations, or the separation into empirical sciences of minds and into rational-axiological sciences of minds.

<c) The Ontology of Values: Formal and Synthetic a Priori in the Axiological Sphere>

Value is a fundamental basic concept for being-formations. It defines a radically new domain in the face of value-free things. So there must after all be an ontology corresponding to it. It is quite remarkable that once formal logic already existed as an analogue, no one ever also tried to develop a formal theory of values, something which nevertheless stood to reason as a necessary postulate once the right foundations were found. One is immediately convinced that there is an a priori, namely, a formal a priori, of the sphere of values by a look at both the principles stated by *Brentano* in his brilliant work *The Origin of Our Knowledge of Right and*

Wrong, which read that what is good and what is known as good is to be preferred over something bad and what is known as bad, and that the existence of a something known as good is to be preferred over its non-existence and, by the same token, the non-existence of something known as bad is to be preferred to its existence. If we set aside the reference to acts, then we obviously more simply have the principles that every good thing is superior to a bad thing, or every value <is superior> to a non-value, furthermore, that the existence of something good is better than its non-existence. *Brentano* further deduced from this that something good on its own is better than the same good mixed with something evil, that the sum of something good and something evil is better than the same evil on its own, that a whole good thing is better than a mere part of it (part as good), that the sum of many good things <is> better that any of them alone, and so forth.[1]

In my Göttingen lectures on the fundamental problems of ethics, I then tried to develop the thought–which even *Brentano* had oddly not grasped–of giving reality to the idea of a formal theory of value in analogy with the formal theory of being.

The formal theory of value is, on the one hand, and first, analo-
294 gous to the synthetic-formal theory of being of nature, to the ontology of nature in terms of space, time, motion, substantiality, causality, on the other hand, however, also analogous to analytic-formal ontology, to analytic logic, provided namely the valuating is, like judging thinking, a founded act. And, as in thinking, higher thought positings with their own forms are founded upon plain positings, so in the area of valuating, higher value-positings, preferring and discounting, as well as the forms of intrinsic valuating and mediatory valuating for the sake of something else, are founded upon simple valuating positings.

However, more needs to be said about the justification of talk of formal axiology. It is formal ontology of values insofar as it leaves out of account the particular nature of the value-matter,

[1] Franz Brentano, *The Origin of Our Knowledge of Right and Wrong*, Abingdon: Routledge, 2009 (1969), pp. 27–28. Originally published as *Vom Ursprung der sittlichen Erkenntnis*, Leipzig: Duncker & Humblot, 1899. (Translator's note).

similar to the way formal analytic ontology deals with objects in general without asking about the specific domains and regions of objects and not at all about their particular species and genera within a region. That does not, however, rule out there being a wealth of a priori knowledge of an axiological kind that lies precisely in what has been valued as such, similarly to the way that the analytic-ontological a priori does not rule out something else, a synthetic a priori. One would like to feel honestly tempted to carry the distinction between "analytic" and "synthetic" over to the axiological field, whereby however its meaning would be somewhat transformed. From the logical point of view, the analytical is what is grounded in the essence of the formal analytic categories (as it were, what is grounded in the empty essence). From the axiological standpoint, the analytical is what is grounded in the essence of the axiological categories "beauty" and "goodness", as well as in the associated relational categories. That is the emptiness in the axiological sense. However, every a priori in the axiological sphere would be synthetic that was grounded in value-things (in the valuation-matter), but naturally not, if they are mere things, but <if they> are value-objects. You will readily understand what is meant by that when I remind you of the values of the sphere of consciousness, where the great headings of reason itself are unconditionally-universal, i.e., therefore, a priori, value-headings, for example, knowledge, namely, knowledge in the concise sense, rationally validating knowledge, to begin with, seeing knowledge as judging which is not only certain of its truth, but is certain in insightful 295
grounding. All such judging is a value, namely, an absolute value, an intrinsic value, and not a mere mediatory value, i.e. a value for the sake of something else. We must thereby distinguish between judging as consciousness and the a priori knowledge pertinent to it that is directed to the essence "judging." But beyond this a priori is the a priori that the proposition expresses. All insightful judging, every insightfully knowing truth is an absolute value. Having value does not constitute the essence of the judgment. Value is not a constituent of essence, of the Idea "judgment" and even less so of the Idea "insightful judgment." But having worth is grounded in the essence of such judging, namely, having positive worth. In my terminology, insightful judging is something beautiful, and not for-

tuitously so, but *a priori* so, consequently, valid in unconditional universality for each possible individual case. And correspondingly, according to an a priori of formal axiology, each actual case of insightful judging is at the same time a good-value. It is a good in itself that is insightfully judged.

To be borne in mind in the process is that a mediatory value can combine with the intrinsic value, that, therefore, for the sake of something else, what is good or beautiful in itself can be bad or disturbing of beauty.

Likewise, just as by its very essence, every insight is beautiful and, posited as an existent, is something good, so every act of reason in general is by its very essence something beautiful, therefore, every perfect perceiving and experiencing, every perfect valuating itself.

How far the material axiological a priori reaches can only be learned from closer study, and this would be very much needed. There is scarcely any talk of a systematic-rigorous science that accomplishes what is needed here. This would involve, for example, the question as to which genera of objectivities can be intrinsic values. Whether, besides the values of the sphere of consciousness and <the> subjective sphere in general, natural objectivity can also be an intrinsic value, or whether not instead all concrete valuation is based on the fact that what is concrete appears, and the way of appearing is the essential value-bearer, or if need be, is based on the fact that, by virtue of empirical consequences of consciousness, or other psychical consequences, what is concrete is valuated as
296 being useful–must in any event be investigated.[1]

In one respect, valuating–and positive and negative being of value–is obviously all-embracing, namely, when we admit both intrinsic valuating and mediatory valuating. *A priori*, all concreteness can really enter into relationship with consciousness, therefore, at least in this regard, have mediatory value.

---

[1] Subsequent note by *Husserl*, "The account is unfortunately very incomplete. I would have had to deal with valuating and willing together from the outset. It is also clear that all value-free existence forms a single main domain, and similarly all valuating and wishing". (Editor's note).

It is likewise to be pondered from the psychical perspective whether states of mind, for example, knowing states of mind, knowledge, mental capacities, habits draw their value from the fact that they are sources of acts themselves having value, and so forth.[1]

After engaging in these reflections, we therefore have a formal and <a> material axiology belonging to the domain of values, and both conceived of as a priori-ontological disciplines. In addition, we then have empirical axiological knowledge. First, <we have> carryovers from a priori matters of value to given ones and then empirical value-influences, the functional dependencies of given existential values, of the adding or removing of other value facts. In all these, the a priori everywhere prevails over the a posteriori.

There may be matters of contingent subjectivity–how this is pleasing to one person, that to another–is valuated, and people or other empirical beings may go to incredible lengths in valuating. But, if we ask how something is not only considered to be of value, but really is of value, and make that clear to ourselves in direct value-givenness, then the thought is not to be dismissed that only what the being-of-value warrants in terms of its essence can in itself be of value, and therein lies the fact that intrinsic value cannot be grounded in the facticity of existence, but only in the a priori essence of what is existing. And it is furthermore certain that all value-dependencies, value-influences, all relationships of intrinsic values and mediatory values and the associated increases and decreases in value are purely grounded in the a priori, namely, in the essence. But, obviously the facticity remains, the fact that precisely exactly this and that is there with its essence and that, through influence, the associated values are then to be determined in terms of it and not in terms of a different value, of a value that *de facto* would precisely be different if the value-grounding matters were different in this factual world. Finally, this is however no 297
miracle and does not distinguish essentially empirical-axiological sciences from empirical natural sciences (sciences of reality). For everywhere, a posteriori knowledge is determined by a priori laws

[1] Continuation 27 <from p. 300, line 18>. Compare the *Textkritischen Anhang* of the German edition, p. 521 (to p. 296, line 9) (Hua XXX) (Editor's note).

at every step, and everywhere the non-rationality of the a posteriori lies in the facticity of existence.

The problems of reason, the problems of axiological noetics, then follow on ontological value problems.

<d) The Relationship Between Axiology and Practice>

In the last lectures, I dealt with the disciplines that relate to the axiological domain. I hope to have awakened in you the conviction that just as apprehending natural existence takes place in natural experiencing through the medium of sensory appearing, so apprehending of being-of-value does in value-experiencing through the medium of value-appearing. We observe thereby that apprehending of value stands in contrast to each and every apprehending of being, whether it is a matter of existential apprehending of physical nature or of mental existence. We know in such a way that we must gather all existence that is in itself fully value-free individual being into a universally united region that is then organized into sub-regions in terms of physical being, individual and social mental being. On the other side, stands the "being-of-value" region. However, even in the latter case, the question arises as to whether we must not proceed to a most universal region.

I have in particular not yet spoken up until now of the willing, acting, engaging in creative activity that differs–which concerns the ways of consciousness–from mere valuating as valuating as beautiful and valuating as good, or differs from pleasure and displeasure, from rejoicing and being sad and all related modes of this group that come into consideration for value-taking. As in all the phenomenological reflections related to the inner sphere as a whole, great difficulties lie here, namely, in the drawing of proper boundaries and in bringing to full analytic clarity the basing of acts upon one another–and correlatively those of objectivities. One is very soon convinced that just as pleasure is a form of consciousness in which something is conscious as being pleasing, so willing
298 is a form of consciousness in which something is conscious as something that practically ought to be. In this sphere, one must also distinguish living-in-practical-consciousness––therefore, willing, deciding, acting, functioning, effectuating–from the wholly dif-

ferent attitude in which we engage in reflection and look at these modes of willing and make them into objects experientially. In deciding, we are conscious of the decision. We are however first conscious of the deciding in reflection. In acting, we are conscious of the action–and especially in creative activity, accomplishing, we are conscious of the accomplishment, of the maturing activity as such.

Now, the objectivities occurring here are also subject to rational assessment–and in a different sense. For example, a decision can be rational or non-rational, may or may not have led to act or deed.

It is clear that deciding presupposes valuating, or is essentially connected with it, in the sense that nothing can be willed that is not valuated. For example, willing preferring presupposes valuating preferring. Something in itself unlovable may even be willed. It can only be so when it is necessarily the lesser evil in a disjunction, between which one must choose, or when overwhelmingly favorable consequences are attached to it, and so on. On the other hand, the willing itself–and the decision made–can be valuated. In the latter respect, as in the former, rational questions naturally arise, therefore, the <question> as to whether the valuating guiding the wills is rational, and then that as to whether the will itself, provided it complies with this valuating, is "motivated" by it, is rational or not. But, in contrast, the question still arises as to whether the willing is also spontaneously subject to yet another rational question, whether therefore the will is rational because it is an object of rational valuating, or whether it is an object of rational valuating because it is rational.

Further difficulties arise with acts, namely, <in the case of> ones in the sense of ποιεῖν and of ones in the sense of πράττειν. Not all acting really sets its goal in a deed that in acting leads to productive creation.

External acts and deeds are, on the one hand, physical objectivities, physical processes or physical things. They are also subject to psychophysical assessment, namely, with respect to facticity, the fact that in the consciousness of the one acting, willing–and inter- 299
woven in certain ways with the willing–a flow of phenomena takes place and that it can at the same time be said that in nature the act

as physical event may occur subsequent to the willing. Lying along other lines, however, is the voluntary rational assessment, the question as to whether the act is a rational one, as to whether its goal, the goal of the willing is in itself rational, or exists for the sake of something else, and whether the means to the end are the right means.

Comprehensive and difficult analyses are needed to clarify all the relevant relationships and to answer the questions about specific willing-meanings and willing-objectivities, which owing to the shortness of time I cannot enter into further. Since, however, the close connection of these questions with axiological ones is manifest from the start. From now on I want always to speak of axiology and practice together.

Every good-value that is conceivable as being the ultimate goal of an act, or for which—as I can also say—an act is conceivable that realizes it, is a practical good. The formal propositions for values, and especially for good-values, automatically carry over to what is good practically, and one is in general very quickly persuaded that a formal a priori also governs in the practical sphere, that therefore just as one is to talk of a formal axiology, so one is to talk of a formal practice, where both disciplines are intertwined with one another in the most intimate way. That is the obvious consequence of the fact that willing is founded in valuating, volition-objectivity in volition-objectivity, will-reason in valuating reason.

In contrast to formal (analytic, in the recently discussed modified sense) axiology and practice, I then speak of material axiology and practice. In relationship to this I want now to provide some more explanations and follow-ups.

<e) The Different Levels of Generality of Axiological-Practical Reflection>

In formal logic, I spoke of the difference between existence-judgments and essence-judgments. Existence thereby had a completely general sense. In the empirical sphere, existence is what is empirical in physical or psychophysical nature. Matter-of-fact[1] judgments in which empirical existence, reality in the ordinary sense, is posited stand opposite the essence judgments, in which such reality positing does not take place. 300

In the axiological and practical sphere we have completely parallel empirical valuating, positing of individually determined valuating and, as a result, empirical value judgments, and in contrast, a priori value judgments that presuppose general, that is, essentially general valuating.

Accordingly, the concept of material value judgment and of material axiology can also be understood in a twofold way. It can thereby be a matter of empirical and of purely a priori disciplines. Most value judgments of ordinary life, and in the practical sphere, all the usual judgments about practicability and impracticality, utility and inutility, provide examples of empirical value judgments. Apperceptions of purpose are embedded from the start in most ordinary concepts. So it is when we speak of a table, of a house, farmland, and so forth. And all that is invariably assessed as good and bad, as practically good and perhaps also as beautiful, whereby practical value is once more connected with beauty.

That gives rise to scientific investigations in different ways. (1) One can–as, for example, economic theory does–start from the fact that certain things are customarily valuated in the factual human community, and possibly also that in certain ways they are actually also values, namely, involve purposiveness upon the assumption of prevailing goals, whereby however in so doing the latter leading setting of goals and valuatings are accepted as pregivens precisely in view of the fact that they are treated by human beings as existing values and existing goals. (2) One can also again valuate these existing purposes and values themselves–as for example in any ethical assessment. From the axiological standpoint, scientific reflection is therefore at times hypothetical, at times absolute or categorical.

---

[1] In English in the original. (Translator's note).

<f) The Possibility of Scientific Constructions of Value-Configurations as Constructions of Ideas of Possible Nature Realized from Axiological Points of View, as well as Possible Consciousnesses and Intellectual Communities>

If we now inquire into the principles to which all valuating–
301 hypothetical as well as categorical–is subject here, then we are led to a priori axiological disciplines, that is, on the one hand, to the formal axiology and practice that establishes all axiological and practical essence-insights that are independent of the particular nature of the makeup of what is valued, and on the hand, to a priori material axiologies that arise out of the fact that, besides the basic concepts of analytics, we draw in the regional <concepts>. Therefore, nature in general can be valuated to start with, and nature can also be seen as a field of possible praxis. Furthermore, people can–and the human community in general can–likewise be valuated, namely, in terms of everything pertinent here: human inner life, human consciousness, and consciousness in general, in terms of all its basic configurations of rational consciousness, and so forth. The level of axiological-practical reflection can thereby be that of differing universality. We can take nature to begin with just as it actually is and within nature then once again the genera of psychical beings and forms of society actually occurring in it, as well as the other forms of culture. From what is actual we can rise up to the Idea. We therefore form the Idea of nature, the Idea of humans, the Idea of human culture. In so doing, we can proceed in different directions from the given and to the Idea and to the axiological-ideal. We take, for instance, a type of societal and public existence and ask *idealiter* how the ideal of proper practical conduct is to be determined for citizens of this kind of country, for a person with this level of culture, under such and such generally to be described conditions. Belonging here are therefore so-called utopias, every blueprint for an ideal state–such as that of the *Platonists*–, the attempt to construct an axiologically and practically perfect social body, to the point of constructing the Idea of a perfect world community of human beings. Such constructions could then be scientific if we had the corresponding scientific elementary laws of an axiological-practical kind at our disposal.

In the analytic-logical sphere, mathematicians construct the Ideas of possible theories, the possible definite manifold forms *a priori* and absolutely exactly on the basis of the exactly developed lower levels of analytics, of apophantics especially. Analogous pos- 302
sibilities of construction arise in the natural sphere.[1] We indeed have at our disposal a perfectly developed ontology of nature as regards some disciplines. The Idea of nature is formally defined by the axioms of this ontology. Furthermore, we can carry the most universal physical principles–for example, the principle of the conservation of energy–over into the material Idea of nature. Defining the forms of these principles, we can then constructively outline different variations of the Idea of given nature, therefore, for example, ask how a nature would be constituted in which one or another basic law of mechanics <would> or would not hold, or <in which> this or that part of the principle of the conservation of energy would be modified, and so forth. We can even modify the formal Idea of nature, namely, by holding on to a certain universality, therefore by that universalizing the Idea of the thing. We can then distinguish within this ideal sphere between Idea of given nature and Idea of possible nature in general in the sense of the universalized concept.

In precisely the same way, we can then proceed farther into the sphere of the individual mind and of <the> communal spirit on the condition that we already have a scientific ontology of spirits at our disposal. The human beings given and the human communities given, the forms of religious, artistic, scientific culture given, and so forth, are facticities, but in terms of their form possibilities selected from spheres of a priori possibilities defined *a priori*. On the basis of such ontologies, one would be able to determine constructively and scientifically multiple forms of possible communities of intelligent, human-like beings in a natural world that are not actualized in actual fact. With all of them we are still in the region of value-free existence.

---

[1] Subsequent note by *Husserl*, "An important observation forces itself on us here. Axiological and practical analytics yields an a priori just as absolute as analytic logic. Is the ontology of nature with respect to space and probably also with respect to causality, different? Yet: Time has a fixed a priori. How is that to be explained?" (Editor's note).

Now, such constructions can, however, be guided by axiological points of view. If we let nature and the realm of the spirit be the
303 factual substrate for value-configurations, then we can naturally also realize constructions in the axiological and practical spheres, namely, constructions of value-configurations, when, besides the ontologies of value-free being we also have the ontology of values and of practical objectivities at our disposal. We can therefore ask: Supposing that in a natural world (which corresponds to the Idea of nature in general as it unfolds scientifically in nature-ontology) there were minds and intellectual communities, supposing that different kinds of cultures developed in it, and supposing that these minds, and so forth, were to conform to the universal, a priori requirements of the ontology of the mind–though they may also differ in the contingent empirical individuations of actual human beings, the Idea of the mind really still leaves many individuations open–then which valuable individuations of human beings and human communities, which valuable culture forms of every kind are then *a priori* possible, and how do things stand with the ordering of these configurations in terms of different and higher values? Are there possibly ready-made configurations there? Are such configurations–which can be considered as the highest, as absolutely insuperable values–*a priori* conceivable? Is a directing of societal willing toward a finite, delimited goal as a fixed boundary-value thus conceivable? Or, would the ideally greatest worth of human or human-like community and community development have to lie in an infinite *progressus* of value configurations that would always be raised anew, so that in ideal striving toward them, humanity and human beings would always have to attain to higher value *ad infinitum*?

It is clear that constructing and axiologically exploiting such possibilities is at the same time tantamount to creating a scientific system of norms up against which the givens of the empirical life of the spirit can be measured, and <that that> is again tantamount to providing the scientifically grounded goals and means for an ideal rational praxis. All ethics and politics in fact proceed constructively. But a scientific ethics and scientific practice of community requires scientificity of such constructions, and for that, as we see with Evidenz, a scientific axiology and practice is necessary–

indeed, going further still, the whole system of scientific ontologies. We obviously still have next to nothing of the kind. That it is here, however, a matter of thoroughly necessary goals of our 304 knowledge and of goals whose setting and step-by-step realization affect the highest interests of humanity, I need not say.

<g) The Call for a Genuine Science with Regard to the Axiological and Practical Spheres of Being. Against a Falsely Delimited and Naturalistically Misinterpreted Idea of Reason>

Humans are rational beings, that means, <1.> they are minds, they engage in acts of consciousness that as such are subject to the rational questions of validity and lack of validity. And that further means 2. They are free rational beings. Ad 1: They are capable of insightful apprehension of goodness and they are capable of the science that determines in scientific objectivity what is true, what is existing, what has value, what is practically good and understands this in terms of its connections. And <ad 2 it means> once more that they are capable of a rational praxis in conducting themselves and shaping their lives and the world, of a rational praxis that is rationally motivated by such intuitive and scientific insights. Of course, lofty values of humanity already lie prior to science and prior to all rational reflection in the organic-natural development of communities and individuals in them. But once humans had benefited from the tree of scientific knowledge, new, incomparably higher and incomparably further reaching and integrating theoretical possibilities, value-possibilities, and practical possibilities opened up for them. Therefore, the time of natural-unscientific living from day to day, of evolving in merely intuitive thought- and value- and willing-outlook is over once and for all. The sovereignty of scientific reason has begun, and implied in its essence is the fact that it wants to become and must acquire universal dominion. And it must above all not be merely dominion in the value-free existential sphere.

One can say that the tremendous conflicts that constitute the misfortune of the modern era have their main source in the antagonism between claims of science and claims of the mind. However,

this antagonism exclusively springs from the fact that up until now genuine, fruitful, rigorous science has just been able to take hold in the region of the natural sphere, of what is value-free. In terms of its aspiration, modern science indeed even extends to the remain-
305 ing spheres of reason and being. But, it is a false aspiration. For, through an understandable failure to recognize the limits of nature, and with the so obvious psychologization of the axiological and practical objectivities in the natural attitude, people believe they can tackle and solve axiological and practical problems with the natural sciences and psychology, which is of course only possible by distorting the problems, thereby at the same time preparing the way for an intolerable axiological skepticism. Precisely this skepticism–by violating the dignity and validity of all ideals of the mind–generates that inner discontent of the modern era, those tremendous tensions sickening it. Pseudo-science must make room for genuine science, and genuine science requires free commitment to the intrinsic meaning of objectivities and objectivity-regions that it is to bring to pure, empirical knowledge.

The Idea of reason, and correlatively, the Idea of veritable being, reaches as far as the sphere of possible, necessary science does. That therefore concerns the axiological and practical sphere of being. Only a human life that has experienced enlightenment about its meaning, determination of its value-realities and value-possibilities–as well as fixed norms for its developments and possibilities of development–from true, genuine science can realize intrinsically relatively highest levels of value in an infinite *progressus* of increase in value. Natural cultural development must (this is the necessary path in the life of humanity from the stage of nature to the stage of reason) change into scientifically substantiated rational activity. Just as in the life of an individual person matters must not be left to rest with natural living from day to day, but rather individual life must assume the character of a work of reason in its overall unity, so it is with respect to human and cultural development as a whole. The valued personality must not simply and plainly exist and become, but it must itself become object, object of creative shaping in the interest of practical reason. Only too narrowly and mistakenly restricted Ideas about reason can discredit such a goal in that in complying they cause the wrong con-

tent to be substituted for the already correctly recognized general requirement. Investigating how self-development and community development are to be carried out in accordance with the highest Ideas of reason, which ideal possibilities general obtain here, and 306
how matters stand with their system of values and their practical feasibility is a matter of rigorous science, first of all, of that highest constructive theory of norms of which I spoke earlier and the empirical-practical theories of the art of based on them. So, it is in fact here a matter of sciences of incomparable worth, the development of which must constitute the site for making possible the highest values of mankind.

<h) The General Schema of the Kantian Transcendental Questions Is Also Applicable Outside the Specific Genus of Epistemological Ideals>

I am here adding something more of great interest. Namely, I want to show that *Kant*'s transcendental-logical way of reflecting upon knowledge, by means of which he believed he could provide a theory of knowledge, also falls within the context of my reflections. *Kant*'s transcendental philosophy is actually quite teleological in nature and is–though this too is not emphasized forcefully enough and was emphasized by *Kant* himself–conducted from an axiological viewpoint.

Let us consider the following. A world is presented to human consciousness through the medium of sensory phenomena, initially, the world of plain experience spatiotemporally boundlessly governed by rough patterns that force themselves on people in their initially only practical attitude and determine their conduct. Then science appears on the scene and, pursuing the empirical coherencies of plain experience and exploring them in thought, people work out a natural world, a world subject to exact laws, as a correlate of natural science. And then one says to oneself that the real world is governed thoroughly by laws. The laws were just not discerned by unscientific human beings. And furthermore, humans are adapted to nature in such a way that the truly existing nature in which they live is not only a world rigorously governed by laws, but is also discernible as such in human natural science, and accordingly can be mastered practically by human scientific technology.

Let us now turn things around, namely, with a view to teleologi-
307 cal reflection. Let us assume there is a consciousness endowed like ours with sensory phenomena. We do not say that in every respect they are sensory phenomena as we actually have them. However, sensory phenomena they are, and as such phenomena, they are phenomena of an existence, of a "natural world". We do not, however, say that it is a natural world in the sense defined by the ontology of nature, but in a more universalized sense. "Natural world" would be a name for a being displaying itself in the phenomena of consciousness that as such, as it were, pretends to be in itself. Now we ask the following question. What must hold *a priori* for an ego-consciousness and its phenomena and combinations of phenomena for what is called the natural world displaying itself in it to be determinable in accordance with strict laws and unambiguously explicable as truly objectively valid being-in-itself? Or, what amounts to the same thing, what must hold for consciousness and the phenomena of consciousness for it to be theoretically possible for consciousness to know its appearing by means of an exact natural science? That is the general schema of the *Kant*ian transcendental question. The extent to which it is in general answerable in this generality, whether one must not make specific assumptions in order to make it scientifically comprehensible and answerable, remains to be seen here. The question is obviously teleological and is conducted from an axiological perspective, that is to say that we place value on the possibility of knowledge. Knowledge is indeed one of the basic values. However, comparing cognitive values, we find that the possibility of scientific and exactly scientific knowledge is of especially high value.

Now, from this axiological viewpoint, it is remarkable that an objectivity of existence displays itself in our consciousness not only in general through phenomena, but that the objectivity called the natural world actually displaying itself has the extraordinary theoretical virtue of being in keeping with the highest interests of knowledge. This is so, namely, provided the natural world consciously displaying itself is such that, in the process of investigating knowledge, it leads to mathematical natural science with all the

wonderful theories, and that all existing individual things can be unambiguously determined and explained in conformity with these theories, whereby every non-rational contingency leaving our drive for knowledge unsatisfied on grounds of principle is ruled out on grounds of principle. Therefore, that is what is won- 308
derful, and that is decisive for our teleological question. Let us leave fact qua fact to one side. Let us rather reflect as to how such a natural world consciously displaying itself must be and how it must display itself in the consciousness so that generally speaking, this virtue obtains. Are perhaps many different kinds of natural worlds, still many others besides the actual one, conceivable that possess the same virtue? Or, are many kinds of ways of displaying and other consciousness-arrangements conceivable–namely, *a priori* conceivable–for the consciousness that knows the natural world if one adheres to the Idea of such a precisely to be determined virtue? Of course, the Idea of this virtue can still be grasped differently. We can, for instance, remove just the unambiguous determinability, the possibility of objectively valid "empirical judgments"; and from there on try to deduce that then all phenomenal objectivity must have spatial form and temporal form, that the law of causation must hold, and so forth. One could then go further and, after the teleological necessity of a mathematical natural science—even though not one with our specific laws of nature—would be deduced, point to further teleologies that become observable by us with the conformation of human knowledge to the natural world, for example, the fact that phenomenal things can be classified according to empirical genera and species, and so on. This procedure therefore characterizes *Kant*ian transcendental philosophy. And, apart from various obscurities and absurdities, it shows us that–starting from what is given and its value-properties of an intellectual kind–scientific-teleological reflections can actually be undertaken that seek to determine the other way around which characteristics objectivity of knowledge must have to be able to be in conformity with certain epistemological ideals, or which characteristics can objectively define a consciousness if it is a nature being-in-itself (displaying itself through phenomena).

Now, it is however clear that we can <also> ask and scientifically deal with just such questions outside the special genus of

epistemological ideals. Therefore, what characteristic must a world have not only to be in perfect conformity with certain higher epistemological ideals, but also with esthetic, ethical, and all possible ideals? How must a world be constituted if it is to possess not only
309 theoretical, but every kind of, virtue–a world good, beautiful in every respect, a most perfect world of all? And this question perhaps initially concerns the natural world, but then further the whole world, the physical-psychical world. Naturally, however, the scientific material and the method of such investigations as scientific lies in all the ontologies presented.

That is the most general sketch of the *Kant*ian transcendental question. The extent to which it is answerable in this generality, whether one must not make further, more specific assumptions, is not under discussion here. The question is teleological. We can say that if we place value on the possibility of knowledge, and if we find especially high value in the possibility of scientific, and even exactly scientific, knowledge, then it can be said that the natural world that is actually there is not only actual, but is such that it is in compliance with our highest epistemological interests. It has, we can say, intellectual virtue. It does us the favor of letting its things to be brought under empirical generic concepts. For example, the organic world orders itself in systems of classification in marvelous ways, physical things likewise. They are analogous to one another, are measurable, and so forth. None of that needed to be, neither on the side of nature, nor on the side of knowing beings.

*Kant*ian transcendental philosophy shows that scientific reflections can be undertaken that–starting from what is given and its teleology–ask the opposite question: How must consciousness with its consciousness-correlates ideally be constituted, which arrangements and forms and categorial laws must be assumed for this for an ideally indicated teleology to be able to be fulfilled?

In this transcendental reflection on the natural world, the investigation is conducted theoretically because the governing ideal adheres to certain theoretical virtues that manifest themselves in theoretical properties.

It is different when we ask how a world must be constituted in order to conform most perfectly to, not only theoretical, but also esthetic, ethical, and all possible value ideals, how a world must be

constituted so that it not only makes possible the most perfect theoretical knowledge–and in infinite *progressus*, so that delight in knowledge is enkindled ever afresh ad infinitum–, but that it also 310
makes values of beauty a reality, and to the highest degree, and not the least values of the will, in short, that it is the most beautiful and best world in every respect, the best of all possible worlds.

Obviously, scientific construction of such systematic ideals can, however, only be possible to the extent that scientific material by way of fundamental principles and primitive theories is on hand. That must, however, be developed in a systematic implementation of analytic axiological disciplines and in a systematic exploration of the primitive material-a priori principles relating to the different domains of being. Where this firm foothold is lacking, there are only teleological novels, but not scientific teleological constructions. The relationship of the analytic theory of theories to basic apophantics and formal ontology is always to be thought of as an analogy. And finally, the systematic implementation of axiological-practical theories of reason is also necessary in all directions, because the investigation ultimately transforms into an investigation of the teleological reason-consciousness.

As regards the Idea, it is just as much a matter of rigorous scientific investigations as in the analytic theory of manifolds. Without a substratum of rigorously scientifically developed ontologies of existence and values, we however produce teleological novels and not teleological disciplines.

With the just completed reflections, I have quite greatly expanded the Idea of logic as theory of science and at the same time created differently defined Ideas of logic. Each of the regional ontologies really presents itself to us as a kind of logic, axiology, as it were, as logic of values, practice as logic of acts, of ethical, of practical deeds, and so forth. Naturally, much research will still be needed to set these disciplines on the right paths. They would include the principles and norms upon which the constitution of empirical sciences of nature and the mind <depends>, which must be grasped in the most comprehensive epistemological interest, not only as existential sciences, but at the same time also as valuating and technical sciences. So the valuation of nature with respect to beauty leads to the constitution of a possible esthetics of nature,

311 which is in turn the foundation of an esthetic nature-practice, because nature can be creatively reshaped in meaning by human Ideas, of which the art of landscape gardening is an example.

The fact that the problems of scientific teleology, initially as pure and then as empirical teleology, lead at once to metaphysics is clear. They lead to it provided the knowledge of a possible actual teleology of the given world of nature and the mind raises the question as to how one is to understand that the absolute being that comes to be known in all empirical sciences purified of all confusions is teleological, in conformity with Ideas of value.

<Chapter 13 311

The Idea of Noetics>

<§ 66. Noetic Reflections in the Sciences as Critical-Methodical Reflections in the Service of the Primary Interest in Things>

I am devoting the last lectures remaining to us to a closer examination of the Idea of noetics that we encountered in all regions opposite to the thing-oriented ontology. Corresponding to every ontology, to analytic ontology, to ontology of the natural world, to the ontology of values, and so forth, is a noetics. Obviously, analytic noetics as counterpart of analytic ontology will also be the most universal of all here and of a universal-science-theoretic nature in a particularly concise sense.

A look at present-day sciences shows that when we in general speak of theory of science and wish to align “logic” with “theory of science”, the sphere of noetic reflections does not at all lie far off. It will be good to go into that somewhat, since in that way we find the best gateway to the clarification of the meaning of the noetic line of research and noetic issues. When we sought to obtain the first Idea of logic, that of formal and to begin with apophantic logic, we directed our attention to the fact that sciences live on objectively in the form of textbooks, treatises, and so forth, in which the scientific theories, the teaching methods find verbal expression. This 312
remarkable objective continued existence is a continued existence of interrelated statements, and these statements have their meanings in judgments and relationships of judgments that for their part refer to the objectivities of the scientific domain, to things of the natural world and natural relationships in the natural sciences, to numbers in arithmetic, to geometrical forms in geometry, and so forth. In all of them, we adopt a thing-approach. The meanings in relation to which we position ourselves in reflecting on meaning and which we call “judgments”, “inferences”, “proofs”, are not, for

E. Husserl, *Logic and General Theory of Science*,
Husserliana: Edmund Husserl – Collected Works 15,
https://doi.org/10.1007/978-3-030-14529-3_13

example, the experiences of judging, of considering-true and considering-probable, of being certain and presuming, but the judgment-contents, the supposed truths, the supposed possibilities, probabilities. While we started from this objective continuance of the sciences, we turned our research just to these ideal meanings and their relationship to the meant objectivities, and so we practiced theory of forms of meanings, apophantic theory of validity, formal ontology, up to the theory of theories.

Precisely this approach is the normal one for research into the present-day sciences. Natural scientists turn to facts, but in so doing they are more and more compelled to turn to meanings. They must direct their attention to truths and relationships of truths, to inferences and proofs. However, not so rarely do they also reflect in other ways. Not rarely, they turn to subjectivity, and this also occasionally finds expression in the scientific statements of writings. We must thereby immediately separate the extra-essential from the essential. We eliminate all the occasional recourse to the historical facticity of the investigation, therefore, where there is, for instance, talk of when and where, and by whom a theorem was discovered or contested, proven, or even exposed as false, likewise, every account by the authors of the motive and reason behind their research. There are, however, other forms of subjective discourse that are devoid of any reference to the contingent empirical persons investigating and substantiating it, to the circle of readers and fellow scholars, and so on, or where at the very least the reference to such empirical-psychological facts is extra-essential. Essential, however, <is> the reference to the acts of position-taking, or rather, to their ideal essence. That is the case
313 in all statements in which there is talk of the fact that certain judging position-takings may be rationally required or unjustified from the standpoint of reason, namely, for instance in reference to the fact that other position-takings may have already been given beforehand or performed. I call to mind expressions such as, "This established, it is certain that now such and such must also further hold"; or, "Assuming this, it is a justified question, can legitimately be doubted, it is a legitimate presumption, it is a possibility worthy of consideration, and so forth." Here, one is not talking about facts and meanings, but about judging–presuming, asking, doubting–

and about its justification. Or, think about such forms of discourse as one after all frequently hears them in all the sciences: One must admit proposition A, because it is a direct Evidenz, an axiom. Or: One must admit it, because it is to be concluded with Evidenz from such and such premises. Since we know that A B etc. is, then the presumption that C is cannot be ruled out. It is easy to see that one is indeed talking of acts here, that however any individuality again specified can all the same be eliminated. Such propositions make a claim that means to be independent of psychological fact. If our knowing that A B is, grants legitimacy to the presumption that C is, then that also holds generally, since we really believe ourselves able to justify such talk before everyone. That holds if in general someone, anyone, knows AB (and has no further knowledge at his or her disposal) on the basis of which the presumption C would be justified. So it is universally. The insight that a conclusion is grounded in its premises, combined with the insight that they are true, justifies not only me, but justifies anyone having such insights, to believe the conclusion to be true.

Naturally, knowing, presuming, questioning, and so forth are acts of an ego. But whether this ego is human or animal, divine or diabolical, is completely indifferent to the meaning of such statements of justification. Whether humans or the devil toss dice, the justified presumptions are the same, provided the preconditions, the die, <the> rules of the game and so on are the same. And anyone who violates the laws here, anyone who presumes differently than what proceeds from the essence of what is the situation with respect to its laws, presumes unjustifiably. Human and diabolical stupidity are completely alike in this respect and are measured by such law-dictated standards. 314

We therefore see that in the sciences it is not exclusively facts and meanings of statements that are judged and pondered, but also thinking itself with respect to its legitimacy. And, it thereby occurs to us from the outset that, for the purpose of assessing legitimacy, there is, at the very least, an aspiration to an independency of the validity of the law from contingent subjectivity.

Admittedly, the extent to which statements about the legitimacy of thought-acts occur in the scientific exposition varies greatly. They do not play a very prominent role in the mathematical sciences. Mathematicians love to set forth theories simply as deductive systems, basic laws and corollaries. To be sure, they occasionally also speak of their methods, but even there in a manner that does not allow precisely what is interesting us, assessing legitimacy, to come to the fore. They say, let one proceed in such and such a manner, let one build discriminants, let one multiply equations by one another, etc. The account is provided in such a way that anyone trained in mathematics can inspect the legitimacy of such a procedure step by step. Once in a while the situation becomes different when the legitimacy of certain fundamental concept-formations becomes doubtful to them as a consequence of emerging contradictions, and they are then inclined to reflect upon the basic concepts and their interrelationship, because that finally compels them to go back to givens.

Reflection on the method and on the ways of justifying thought processes has been necessary in other sciences to a much greater degree. The natural sciences, for example, indeed use deduction and mathematization to a great extent. We really have the vast deductive disciplines of theoretical physics: mechanics, acoustics, optics, etc. in them, but the basic principles of the deduction there are not axioms whose truth can be grasped in an evident way without further ado. One cannot set forth the basic principles there without further wasting a word, simply trusting in the fact that everyone could become conscious of the legitimacy of such an arrangement through recourse to Evidenz without further ado. The basic principles spring from experience, but not from direct experience, as if it were enough to perceive and make perception-judgments, but from long processes of methodical processing of thoughts. And this processing, which is included under the rather
315 vague heading of "inductive thinking, inductive methods", is not a chain of such directly obvious steps in thinking, as in the case of deduction, where in each step of inferring one follows the apodictic Evidenz and in reflecting on the necessary validity can become aware of the necessary legitimacy of the procedure. On the contrary, although it is also borne by a general consciousness of legitimacy, here the method harbors all sorts of puzzles.

The average natural scientist does not notice this. Admittedly, natural scientists also once in a while reflect about their methods in order to ensure the correctness of their procedures. However, this correctness is not gauged by real insight, but rather by agreement with generally respected methodological precepts and with a certain consciousness of correctness deriving from schooling and tradition. One reflects and finds confirmation. One must proceed in this way. One does so in all similar cases, and so forth. Competent natural scientists can proceed correctly without in the least being able to provide an account of the inherent legitimacy of their proceeding, because the fact that one is accustomed to proceeding in this way in similar cases in natural science and in so doing fares well is no genuine inherent justification, any more than it would be a justification for calculating logarithms from the given tables to say that everyone proceeds this way, and in doing so they fare well. In hindsight, this is right. Logarithmic tables have indeed not fallen down from heaven any more than the methods of natural science have. Methods are creations of great thinkers. Pioneers of method are hence also those of real science. Creative thinkers had to gain each methodological step by strenuous painstaking effort. They, who started from the pre-scientific standpoint or from the initial methodologically imperfect beginnings of science constantly reflected on the legitimacy of their empirical methods. They engaged in perception and other experience, empirical generalization and formation of hypotheses, natural scientific concept formation, participation of thinking in forming objectively valid empirical judgments, and so forth, an ever more deeply penetrating critique, and they fought a hard fight with trickery and sham, i.e., with everything vitiating the legitimate value of cognitive acts.

They repaired precisely these critical-methodical reflections in order to devise new methods that could satisfy the standards of 316
legitimacy they beheld. Of course, they were primarily interested in things, not in the cognitive acts. But how is one to grasp and objectively validly determine the things? How are the things given? How do they reveal their properties, relations, laws? In plain experience they seem <to be> given without further ado. But plain experience alone does not do that. Experience can also be deceptive experience. When, therefore, can perception and experience justifiably say that the things may really exist? And what can it at best accomplish for the grounding of scientific judgments? No

plain empirical judgment is already a scientific judgment. Why is it not? Why and how must we advance from experience to experience? In accordance with which methodological guidelines must we connect them? How do we grasp them in thought and process them in order to determine scientifically-objectively the real thing displaying itself in them? Those are the questions. All these cognitive processes are subjective acts. In the contexts of cognitive consciousness itself, everything that makes up its claim to legitimacy, and what, if it is genuine knowledge, intelligibly grounds it, must be discoverable and exposable. Creative natural scientists therefore engaged in such reflections about what lends legitimacy in cognitive procedures, while the method that was developed, and then over generations became the routine practice of average researchers, was used without insight into the reasons for it–in the way craftspeople practice the tricks of their trade on the basis of theory and tradition and without theoretical insight.

<§ 67. General Noetics as Systematically Formal Theory of Justification of Knowledge>

However, though it serves science, the kind of insight possessed by the great pioneers of true methods is not itself scientific insight. The search for justification of knowledge clings to contingent motives there, to methodological needs that have grown out of a particular science and particular problems. It is not radical enough and it is in need of systematic research pursuing the relations of justification of all kinds of knowledge in terms of their general relationships. It is, moreover, to be noted that creative minds are
317 also to a great extent guided by instinct and feel and by testing through subsequent scientific successes, and so genuine insight–which, however, in many cases does not inquire into the systematics of the ultimate grounds of justification–is here and there combined. In any case, in view of the problems of method, the matter cannot rest with such more contingent procedures clinging to specialized scientific interest.

The specific scientific method is the property of the specialized science. But it is only a particular grouping of methodological elements that point beyond the specialized science and are partly common to all sciences in general, or to all sciences of one and the same region. Everything that we call method, every tightly ordered procedure for substantiating knowledge possessing a claim to legitimacy, leads us by means of analysis back to the manifold basic kinds of acts that we include under the very multifold word "knowledge". All these acts must be examined thoroughly, therefore, not only the specifically intellectual position-takings we call acts of judgment, but all acts founding them, proving or disproving them. They are all linked to one another by essential relationships. All claim to be legitimate, or their specific legitimation functions come under the heading "proving legitimacy" or "disproving legitimacy". Judgments draw justification from perceptions (*Wahrnehmungen*)[1] or value-takings (*Wertnehmungen*), and so these acts become legitimizing acts. Perceptions come into conflict with perceptions. They challenge each other's legitimacy, and nullified perception counts as pseudo-perception (*Scheinwahrnehmung*). Judgments can also legitimate other judgments, whether with respect to judgment certainty, or with respect to probability. Questions can find rational grounds in presumptions, and so forth.

It is clear that insofar as every science establishes truth or probability, and correlatively every science states, thinks, judges, substantiates judgingly, and so forth, a broader range of knowing acts and ways of proving legitimacy must be common to all possible sciences. And, so it is apparent as a further consequence that there must be a scientific discipline that has the character of a systematic theory of justification of knowledge and is directed at the problems of justification common to all sciences as such. With respect to cognitive acts and the questions of legitimacy relevant to them, this science would therefore have validity for all actual and 318
possible sciences that is just as universal as that of formal logic with respect to problems of meaning and objectivity. In other words, standing opposite to formal logic, analytic ontology and theory of meaning must be, as it were, a formal noetics, a formal theory of justification of knowledge, and the latter will be attached to the former in the most intimate way. Corresponding to every

[1] Once again, the German word for perception is "*Wahrnehmung*", which literally means true-taking (Translator's note).

judgment-form in the meaning-theoretical sense is a mode of acts of judgment, namely, of such acts as they could occur in every particular thought-sphere, because the generality of form in the meaning has its correlate in the generality of the make-up of acts of judgment, which rules out any tie with special cognitive material. As, however, concerns the laws pertaining to the legitimacy of the judging, every formal law of truth and probability–for example, every syllogistic law, in certain ways a rule of justification–can be turned around into a law of justification of deductive knowledge. All thinking that infers from two premises of the forms "All A are B" and "All B are C" that All A are C is legitimate, is to be performed with possible apodictic Evidenz, and can prove its legitimacy in that. Of course, one must not for that reason believe that the entire formal theory of justification is merely a turning around of ont<ological> analytics, as if it there were nothing more to explore here. And one must not identify without further ado theory of justification of thinking in analytics and formal theory of justification of knowledge in the sense of my formal noetics. The theory of justification of formal logical thinking fits in it, but is not everything. In addition, the problems of general formal noetics follow on the problems proper to the essential regions of objectivities and ontologies.

Formal noetics can answer the question as to how judgments are substantiated by pointing to the difference between judgments to be substantiated indirectly and directly where, in the sense of this discipline, judgment is now always understood as act and not as supposed truth. With the generality of its approach, it can then only answer the question as to what direct substantiation signifies, how the difference between blind and insightful judging is to be understood, by pointing back to underlying straightforwardly presentive acts. But, since the fundamental differences between
319 regions and domains that determine the fundamental classes of the sciences come into consideration here, general noetics cannot deal decisively with the particular make-up of these acts and the manner in which they in particular provide foundations and make objective validity possible in special spheres of being. And, thus we overlook the fact that investigation then continues in disciplines that stand correlatively opposite the regional ontologies as the

noetic disciplines corresponding to them. Besides a science of justification of analytic knowledge and of the most general justification of analytic knowledge and of the most general justification of knowledge in general (provided it has not yet been split into regions), we encounter sciences of the justification of empirical knowledge, of the justification of valuating knowledge, of the justification of knowledge with respect to rational praxis. A complete theory of justification of knowledge, a comprehensive noetics, which by virtue of the essential relationships of the acts must be a unit, would therefore have to embrace both the most general questions of justification common to all sciences–among them naturally the theory of justification of analytic knowledge or the theory of analytic reason–<and also> then, in addition, the disciplines of experiencing reason that knows real existence, likewise those of axiological and practical reason, namely, from a formal and material perspective.

In realizing it, one will naturally begin with general noetics and in so doing again with consideration of the most general questions of all. The point of departure will thereby be the natural one, precisely the one prevailing in the naturally developed sciences standing before our eyes as exemplary. It therefore goes without saying that there are many sorts of objectivities about which we humans can have knowledge, and paramount is systematic science. Knowing as a state of mind thereby leads back to actual knowing, therefore to judging. It thereby further goes without saying that not every instance of judging is already knowing, that not every one has, as people also say, cognitive value. Rather that would only hold of substantiated judging. All scientific thinking would be substantiating thinking. As a matter of principle, science does not want to settle for the ordinary blind, instinctive, routine judging acquired from authorities and traditions, but systematically to substantiate judgments concerning its subject area and to advance in well-ordered *progressus* from legitimately substantiated judging 320
to other such judging in order to make sure of the systematic ordering of the truths relating to objects by this means. Therefore, the first questions of noetics, the most general of all, will be: What distinguishes judging in general from legitimately substantiated judging proving its alleged legitimacy? What does this legitimate

substantiation look like? How is the meaning of such an achievement to be understood? Underlying the constant criticism that scholars level in their judging aimed at proving legitimacy and perhaps denying legitimacy, is the conviction that every judgment that is not already substantiated can be substantiated, or more concretely, that its legitimacy can either be substantiated or disproved. How do things stand with this conviction itself, and how do these contrasts between substantiating and disproving stand in relationship to one another? In addition, the very judgment that was blind at first acquires a certain precedence due to substantiation. It shows its legitimacy. But does it thereby take on a quality of legitimacy? A new intellectual quality that it did not have before? Well, <we> must then, however, ask how it happens that after the substantiation we say that our judgment was already, when we had not yet substantiated it, justified beforehand, and why we do not say that was only now, and only at this time, only so long as <it> is laden with the quality of legitimacy, precisely a legitimate one. Every judging is a "supposing to be true". Each one makes a "claim to legitimacy". That is already a curious, therefore, problematic thing. While we are judging, the need awakens to prove this claim to legitimacy. We ask ourselves, Is it really so? Is our judging genuinely valid? And afterward we say, Yes, our judging was valid, because "it turns out to be substantiated". But what does the quality of substantiation that subsequently ensues have to do with the earlier judging? And what does it have to do with future judging, our own and that of our fellow human beings? Curiously, we really believe that the quality of legitimacy has this wondrous transcendence: Once substantiated, a judgment A is quite simply substantiated; whoever judges A, judges legitimately, and never ever can another A prove not to be legitimate if we have recognized the same A as legitimate. What sort of odd claims are those? What does the facticity of the world, with all its human beings and all the judgments made by people, care about the eventuality that *hic et nunc* something called "quality of legitimacy" is associated with
321 judgment A in something called a substantiation process? Does our momentary substantiating govern the course of the whole psychophysical natural world, our own future psychical nature as well as the totality of nature with all thinking beings?

<§ 68. The Radical Problem of Evidenz and of Givenness>

<a) In What Respect Does Purely Immanently Evident Judging Differ from Non-evident Judging? What Does Givenness for the Different Regions and Categories of Objectivities Look Like?>

In all reflections about the justification of knowledge, besides the words "legitimacy", "validity", "truth" and their negations, the word "Evidenz", along with the antonym "absurdity", also plays an uncomfortable role. One perhaps says "legitimacy" means the same as validity or truth. Legitimacy is to be distinguished from consciousness of legitimacy, from Evidenz. In substantiation we become conscious of legitimacy. The judgment lacking Evidenz turns into an evident one. As a result, it acquires a remarkable intellectual quality. In disproving, we become conscious of the negative legitimacy, of the lack of cognitive value. The judgment acquires a contrasting intellectual quality, that of "absurdity", that of obvious nullity. But, how is this being-conscious of legitimacy, of validity to be understood—especially since this consciousness itself can, however, in turn be legitimate or merely supposed? The previously indicated fatal problem naturally recurs, <namely,> how the subjective, factual quality of Evidenz can claim to guarantee "truth", "objective validity", independently of the subject and the contingent moment. One does not speak of Evidenz only with respect to the thought-acts that we call acts of judgment. In that they grasp truth, judgments ought to refer to objectivities in the right ways, namely, to the objectivities about which the truths say something true. Corresponding to the Evidenz of the judgment ought then be the Evidenz of the givenness of the object. And since the first problem with respect to judgment-Evidenz must be to clarify what evident judgments actually look like, in what respect they purely immanently differ from judgments that are not evident, the parallel question arises with respect to objective givenness, therefore, the question as to how what is called, and in 322
the most concise sense is called, givenness of what is objective differs from an object-consciousness that we designate <as> not-presentive and that is not to contain any guarantee at all regarding

the being of the object meant. And, it goes without saying that this question must first be considered from a descriptive perspective, through immanent reflection on the ways of consciousness itself, apart from those broader, in any case probably dependent, questions about the meaning of the claim to legitimacy reaching beyond the contingent act and its subject.

The general question is differentiated naturally. What does givenness of an object look like in all the different regions and categories of objectivities? The formal ontological forming is also there and to be inquired into above all else. What does givenness look like for an object and what does it look like for a state-of-affairs, for an individual object, for a genus and species, for a formal category? Moreover, what <does givenness look like> for a truth, for a possibility, a probability, namely, as such, initially precisely in formal universality. What <does it look like> for a fact? How <does it look> for a fact and for a law? Likewise then, <What does it look like> for the ontological forms "thing", "property", "causal property", "natural law", etc. Again in the axiological sphere–what does the Evidenz for a value, for a value-state, for a beauty-value and a good-value, and so forth, look like?

<b) How Can Evidenz Secure Truth for Us?>

In the last lecture, I began to discuss the curious problems of Evidenz and givenness. From the little I indicated, you have become conscious of the radical meaning of these problems for all our knowledge–though it may also carve out ever so great, exact sciences for us. As long as we are within the sciences and practice them naïvely, we dwell in the Evidenz of their procedures, the intrinsic rationality of which overmasters us. Step by step we see, "That is how it is and it must be so", and the theories built one upon the other delight us. But, as soon as we begin to reflect, the puzzle is there as to how knowledge in general may be possible, how its claim to legitimacy may in principle be understood. Scientifically
323 judging is not judging of a random kind. It is evident judging. Evidenz, one says, is what necessarily assures us that we are in the possession of truth. We can judge without seeing, blindly, routinely, and so on. We can, however, also judge insightfully–only if we do

that does our judging have objective cognitive value. Were we not capable of Evidenz, then all talk of truth and science would not make any rational sense. Quite so, we said, but we want to understand how Evidenz has the wonderful quality of assuring us of truth. How can an immanent, positive difference in judging have such immense import? Are not things in themselves undisturbed in their being and essence by our contingent psychical experiences? How does our evident judging, our evident substantiating, our proving conducted in such and such forms, our evident theorizing for this purpose come to prescribe laws for the world existing-in-itself? The inner course of our evident theorizing ought to have results conforming to the course of the world existing-in-itself. But what does the world existing-in-itself care about our psychical processes? Evidenz is supposed to guarantee truth, but who guarantees for us that it achieves this? Could not, or so we may ask with *Descartes*, an almighty God, or a powerful tricky demon have created us in such a way that we would have to err consistently precisely in Evidenz? Ought pre-established harmony exist between our Evidenz and the being-in-itself? Now one dimly senses the *Widersinn* of such possibilities and everything presupposed in such questions. But how is one to clarify things? How is one to save oneself from the skepticism that threatens to quash any possibility of knowledge?

<c) Triumphing Over the Psychological Feeling and Index Theory of Evidenz>

How little modern times have done justice to those problems in this respect–however great a role the critique of reason may be playing in it–is shown by a look at the prevailing psychological theory of knowledge in which, moreover, many philosophical schools of thought calling themselves idealistic remained caught up. To the question as to what Evidenz may be, one replies there that it is a peculiar pleasurable feeling, otherwise undefinable, but well-known to everyone through inner experience. It is also called 324
a feeling of thought-necessity or judgment-necessity. Like any feeling, it may have its negative correlate, namely, in the quality of thought-impossibility, of negative Evidenz, as it were. But, calling

Evidenz a feeling, and a feeling of thought-necessity or of thought-impossibility does not mean the question has been answered as to how it is to be understood that if our subjective experience of thinking is laden with this quality–that of thought-necessity–what is thought must really be. What does Being care about thinking and its feeling-qualities. If one positions oneself on psychological ground, then the theory of knowledge, of the various kinds of substantiation, of the various acts of grasping objects and determining objects reduces down to a psychology of Evidenz-feelings. Noetics turns into a psychological discipline. In experience, we study human consciousness, human perceiving and other experiencing, the judging building upon that in human beings in this or that manner, and afterward see when Evidenz-feelings associate with that and which empirical laws for the occurrence of Evidenz-feelings emerge there. It then emerges that general conditions of the possibility of Evidenz lie in the judgment-contents, in the logical propositions, and more precisely in their forms. It then becomes apparent that Evidenz cannot affix itself to every arbitrary judgment. One would have to explain that following the stating of psychological experience, every judgment has a judgment-content, a judged something. On its form depends whether the Evidenz-feeling can empirically set in or not by virtue of laws. Here, it is consistent with the psychological outlook for the judgment-content to be viewed as an experiential moment of judging. For the psychological approach, all formal logical laws, those of our analytics, turn into noetic laws, namely, into psychological laws of possible judgment-Evidenz. And, <it> must thereby then also be psychological experience that teaches–or for the purpose of this approach would have to teach–that the feeling of positive Evidenz is incompatible with that of negative Evidenz for the same judgment-content, that is, also in the sense that it can never happen that if a judgment has in one case been characterized as a necessary thought, it can in another case be characterized as an impos-
325 sible thought, and so forth.

If one pursues this approach in a critical manner–however appealing it may at first seem to be–then one lands in the greatest absurdities of the kind I discussed in my *Prolegomena to Pure Logic*, and which I ask you to consult now and again. If everything is reduced to the factual constitution of the human mind and to empirical laws governing it, then one has to say that a change in human beings and a change in the factual constitution of their minds could also bring other laws with it. In the struggle for existence, human beings perhaps change, so that Evidenz-feelings arrange themselves differently, that every contradictory human or suprahuman judging with the quality of thought-necessity, every unambiguous one as a thought-possibility, would be there, that the law of contradiction would figure there as evidently false, that $2 \times 2 = 5$, as evidently true, and so forth. One is then easily convinced that the psychological theory cancels itself out. For the purpose of its stating, it presupposes precisely what this meaning as <meaning of the> stating, as <meaning of the> judging requires, but *in thesi* it teaches that precisely this would not possibly hold, and so forth. One also shortly realizes that, like analytics, all ontologies would have to dissolve in empirical psychology, and ultimately all science in general. Psychology as empirical science would in one case presuppose the existence of humans in the natural world, and so non-psychological natural science, and, on the other hand, would nevertheless again reduce every non-psychological science to coincidences of human acts. In short, a multitude of absurdities intertwined with one another emerges as a consequence of psychologism. They remain concealed from it only due to the fact that it is careful enough not to think its theories through to the end and, on decisive points, to take the edge off of the result through veiled borrowings from idealism.

The entire situation changes when one makes it clear to oneself from the start that the difference between non-evident and evident judgment of the same judgment-content does not consist of the fact that some intellectual quality adds on as a puzzling and actually fictive index *veri et falsi* to the non-evident, otherwise remaining unchanged, judgment-experience, but of the fact that the entire judgment-experience thereby radically changes through mere holding on to a general moment, precisely the one for whose
sake we are in the habit of speaking about a judgment of the same 326
content or, in short, of the same judgment. The same judgment $2 \times 2 = 4$ is in no way phenomenologically constantly the same. On one occasion, we judge completely vaguely, as it were verbally, as

when we emptily reproduce one times one ($1 \times 1 = 1$) (with that familiar vague consciousness of understanding). On another occasion, we carry out the judgment in a completely different way. More or less clear intuitions also combine with the words, and perhaps such a combining and relating in thought, not to be described more precisely here, builds upon the underlying intuitions that, instead of the vague consciousness of understanding, we have insight. What previously was, so to speak, a mere bill of exchange, a mere notification, is now the clean cash payment that redeems the bill of exchange. We are judging not only, believing not merely, in the manner of a vague δόξα. What we are believing is really given to us, stands there itself, clearly and, so to speak, in person, and in such a way that it would be senseless to look for anything further behind this given. If we compare the forms of the non-evident and evident judging, then we therefore see how ridiculous the approach is that sees Evidenz as a merely appended index, a feeling that, adding on to the otherwise identical judgment-experience, is supposed to be a notification of truth. Only those can talk in this way who do not themselves study the forms, rather have concocted a theory for themselves for extraneous reasons.

<d) Evidenz-Consciousness as Consciousness of Givenness of the State-of-Affairs Itself. All Regions of Objectivities Need Their Own Investigations into the Possibility of Their Givenness>

If one delves into Evidenz-consciousness and comes to realize what it truly is, namely, self-presentive consciousness in contrast to mere believing, without itself grasping, <in contrast> to empty intention without fulfillment, if one, I say, comes to realize this difference, then one sees that obviously self-presentive consciousness of this kind in can indeed solve puzzles, but not itself contain more puzzles. Everything puzzling, everything problematic lies on the side of mere believing. Wanting to deal with seeing self-grasping, self-having as a puzzle means not understanding oneself. It means philosophizing about Evidenz from on high instead of having a look at the Evidenz itself, of bringing it itself to Evidenz.

327 A second thing that one must make clear to oneself beforehand is that, just like blind judgment-consciousness, Evidenz-

consciousness is in that case indeed a psychic experience, and as such is roughly connected to my momentary inner life, that that, however, does not change anything, that, absolutely beheld and conceptualized, what is given there is nothing psychical, but just the state-of-affairs $2 \times 2 = 4$, or the true being of it, that $a + 1 = 1 + a$, that if $a > b$, $b < a$, and so forth. And, in this manner, one can convince oneself that, especially, as in such examples, what is universal, the relationship of universalities, of contentual or formal essentialities, are given absolutely, and that their givenness signifies that they are directly and themselves grasped in their being, and that it makes no sense to doubt their being, or to view their being as psychological being. One must come to realize that, just as in reflection, momentary Evidenz-consciousness can be brought to givenness, the essence of Evidenz-consciousness can too, and that everything that can be said of Evidenz-consciousness is not said of it as of the contingent psychological fact, but <is said> of it by virtue of its essence. From here on, the path leads into the phenomenology of consciousness as an essence-theory of acts of judgment and Evidenz-acts and all kinds of consciousness related to them. All essence-relations of this sphere are explorable in absolute Evidenz, in an Evidenz that brings the essence-relations of acts to givenness.

If one is ready, then one can solve all the particular problems of Evidenz, because with the general clarification of psychological follies and with the general finding that direct Evidenz is by essence presentive consciousness that leaves open no question about the whether and how of the givenness, not everything is achieved that must be achieved if knowledge is to shed its puzzles regarding its possible unconditional validity. The general puzzle as to how Evidenz can grasp objectivity initially appears solved if using examples we realize that something being in itself, such as a mathematical state-of-affairs like $a + 1 = 1 + a$, can be completely given, and that that givenness does not mean having some kind of magical psychical index, but just having the state-of-affairs itself. But, for all basic kinds, for all regions of objectivities, big problems, 328
<questions> result regarding the possibility of their givenness, and everywhere they call for their own exceedingly difficult investigations. Were Evidenz everywhere plain direct beholding, in which

what is objective were plainly and directly given, then general clarification would be sufficient. But not all Evidenz is direct Evidenz, and not all objectivities are to be brought to direct beholding as those axiomatic universalities are. For example, natural objects, things, are not completely and indubitably given in direct Evidenz, namely, in their full reality. Things are given to us through phenomena, and givenness through phenomena leaves open the fact that the phenomena may perhaps not exist at all. We speak of hallucination, of illusion. Is that then an accident of human knowledge? Or, is it part of the essence of thingness that it is in principle only given through phenomena and, then, in incomplete ways. And what does “incompletely given” mean, since one is thereby, however, still talking about “given”. How is it to be understood that such direct givenness has cognitive value all the same and can be a foundation for scientific knowledge, even natural knowledge, in the form of exact science?

<§ 69. The Solution of the Problem of Reason Requires a Systematic Phenomenology of Consciousness and Its Consciousness-Correlates. Transformation of All Knowledge into Absolute, Metaphysical Knowledge. Noetics as Theory of Science in the Highest Sense>

All natural scientific knowledge is indirect knowledge. Only in certain methodical empirical methods, which are called inductive there, does the being and being-so of natural objects come to adequate, as it were, scientific, givenness. How is this knowledge and how are the justifying Evidenzen to be described and all their steps and the principles governing them to be elucidated? Problems arise in a similar manner in all possible spheres of knowledge, even in the sphere of analytic thinking. So there we have the difference between direct knowledge, in which universal forms and laws come to us in direct, evident givenness, and indirect
329 knowledge that in certain ordered processes leads from direct givens to new ones, which by their nature are perhaps not to be had directly. And the same is so in formal and material axiology and practice. Things are given through experience, and empirical givenness is givenness through the peculiar medium of phenomena.

Values and what is practically good require a different givenness. Experience may also underlie them. Needed there are inner experiences, <experiences> of pleasure, rejoicing, of wishing, of willing. Anyone who dispensed with such acts would not understand what is meant by something beautiful, something good, an achievement, an ethical deed, and so on. And even there we have various differences of indirectness and directness. General talk of Evidenz therefore says little. Objectivity and consciousness with its consciousness-correlates, its opinions, phenomena and so on display remarkable affinity. While, practicing sciences of the different objectivities, we dwell in Evidenz, engage in reason-consciousness with the content that precisely knowledge of such objectivities requires, this consciousness is not our object. We experience it, but we do not know it. If, however, we want to understand how this wonderful fact of knowledge actually looks, how it achieves and can achieve what is put down before us as objectively valid science, then we must inquire into the correlations among consciousness, consciousness-meaning, consciousness-phenomenon, and consciously supposed objectivity. We must follow the guiding thread of all basic kinds of objectivities and of the ontological principles to answer the questions as to how objectivity may come to givenness, how the being and being-so of objectivity may reveal itself to the consciousness, how–in which systems of "phenomena"–objectivity may necessarily display itself, how thinking may conform to experiences, how it may be understood that thinking operating in accordance with certain methods could be able to reach beyond the phenomenon and prescribe the rules for the future course of phenomena,–and for every other sphere, just as for the empirical sphere.

Only a systematic analysis, namely, an essence-analysis of consciousness in terms of all its basic specifics, in terms of all its essence-correlates can solve the problems of the critique of reason, can realize an unambiguous and fully satisfactory noetics. The fact that it belongs to the essence <of> every kind of consciousness to be consciousness of something is the point of departure. 330
But, for every basic kind of objectivity, consciousness, and more precisely the presentive consciousness, is characterized differently. It has different essence-formations that are connected *a priori*

with the essence of the objectivities concerned. The question as to how objectively valid science may be possible as a system of knowledge in which the content of science comes to insightful substantiation therefore leads back to a systematic phenomenology of consciousness and its consciousness-correlates, to a study of the essence-relations that link the different basic formations of consciousness to certain basic formations of associated meanings and associated meant objectivities. Only when this phenomenological theory of reason is carried out do we thoroughly understand the meaning of the scientific knowledge of each region, and only then do we have definitive, absolutely clarified sciences free of puzzles in every way instead of naïve sciences unclarified as regards their meaning. Only then does everything problematic presented to us by the different spheres of being disappear. All false interpretations attaching themselves to scientifically known being shed their illusion, and the naïve unclarified knowledge of being is transformed into absolute knowledge of being–and that means into philosophical knowledge.

From the natural sciences, the way therefore leads to the ontologies, up to the most universal ontology (to analytics), then further to the noetics following all these ontologies, elucidating their principles and methods, and all particular scientific knowledge is then elucidated through noetics. And all knowledge is then transformed into absolute knowledge, into metaphysical knowledge. Thus noetics is theory of science in the highest sense, and at the same time that discipline that makes possible an ultimate and highest fulfillment of our epistemological needs, because only absolute knowledge can afford ultimate epistemological satisfaction. Above all knowledge–this is part of its essence–hovers the Idea of the absolute as its guiding star. If philosophy is the name for every kind of scientific investigation that seeks to serve our striving for absolute knowledge over and above the depths of scientific work, then all logical disciplines, and noetics foremost, deserve to be called philosophical disciplines.

Appendix I (to Sections I–III): <For the 1910/11 Lecture. Concluding Remarks and Plans>[1]

...To what extent can all that be conveyed to beginners? I can probably only give them a faint idea of what is to be accomplished and probably possible here. The concepts of a priori and a posteriori must be thoroughly discussed precisely in the introduction, while the difference between analytic and synthetic, what is merely formally and materially a priori valid is to be given in addition to formal logic.

5) The next step would probably at first lead back to formal logic and from there on to the formally restricted noetic theory of norms (and methodology) and from there to the idea of an a priori founded noetic theory of norms and methodology in general that makes use of all a priori ontologies and axiologies and, when these are complete, would also have to provide all complete methods–and all in any way possible.

6) Is anything still missing? Now the critique of reason would come. Can one seriously distinguish the critique of reason from noetics? In the lectures, I distinguish between a noetics of what is superficial and a noetics of what is profound, the latter the genuine theory of knowledge. Is this distinction to be implemented? In any event, a theory of justification of knowledge can be drafted up to a certain degree without becoming involved in the ultimate epistemological difficulties.

[1] Probably beginning of 1911. (Editor's note)

E. Husserl, *Logic and General Theory of Science*,
Husserliana: Edmund Husserl – Collected Works 15,
https://doi.org/10.1007/978-3-030-14529-3

The foregoing pages were used a great deal for the lectures, namely, for the Winter Lectures 1910/11 (Logic–Theory of Knowledge).

The investigation was not completed in the lecture. So many things remain in abeyance. The difference in the position of the formal disciplines of axiology–which however occupy another position vis-à-vis formal analytics–and of the ontology of nature, which is not however a formal discipline in the genuine sense, has not become completely clear to me.

All the pages on theory of science must be reworked, and everything judgment-theoretical that has been mixed in there
334 must be eliminated. Does that not yield a treatise in its own right?

Plans: Introduction to theory of science as "classification of sciences", as theory of the domains and regions of possible knowledge in general: theory of categories. Whether the writings of Jonas *Cohn* and Emil *Lask* should be able to help somewhat there?

General clarification of the Ideas "meaning", "judgment" etc., within the context of a general theory of consciousness. That is the most burning desideratum. And I shall have to go into that first. From there <on out> transition to a systematic phenomenology of consciousness with all the fundamental distinctions: impression and idea, etc.

To begin with, however, I have to attend to the printing of the second edition of the *Logical Investigations*. That will give me a good bit of work. The theory of meaning ought to remain unchanged in the process.

Appendix II (to §1) <Logic as Philosophical Science of Knowledge, as First Philosophy>[1]

For scientific specialists who have devoted their lives to the systematic examination of logic, it is now and then truly entertaining to observe how unsuspectingly even well-established masters of theoretical science take no notice of the genuine problems and deeper lying difficulties when they indulge inclinations to philosophical reflection, but disdain to derive the benefit of the scientific work of thousands of years in this field. In our time, newly

[1] From the lecture course of 1910/11 WS. (Editor's note)

awakened philosophical interests are indeed very popular in all sciences, and not the least in highly developed natural scientific science-theoretical reflections. For the most part, they are unfortunately of a very naïve kind. There is thus quite the same gap between them and the heights of true science as, for instance, between the practically wise peasant's observations of nature and the theoretical natural knowledge of physicists or astronomers. I believe I may state this judgment without exaggeration, although it is certain that the science or set of sciences pertaining to the understanding has not remotely attained the degree of theoretical sophistication that modern mathematics, the modern natural sciences, and not least also the philosophical-historical disciplines of the modern era honors so gloriously. In spite of all the efforts that the greatest minds have applied to the analytic elucidation of the sphere of understanding over thousands of years, the number of firmly secured insights and theories there is disgracefully small in comparison, in particular, to the younger and so very successful disciplines I have just named. But, for that reason, what I maintained earlier however holds good.

The greatest philosophers, who for the most part were also the creative minds of the beginnings in the exact sciences, did not succeed in gaining as steady a foothold and in ultimately conquering vast areas in the field of philosophy as in the other sciences. So, the 335
endless efforts did not, however, remain misspent. In order just to close in on the problems, it was in particular necessary for them to circumscribe them into groups, to test the major points for their possible solution, and to develop the unique methods that the incomparable uniqueness of these problems precisely requires. The temple of philosophy dwelt, as it were, on steep, distant far-off heights. We must first struggle to scale them before we can set foot in it and fathom its secrets. Many, and great, things have now been accomplished by way of preparatory work, and the number of solving and really compelling theories not much greater than in the specialized sciences erected upon a more accessible basis. So the gap between what has already been accomplished in forms of science and what is to see and grasp in the naïve standpoint is in fact not less, indeed, undoubtedly actually greater than, for instance, that between the sciences of nature and the empirical life of the spirit.

My task will then have to be to raise you above this naïve standpoint and show you the way to the mysterious solitudes in which one day the sphinx of knowledge must unveil its riddle. I could also call it the way to the "mothers" of knowledge, to the essence-principles of knowledge in terms of its ultimate origins.

Regarding these mothers, I love to turn to Mephistopheles' words, "Enthroned sublime in solitude are goddesses. Around them is no place, still less any time." And unfortunately, it is stated no less significantly, "To speak of them is embarrassment." As you will remember, Mephistopheles is keen to dissuade the aspirant from taking the way "into the untrodden–the not to be trodden" and paints the solitudes in a ghastly enough way, "Nothing will you see in interminably empty farness, the step you take, you will not hear, nothing firm find where you rest." We must, though, not allow ourselves to be frightened, and with Faust answer, "Just keep on, we want to fathom it. In your nothing, I hope to find the universe."[1]

It will naturally also be my task to clarify for you the meaning of the problems arising here for all human knowledge, so that you understand that in the case of logic as philosophical science of knowledge, it is not a matter of subordinate specialized fields in the realm of manmade sciences, which like any others find their enthusiasts, which though someone aspiring to deeper learning could very well do without. Ultimately, learning really cannot look for its goal in learning about anything and everything, but only in what is ultimately meaningful, what is deepening, indeed enabling, understanding of self and the world.

It will, however, become apparent that nothing is to be expected of an ultimate understanding of the world as long as one has not raised oneself up to the set of problems of the analysis of the understanding and reason and adopted a position with regard to them. If, therefore, philosophy is the word for the loftiest goals of knowledge and the sciences directly oriented toward them, then the discipline we want to devote ourselves to here is to be labelled first philosophy in the most concise sense of the word, for it is a
336 matter of investigations that must go before and must be attended to before further philosophizing can seriously be contemplated.

---

[1] Johann Wolfgang von Goethe, *Faust*, Section II, Act I. (Translator's note)

The imperfect state of our knowledge of the understanding therefore also sufficiently explains the imperfect development of philosophy in general and in turn explains why a dogmatic-didactic presentation of philosophy that ought to be able to lay claim to scientificity provisionally yields nothing and cannot yield anything. Kant already loved to say: One cannot learn philosophy, but only to philosophize.[1] One can learn mechanics. One can learn acoustics, optics, chemistry, geometry, and so all branches of exact and well-developed sciences. Of course, none of these disciplines, as marvelous as the things they have been able to do are, is finalized. Everywhere there is an abundance of unsolved problems. Their horizon is infinite. It recedes with each step and remains infinite. But, within the sphere of explored and surveyable finitude, we have found our bearings on all sides. The facts are established through observation and experimentation, through various expertly developed methods precisely suited to the different fields. In terms of laws, everything is in order, brought under verified theories. And, these facts and theories are then expounded dogmatically in systematically ordered coherency. One can and must learn them. So, dogmatic expositions are never lacking in philosophy. But, their execution and meaning are different there. There is indeed no question of there being as soundly, thoroughly formed and securely founded descriptions and theories as, for instance, in physics. And, consequently, they are at most significant as rough drafts, as theoretical possibilities which one never accepts as confirmed truth and which, above all, one must not let influence one simply didactically unless one wants to engage in dogmatic theorizing in a bad, scientifically objectionable sense. Therefore, one cannot learn philosophy.

Though they may have ever so much to say to scholars, dogmatic expositions are a danger for novices. They only too easily deaden their sense of genuine scientificity. They cover up the difficulties from them, like chasms covered over with greenery. If there is stuff of facts and theories firmly grounded only to the most modest extent here, then the sole justified pedagogical task only

[1] Immanuel Kant, *Critique of Pure Reason*, II. Transcendental Doctrine of Method, Chapter 3, The Architectonic of Pure Reason. (Translator's note)

lies in training novices to philosophize. That, however, happens through the essential meaning of thousand-year-long efforts having been laid bare to them in accordance with the state of things at the time and its insights, through their having been raised up to an understanding of the problems, goals, methods–and finally, through the deeper meaning of the theoretical attempts and rough drafts in which the greatness, grandeur, and force of philosophical science lies having been made accessible to them. And, this is the way in which I wish to introduce you to logic and critique of knowledge. You ought, as it were, to sense the inmost spirit of their intentions and be most inwardly concerned by the fact that the difficulties to be overcome here are not a question of some scientific sport, but are a serious matter for all those interested in ultimate truth and philosophy.

One experiences nothing of that through the usual expositions
337 of logic accessible to beginners and intended for them. One is still far from having understood anything of logic if one has read such an exposition, and perhaps even if one has written it. If, however, you have been introduced to the inmost essence of work in logic and critique of knowledge, to the nature of their problems and methods, then you are sufficiently prepared to be able to derive benefit and excitement from reading the important expositions of the discipline.

That is therefore to say that these lectures seek to provide and will provide something entirely different from what you could learn from books, since I in general believe that lectures are not there for the purpose of replacing books, let alone that they ought to be spoken books or excerpts from books.

Appendix III (§6): <Distinction Between Explanatory Theory and Theory of Knowledge>[1]

To begin with, I would like, however, to say a few words here to keep you from developing false expectations that the word theory might have awakened in you with respect to the theory of knowledge that is our subject. You should by no means think of the dis-

[1] From the lecture course of WS 1910/11. (Editor's note)

cipline as a theoretical discipline in the sense familiar to you from your work with non-philosophical disciplines, with theoretical mathematics, theoretical physics, and so forth. It lies in the distinctive nature of philosophical knowledge as opposed to all other knowledge that its goal is in no way, and can in no way be, to provide "theories" in the usual sense of explanatory theories, and that for this reason it cannot conform to the expectations with which scientific beginners will come to it.

More precise explanation. Other sciences that do not fall under the heading of philosophy have their prescribed fields in which obvious factual cohesion reigns, and in reference to it, they everywhere set themselves the task of exploring the facts precisely in their factual contexts and in accordance with the laws prevailing in these. In a specific sense we call such sciences theoretical sciences, also nomological and nomologically explanatory sciences. The mathematical sciences and the sciences of physical and psychical nature therefore belong there. In all of them, the ideal is this: To carry out theory in the specific sense of explanatory theory, therefore, to accomplish something like what *Euclidean* geometry already has in the area of spatial configurations, or as recent physics is accomplishing in its exact mathematical disciplines as deductively explanatory theories.

Every field has its multiform conformity to laws. Everything singular, in nature every individual-concrete thing falls under what is universal, and the universality everywhere leads to conformity
to laws. Nothing is isolated in the sphere of numbers, in the spatial 338
sphere—in short, in any mathematical sphere. And in turn—in all of nature, no particular formation is isolated. Every particular formation of being is a formation conforming to laws.

And, then, the overall aim is that of reducing the vast abundance of particular formations to a smallest possible, and in any case finite, number of what are called basic laws by which those particularities can be explained as merely necessary consequences, hence, deductively. If the basic laws are given, then it is therefore a mere fact of purely deductive unfolding to derive, not only all already known particular laws in the form of ordered systems of inferences and proofs, but also a host of other, still unknown, ones. And, that is precisely what explanatory theories accomplish. In the purely mathematical disciplines, these are based upon directly obvious, absolutely valid basic laws called axioms, only in the natural sciences, however, upon hypothetical and inductively obtained propositions that refer back to experience and only have probable validity. Such are the basic laws of mechanics, energetics, and so forth.

One ordinarily distinguishes between concrete and abstract natural sciences and calls the former descriptive, the others explanatory. But, upon close inspection, both are explanatory, each in its manner. Concrete sciences like geology, mineralogy, and so on, first aim at the description of individual and typical individuations of earthly and heavenly existence. In them, however, the goal is ultimately to bring the particular under the law-concepts of what are called the abstract sciences precisely for purposes of explanation. These abstract sciences are no longer concerned with what is concrete, but exclusively with existential formations that conform to laws. What is concrete is only used as an example and only as a springboard in order to make general findings. Physics and chemistry do not deal with the earth, with specific heavenly bodies, not with kinds of minerals, but in general ways with the qualities of matter in general that conform to laws. They then, however, further try to bring the various laws that they find under a theory, to investigate in terms of primitive properties and primitive basic laws out of which the vast abundance of physical laws, as every particular physical theorem states them, can be explained deductively as necessary particularities. If abstract natural science is therefore there and developed to a great enough extent, then all aiming at explanation has attained its goal in the field of the concrete, therefore, of the concrete natural sciences, whenever observation–and perhaps experimental analysis–of what is concrete suffices for subsumption of the concrete under the relevant law-concepts of the abstract sciences. With that, the general theory becomes applicable to the concrete case, and thus the concrete becomes explainable. It becomes understandable from the abstract laws. Theoretical physics and chemistry become, for example, applicable to everything occurring in the area of mineralogy pre-
339 cisely if the minerals are exactly analyzed and described in accordance with physical-chemical methods. Abstract science is the field

of pure theory, concrete science the field of what is individual and concrete that is explained by means of this theory.

(The word explanation naturally has a twofold meaning if on one occasion one speaks of explanation with regard to what is concrete and its subsumption under the concepts of the abstract sciences, and on the other hand, says in the abstract sciences themselves within the context of the theory that through subsumption of the particular laws under the basic laws, the former receive their explanation, as for example *Kepler*'s Laws receive their explanation from the basic law of universal gravitation and the basic laws of mechanics. On the other hand, the one is connected with the other in easily understandable ways, for the drive to conceive of the concrete as not merely factual, but necessarily valid, leads us to the universal and laws, requires the development of pure sciences of laws, and within these, we in turn see ourselves led upward in the hierarchical structure of the law-governed dependencies to the basic laws upon which all full, ultimate explanation is based, therefore, for example, to the basic laws of mechanics, the basic laws of energetics.)

What has been discussed above probably suffices to make the meaning of explanatory theory and theoretically explanatory science clear to you. I now ask: Must then every science of the type described, each one, be oriented toward theoretical explanation? Perhaps, precisely philosophy, and especially the philosophical theory of knowledge, is a discipline, which by its very nature cannot at all be oriented toward theorization in the sense described (deductive theorization from basic laws). And, perhaps, just reflections of the kind we have undertaken for marking pure logic off from the empirical theory of the art of knowledge and from cognitive psychology then constitute a main thing for them. Obviously, systematic investigations, the point of which is to separate on grounds of principle clusters of problems which pertain to the Idea of the understanding, and which consequently also seek to mark off the scientific disciplines required by this Idea on grounds of principle, are also scientific in nature, actually belong together in their own discipline. And yet, what is to be accomplished here is not theoretical explanation in the sense of mathematics and mathematical physics.

These reflections ought to convey the right approach to you. They must not by any means let the feeling grow in you that has perhaps already crept over you that we would, however, spend too much of our time on complicated preparations for things that should always come first. Why not begin right away with the things themselves?

Exclusive concern with theoretically explanatory sciences conceptually leaves you with the impression that it may also be a matter of something of the kind in philosophy, therefore, that also there one would have to proceed as, for example, in mathematics, that is, begin with axioms and continue with deductions–or as in physics, begin with suitable observations and experiments, proceed to formulations of the simplest laws, and then rise to mathe-
340 matical theory. In these sciences one does not engage in much reflection. One begins and carries on. Listeners react impatiently to long-winded reflections in the context of a mathematics or physics lecture. They say to themselves: I did not come here to reflect, but to learn about things.

This is precisely the approach you really must not adopt here. Philosophy is to a great extent a sphere of most rigorous findings, but it is not a field of explanatory theories. In its case, reflections are to a great extent the thing itself, simply because what it must above all secure–the fundamental demarcations in the field of reason–is only to be secured through reflection. Any of you who have at one point had *Kant*'s critique of reason handy–and unfortunately beginners really preferably reach for this difficult book which is completely unfathomable to them–will have already noticed that this basic philosophical work is entirely different in nature from any basic work of the theoretical sciences. And yet, it is a work having the most rigorous scientific intentions. It thoroughly consists of critical reflections from which exact findings, but by no means nomologically explanatory theories ought to proceed.

Accordingly, I ask you to accept that what we have already thought through up to this point, and thus no less everything else, ought not by any means be mere preparations, but already pertains to the systematic structure of the things themselves that make up philosophical theory of knowledge. Naturally, the import of findings made in the beginning can—and especially with respect to the loftiest philosophical problems upon which we have set our hearts—be unclear to beginners. In the course of our work together this clarity will, however, automatically set in.

After these very necessary parenthetical remarks, let us return back to our subject. I had sufficiently delineated a general Idea of pure logic from the angles of psychology and psychologically founded theory of the art of knowledge. I do not by any means have a scientifically determined conception of the clusters of problems satisfying this general framework, or of the different directions in which the Idea of the understanding, or the Idea of knowledge, gives a reason for investigating and establishing different disciplines. It is necessary to remedy this shortcoming now.

Appendix IV (to §9): <Supplement to the General Characterization of the Noetic Theory of Justification. The Ideas of Judgment-Events Relevant from the Viewpoint of the Theory of Knowledge>[1]

I must still emphasize that grasping Ideas about judgment-experiences can run in very different directions. I have up until 341
now favored the Idea that underlies talk of a judgment (for instance, "All men are mortal") where this Idea is impervious to differences of the possible liveliness of our conviction in so judging, or to differently oriented differences of clarity and unclarity, and so on. So, the difference of the Evidenz, the insight, with which what is judged stands before our eyes as factual truth–as opposed to the blindness with which we merely believe without reasons to understand–is also irrelevant for the Idea of judgment under discussion. In certain areas of logic, precisely this Idea of judgment is the necessary and governing one. We can, however, also form other Ideas, for example, the Idea of insightful judgment, the Idea of lively conviction or of conviction with varying degrees of liveliness, and so forth. We can also grasp these Ideas in purity, not therefore speak, for instance, of human or animal judging and posit the existence of a world with people and judging beings. It is just as in the

[1] From the lecture course of WS 1910/11 (Editor's note).

sphere of sounds, where we can attach the pure quality c as an Idea, but can also take along a certain intensity, a certain timbre with it in the pure Idea, and so on.

If I have formed the different Ideas of judgment-events, then I can state in purely ideal generality knowledge that refers to corresponding ideal relationships. There ensues–something of which it is especially a question for normative purposes–ideal relationships between the jurisdiction of truth and the Idea of insightfulness, and in the Idea this in turn relates to what the judgment judges.

Appendix V (to §12): <The Enigma of "Intentionality">[1]

Because the miracle of consciousness is the miracle of what is called intentionality. For the philosophically naïve the most obvious thing of all is the fact that in subjective experiencing, called presenting, judging, valuating, and so forth, something can be intended that is not itself an experience, but lies beyond the experience, and that in relation to such experiences, which are called "consciousness-experiences", subjects can be certain of the objective validity of their intending. For naïve people that is completely obvious–so much so that they do not feel the slightest reason to reflect upon the fact that objects exist in their own right, states-of-affairs obtain in their own right (things change, events take place, natural laws and mathematical laws hold), and that in their experiences of the consciousness, subjects can perceive the objects existing in their own right, determine them thinkingly in valid ways, valuate them in valid ways, and likewise can subjectively know every kind of state-of-affairs, every kind of law obtaining in its own right. This obviousness (and already the most primitive per-
342 ception of a thing of all) is the enigma of all enigmas. And, the first step of the inquiry–provided it first of all only seeks to ascertain what is immanently present there–already shows that the original enigma hinges on the original fact that makes up the essential particularity of each consciousness as such, namely, that in itself each experience that we call a consciousness has a meaning. But hinging on meaning (*Bedeuten*) and the meaning (*Bedeutung*) is the enigmatic distinction of rational, or objective validity, or non-

[1] From the lecture course of WS 1910/11 (Editor's note).

validity that ought to signify "soundness"; an encounter of the consciousness with something beyond it, something transcendent: what consciousness intends as being and what is not itself consciousness actually ought to be.

This, only so that you can sense the importance of the present investigation for the loftiest epistemological interests. It is highly important for the establishment of critique of knowledge, and at the same time it carves out for us the basic concepts with whose delimitation the logical disciplines we are driving at secure their own field of work.

However, in the short digression, I have already heralded a certain extension of the concept of meaning that is indispensable and that I must now explicate more in greater detail.

Appendix VI (to Section II): <On the Content of the Theory of Forms>[1]

For 1910/11

For the conception of natural process:

1) I have not entirely brought the theory of forms to completion. In any case, everything that is found in the theory of inference from 166 on with regard to mere (to be severed from the Idea of validity) inference-forms is to be taken out and to be placed in relationship to the part of the theory of hypothetical and inferring judgments, for example, the differences between simple and compound inferences.
2) Belonging within this context is also the Idea of apophantic-analytic law, of apophantic law of inference (inference-form), of law of proof ("proof-form"), indeed even the Idea of apophantic theory-form. For, a false law of inference, a false law of proof, a false analytic law is also in itself characterized as ap<ophantisch>-analytic. Consequently, much of what is propounded within the context of the "transition" sphere also belongs in the pure theory of forms, especially 160ff.

   An analytic inference itself, proof itself, etc. is also characterized as analytic as a result of the fact that its terms are precisely

[1] Probably Fall 1910. (Editor's note)

nominal terms etc. and its forms "purely" analytic forms. Therefore, that also belongs in there.

3) Also, the difference between a posteriori propositions and a
343 priori propositions, pure concept-propositions, etc.
4) Likewise, the difference between mathematical and non-mathematical propositions. And the difference between the analytic as apophantically analytic and the mathematical of the broader sphere. Naturally, also analytic proof in the narrower sense and higher-level mathematical proof. Therefore, rather everything in the transition area, with a little bit of elimination.

In the theory of forms, it must at least be briefly said that a theory of forms of relation-judgments, or of relat<ions> formations is needed, moreover, of the theory of forms of possible number formations (taking no account of validity), namely, of the configurations of number-formations, both in the theory of cardinal numbers and in the theory of ordinal numbers. And thus everywhere.

Likewise, a theory of forms of the formations of aggregates (set-formations) is necessary. Namely, the theory of combinations and the theory of permutations belong here.

The relationship between mathematics and ontology as formal ontology is a cross I bear. After I have, however, situated the Idea of "factualness" of reality in the realm of meaning, then I really secure the Idea of what is not-factual, and that is what is ont<ological> in the formal sense.

Then, however, relation-concepts such as "whole" and "part" also occur, concepts such as "conjunction" and so on, namely, as apophantic form-concepts inasmuch as we must incorporate these concepts into the judgment-forms as well. The being-in-something indeed occurs in the existential sphere, or as existence-concept.

Appendix VII (to §20): <Independent and Dependent Ideas>[1]

We say that the Idea of a sound-concretum is independent. We imagine a sound. We take it in terms of its entire content-store, apart from its *hic et nunc*. Then, the Idea of this (no longer individual) sound is an independent Idea.

---

[1] November 26, 1910.

We take the Idea of a sound-quality, of this specific sound-quality. Thus it is dependent, just as is the Idea of specific sound-intensity. The generic Ideas "sound-quality" and "sound-intensity" are likewise dependent. How are we to ascertain and elucidate this difference?

One could try this way: the full sound is something in its own right. Namely, if we pretended (of course, an impossibility, indeed an abs<olute> one) that nothing else existed, then that would not change anything about the being of the sound. If, however, we pretended that nothing existed but a specific Idea "sound-quality" (of the singular Idea), then it is evident that this quality cannot exist either. Its being is dependent being. Indeed, one can say of it that its "*conceptus non indiget conc. alterius rei* a *quo formari debeat*", but in another sense, it certainly needs the *conc. alt. r.*, namely, another Idea. It is not an Idea in its own right, but a necessarily 344
interdependent Idea. It is a fully determined Idea only in conjunction with another Idea, and if it is regarded in its own right, as if it were a thing-in-itself. So it brings indeterminacy along with it. It is indeed determined as a singular sound-quality Idea, in its own ideal essence. But because of this "ownness", it has another relationship to otherness with itself, and this remains undetermined. Thought in this quasi-independent way, this singular sound-quality as Idea brings along with it the Idea of "any" sound-intensity and timbre in general, or the Idea of a conjoined something in general that is only determinable by the general Ideas "sound-intensity" and "timbre". We could also say that the sound c is a sound completely determined "in terms of quality". It is not a genus and does not bring any generic generality along with it. But it brings another generality along with it, not an inner one, but an external one, and yet an ideally delimited one, delimited by a generic Idea, but by an extrinsic one. This extrinsic Idea is conjoined with the Idea "sound c" or the genus "sound-quality" and the genus "sound-intensity" are allied, or conjoined, genera. They necessarily belong together with regard to a reification, or to an extension of reifications. The conjunction of genera yields a unit that is not a pure genus, but is a mixed (composite) genus. Their differences are not pure differences, but composite differences based on the coalition of the genus.

In what, then, does the dependence of the allied genera, and in further consequence, of the allied differences, consist? The genus brings with it an indeterminacy that is determinability through an extrinsic genus. The difference brings with it indeterminacy that is determinability by any of the differences of the extrinsic genus.

What does the genus brings indeterminacy with it mean? That it can only be thought of as an Idea that has a certain kind of generality? It can only legitimately be thought of thus–and in turn it can only be thus. It has, on the one hand, its generic generality and, on the other hand, its allied generality, and this implies that it is what it is only in the coalition. We can also say that what cannot be legitimately thought without one another, or what is the same, what cannot be without one another, can also not be given without one another.

But what does "given" mean there? If we take sound-reification, the genus "sound" can be given without something else having had to be given. The genus "sound-quality" can, however, only be given when a specific sound-quality is given, and specific sound-quality only when specific sound-intensity is given, but any one. If, however, I bring sound-quality to givenness for myself, then I do not nevertheless bring the specific quality c or some other specific thing to givenness for myself, and in the same way, the Idea "intensity" and the specific intensity to givenness. I do <not>, however, need to engage in all these ideations (the presentive acts), and yet all these Ideas are "co-"given. The givenness of the Idea "sound-quality" is *a priori* of such a nature that the other Ideas are to be taken out of it by shifting one's gaze. A dependent Idea cannot be
345 given without the allied Ideas being given with it, and if the latter become givens, then the former is given with them. They are "there with one another". Beholding grasp of the one necessarily contains everything making the grasping of all coalitions possible.

If, though, we have an independent Idea, then nothing is implied in it itself that points beyond it. What made it into a component of a whole is not implied in it. If it is given, then no other Idea is given with it, except those that compose it, or that are "necessarily determined" in those.

An allied Idea is not to be deduced from the allied one, or not determined in it, i.e., it is not implied in its own essence. The generic

Idea "sound-quality" has its content. Indeed, this content is precisely sound-quality, the quality purely in itself as αὐτὸ κατ' αὐτό. But it is bound, bound to another Idea. It is allied with it. It is not alongside it. It also undergoes something through this alliance. It acquires an extension through it, since it itself in itself has its "extension", its "generality" as genus. But, in it itself, in its own content, nothing is implied of the other Idea. It is not intrinsically determined in it, or even more clearly, the genus is determined in the species. In the species' own content the genus is implied as content.

We also say that the conclusion is determined in the premises. But, the conclusion is not really (*reel*) implied in the content of the premises as a part. But, the S "lies determined" in the conjoined linking of the premises, i.e., in the unit of the overall meaning embracing them, in the unit of the judgment-meaning P_1 and P_2. So, that is a part-relationship. It does not say, "I have judged S, if I have judged P_1 and P_2". But, even when I have given, or posited, the species, for instance, sound-quality, I have not posited and actually given the genus-quality in general. But I can find the genus in the species. It is implied in it. I can also find what has been determined analytically in the judgment of the premises as a whole (in the meaning unit) if I bring this unit to givenness. In contrast, I cannot find the content of the allied Idea in the content of an Idea. In contrast, with the Idea, I can in each case find the allied one–provided it is in general given.

In the perception of a whole, the parts (provided they figure in the perception) are co-given. In the case of perception of a thing a backdrop is co-given, and so forth. Only when the whole is perceived is the component of the whole for this reason not yet perceived in its own right, and so in the genuine sense, by the same token, a moment, a coloration etc. But, even if not made into an object in its own right, it is nonetheless co-apperceived in the overall apperception. The whole appears and the part appears in the overall phenomenon, on it the backdrop also. Here, we are talking about phenomena, i.e. individual-phenomenon, individual "self-presence", apart from the thinking-out, the thinking in its own right etc., likewise for Ideas.

If the Idea "sound-quality c" is self-given, if it figures as "self-present", then a sound-concretum is necessarily self-given as Idea. It is thereby with the Idea "sound-intensity".

346 If the Idea "sound-quality in general" is given, then the Idea "sound-intensity in general" is co-given. "Idea" is certainly given in the Ideas-approach, and what is co-given is also given as Idea, but I am thinking about the one Idea. The other Idea is "there". But, I must first single it out, first make it into an object.

With that, the same thing is seemingly being said as before in the earlier pages. However, the new thing I am seeking to make clear is how individual grasping, a more or less far-reaching and complex thing-phenomenon, is to be distinguished from the genuinely objectifying thinking-out. Positing-for-itself, and since various things appear, and appear there inclusively, that can be "singled out" in different objectivations, I think, the same thing holds for ideations. The Idea is given, i.e. it appears, and if an Idea is thought, another Idea can "appear" in connection with it, namely, in a self-phenomenon.

Appendix VIII (to §26): <On the Theory of Primitive Full-Nuclei>[1]

I now have to provide another important supplement to the theory of primitive presentations or primitive nuclei. It concerns the general theory of full-nuclei. –Specific features of full-nuclei, such as they, for instance, determine the difference between red and green, do not belong in the general logic of meaning. But, one can probably ascribe differences to it that are connected with the most general differences of all in the sphere of objects in general and reflect them within the nuclei.

I came to the concept of the full-nucleus by contrasting presentations such as "similar" and "similarity", "redness" and "red". Primitive presentations of different syntactical categories coincide, as it were, in terms of a content, the nucleus, to be precise. The same nucleus is formed nominatively on one occasion, not nomi-

[1] From the lecture course of WS 1910/11. (Editor's note)

natively on another, adjectivally or as relat<ional> within the context of a comprehensive predicate. But, I have already pointed out that not every nucleus can, for instance, arbitrarily change its syntactical category. It only holds that each non-nominatively formed nucleus can be nominalized. Inversely, not every nominal nucleus can be de-nominalized though. Every proper noun of an individual thing provides us with an example here. Corresponding to the word Napoleon is an individual proper-presentation that is nominal and can never be made into an adjective. We indeed say “Napoleonic era”, and so on. But, you readily see <how> this adjective does not have the nucleus in common with the nominal presentation Napoleon, as if the same nucleus had only changed its syntactical form. In the place of the simple nucleus of the proper-presentation is now a very complex one: an era, upon which the great spirit of Napoleon left its imprint, and the like. Likewise, it is clear that when the nominal presentation Napoleon stands alone as predicate, it is not of the nature of an adjectival presentation. 347
“This is Napoleon” means this human being is identical with Napoleon, is the same as Napoleon. The proper noun indeed has a different function, but noun remains noun. By proper-presentations in general, <I> understand nominal presentations that include nothing of attributions. An attributive presentation is “an A”, or “this A”, or “a green house”, or “the philosopher Socrates”, and so on. A proper-presentation is “Socrates”, “two”, “redness”. Now, what is characteristic of individual proper-presentations is that their nuclei can only have a nominal form. And if we go back to primitive individual presentations (in contrast to, for example, conjunctions “Socrates and Plato”), then apparent in this particularity of presentations is that of their nuclei. Among primitive nuclei, individual nuclei form a closed class. The general-nuclei stand in contrast to them. And this difference carries over to the sphere of complex nuclei.

If we accord the ambiguous word “concept” (or presentation-content” in the meaning-theoretical sense) a fixed meaning by means of the concept “nucleus”, then we could say that concepts break down into individual concepts and general concepts (universal concepts). The latter further break down again into concrete object-concepts–such as “house”–property concepts, and relation-concepts. Regarding the latter nuclei, that is a fundamental difference that cannot be abolished. We can also really never bring the concept "green" into the function of a relation-concept such as "similar"; and vice versa. The nominalization of non-nominal nuclei is therefore actually the only syntactical operation (namely, <the> direct one) that we can perform in the sphere of nuclei. Moreover, the nuclei in themselves are, however, fundamentally different, each one bound to specific functions. The individual proper-nuclei are pure full-nuclei and do not carry any indeterminacy at all. Where they stand in a fully determinate relationship, the property-nuclei and relation-nuclei do not carry any indeterminacy at all either. For example, "The bench is red here"; "She is standing beside that bench"; and so forth. If, however, we wish to nominalize such nuclei and form a nominal presentation out of them alone, then indeterminacies enter into the context as well, and <namely>, inseparable ones. The presentation "similarity" is actually the presentation "similarity of something to something." The presentation "redness" is actually "redness of something." Belonging to the essence of the property-nucleus is the fact that it points back to a subject, to the essence of the relation-nucleus, that it points back to two nominal components. One or two empty places are then necessary in the nominalization.

It results there however that the nominalizations of property-nuclei and relation-nuclei are actually no longer simple presentations, but already complex ones of an otherwise distinct type. If we remain in the primitive sphere, then we therefore have three irreducible kinds of primitive nuclei.

The secondary nature of those nominalizations also becomes
348 apparent in the fact that fully elucidating, or even making clearly intuitive, leads back from such nominalizations to the non-nominal forms. Nominalization is an operation that, in virtue of a context in which precisely the property functions adjectivally, just brings about a change whose meaning constantly points back to the original form. Similarly, the original function-form of the relation-nucleus is predicative where, as a component part of an adjectival predicate, it occurs in conjunction with a noun.

Appendix IX (to §30): <Plain Predicating and Actually Identifying Predicating>[1]

"This is Heinrich." "The third man there in the line is Franz." "This is the newly appointed mayor." "Schulz is the newly appointed mayor." "This is the midpoint equation of an ellipse." "This is what is called Cardano's formula." There we shall not say "one and the same." For example, in relation to the person entering and not at first identified, "This is one and the same as Heinrich." We find two predicate forms here: proper-identifying ones–a proper name figures in the predicate position–and concept-determining ones. (In addition, mixed ones, "This is the newly appointed mayor Schulz."). Here, however, attributions are present everywhere: this is the one who has recently been appointed mayor and is named Schulz–or is Schulz.

The form with "the same": <The> identifying linkage points back <to> a peculiar linkage of identification by means of which subjects that were determined in different judgments are posited <as> objectively one by means of predicates. Someone is not aware that the algorithm A which he or she is dealing with and which merely presented in a slightly altered form is the midpoint equation of an ellipse is the very same thing. As soon as that is noticed, it is said that "This is indeed one with the midpoint equation!" There, one can however also say, "That is indeed the midpoint equation."

Is it not there a matter of the meaning of "is one with = is none other than," therefore <of> cancelling an originally supposed negation, or of a belief that would be equivalent to a negation? –Or, someone speaks of Alexander and of Paris as if they were different people. Then we say, "Alexander is the same as Paris." But is that not saying Alexander is not not Paris? The victor of Jena is the same as the loser of Waterloo (they are one and the same). The victor of Jena is not the same <as the loser of Waterloo>, is someone else. A is α, the same one is β.

In any case, we must differentiate plain predicating with a nominal proper-presentation or <a> generally determining nominal

---

[1] Probably from the time of the lecture course of WS 1910/11. (Editor's note)

presentation (a conceptual one) and the actually identifying predicating, where the "concept" of identity occurs as relation-nucleus-concept. In one case, the identification occurs as a
349 bringing-to-coincide of the nominal predicate and the nominal subject in the form of the mere Is with respect to the object, where the Is is of course the Is of identification, but does not include the concept of identity. In the other case, the concept of identity occurs.

Identification as relation: What is two is made into what is one, A and B presented in "separation" as being two are identified, made into a unit. What was presented in separation shifts over to what was separate and coincides with it.

Appendix X (to §34): <The Conjunction of a Nominal Presentation with the Negation of Another Nominal Presentation>[1]

In reference to the principle that the And is automatically to have the power to have conjunctively connected nominal presentations only yield a meaning unit when they have a different objective orientation, Miss Hoffa asks how things then stand with constructions such as "Socrates and only Socrates (was the conqueror of Sophism)". That says, "Socrates and none other than Socrates was the conqueror of Sophism".

It is clear that we do not have two conjunctively connected nouns here. When a negation directed toward a noun is applied to it, then the whole is no longer a noun, namely, noun in the form of negation. I have passed over the forms resulting from the introduction of negation. I have exclusively dealt with plural judgments in which conjunctive concatenations of names occur, and that always yields a possible plural subject, it seems to me, when one conjoins names themselves. The conjunctions "Socrates and not Plato", "a man and not a woman" <are> not nominal conjunctions either, but always conjunctions of a nominal presentation and of the negation of another nominal presentation. In contrast, we would have conjoined nominal presentations if we had said, "A man and someone who is not a woman", "An a and a not-b".

[1] From the time of the lecture course of WS 1910/11. (Editor's note)

It is, however, thereby to be noted that also in the cases in which a negation is conjoined with a nominal component, a duality, a dual ray is given in the predication, but we then have a ray of positive and one of negative predication, provided indeed that the negation of the subject-nouns does not also affect the whole manner of predication. "Not Socrates was the teacher of Aristotle" really does not say that we are adding the predicate to the object called not-Socrates. Admittedly, somebody will be co-thought who was in fact Aristotle's teacher, for instance, Plato. But, that is not stated here. It is in certain ways implied, although it is questionable whether it is necessarily <implied> in this form. So, when we say, "This is not green", we really also as a rule have in mind another property of the This that is assigned to the This "in place" of the green, with which the green now conflicts. But it is not stated that this may have a 350
different property of this kind. In addition to that, we would have to say this may have a property different from green.

On the other hand, when we direct the negation toward "Socrates", then we nevertheless let the predicate, so to speak, obtain, and implied there however is "Someone probably, but not Socrates, was the teacher...." We do not, however, say it this way. But, must that then be the thought? Can we not think: Socrates, the teacher of Aristotle? Socrates–no. In any event, not Socrates was the teacher of Aristotle. Nothing then needs to be said about the fact that Aristotle had a "teacher". Perhaps, he was his own teacher, was not anyone's "student".

Appendix XI (to §40): <Old Pages on the Problems of Propositionally Simple Judgments>[1]

<a) On the Meaning of Existential Judgments. The Difference Between Act-Analytical and Meaning-Analytical Investigations>

On the other hand, it is clear that we indeed have equivalents for existential propositions here, but find their original meaning as reality statements changed. If we state existence or non-existence

[1] Probably WS 1908/09 (appended to the lecture course of WS 1910/11 as a supplement). (Editor's note)

of an A, then the "an A" indeed figures in the subject position, and not on the predicate side as in those indeterminacy-judgments (therefore, for example, "Something is an A"–there, I am saying that the "an A" obviously has a different position and function than in the proposition "An A exists"). Things are more promising when, the other way around, we form "An A is something"; "An A is not something", namely, in the sense of 'in general not'. Here, we would have a particular judgment with two terms of particularity for the apparently indefinite existential judgment. But, *nota bene*! It is not the particular judgment. We would namely have to understand it in such a way that on the subject side, the "an A" is free of any existential status, while on the predicate side, the "something" acts as a normal term with such a positing status. Of course, the "is" is not hereby equivalent to the ordinary "is identical", although a nominal term indeed functions on the predicate side. Because the identity judgment also lies on the subject side and in the same way as on the predicate side. By introducing totality-presentations, we can equivalently substitute propositions for these propositions "An A" (or "A's") "is found among things, there is among the things", "an A" (or "A's") "there is not among the things". ("Thing" can be taken in the broadest sense of object.) Naturally, these are statements modified in meaning, but they function practically in thinking as a substitutes for those primitive, particular propositions in the ways described earlier. And, as earlier, the "there is"-
351 thought attaches on to these forms, whereby then the carrying over of talk of "there is"–and related talk–to the reality-judgment becomes explicable.

It is of course doubtful whether this interpretation of existential statements is accurate (I mean now of reality statements), and whether I have not established a mere equivalence again here too, though one of value. I am inclined to the view that, understood as reality judgment, the existential judgment is a basic form that can no longer be reduced. This is not to say that reality judgments do not give grounds for any further investigations–apart from difficult investigations aiming to separate them from different judgments closely united to them through equivalence and to accord them their rightful position in the theory of meaning of judgments. But, if we <abstain> from investigations concerning the structure of the formal logic of validity (therefore, <that> concern the laws of validity, which pertain to the different judgment-forms in accordance with their relationships of validity), these are investigations into origins, phenomenological and epistemological investigations along entirely different lines that must be carefully divorced from formal logic. The question as to how the different judgment-forms are consciously realized, the question as to how the acts, as it were, look in which the judgment meanings form the What of the thinking, which different modes of acts have the same judgment as content, how especially the experiences look in which the judgment is content of an Evidenz, and in which, consequently, the judged state-of-affairs is adequately given—are all extremely important investigations, but they do not concern the theory of meaning itself. Also pertaining to them are the questions about the "origin" of concepts that as formal meaning categories correspond to the form-giving moments of the judgments, as well as <about> the concepts related to this a priori, therefore, concepts such as unit, multiplicity, generality, individuality, having properties, non-existence, and so forth. And pertaining to this also is the concept of reality, of actual being, and on the other hand also <that of> what has been presented as such. In the case of meaning analyses, one will admittedly occasionally not be able to find it good to connect up with subjective experiences and to make any use of the phenomenological distinctions relating to them, as I have especially done in the analysis of the reality judgment. But, the goal there was merely one of putting <myself> into the state of consciousness that must be realized in a lively way for the meaning of the existential judgment that is to be differentiated and, within that, the meaning of the existential predicate to come to life. In the correlation obtaining between acts and their ideal contents, one can in general conduct meaning-analytical and act-analytical investigations in parallel fashion, and one is also compelled to do that for purposes of the deepest knowledge of this correlation. But, at the same time, a logician of meaning is left with the fact that it is not a matter of an objective discovery of the differences between acts and correlations, but of those of meanings and their essential forms, and of no more than that.

352 <b) Quarrel with Assumptions of the Traditional Theory of Judgment>

Earlier investigations into what is called theory of judgment all failed because people were not yet in a position to separate the fundamentally different problems present there. That holds of the theories of judgment of thinkers as eminent as *Brentano*, *Sigwart*, *Bergmann*, and *Erdmann*. Their theories suffer from a faulty psychologism, from a mixing of psychological and meaning-theoretical investigations, the otherwise understandable deep-rooted ailment of logic since *Aristotle*. Despite this mixing already accompanying the oldest logic, up until the blossoming of psychological empiricism since the 1860s this traditional logic was primarily objectively oriented. Only afterward did people fully consciously set about establishing mere psychological analyses through judgment-experiences and establishing how people think only scientifically. In his *Psychology*, the brilliant *Brentano*, for example, aims to establish the descriptive differences of psychic experiences, and among them the intellectual experiences, that really obviously must form the basis of a genetic psychology. He divorces mere acts of presentation from acts of judgment, the latter distinguished by the descriptive trait of conviction, and this moment of conviction, as *Brentano* finds, is either approval or rejection. Judging is believing, and believing is believing that is or is not. And *Brentano* immediately believes the true sense, the meaning of the statement in which we express ourselves judgingly, established in new ways and to have established it as an existential proposition. In order verbally to express a believing, a being convinced, two different things are needed: on the one hand, a sign for what we are convinced of, and, on the other, a sign for the conviction itself, for approval or rejection. Therefore, "A is" and "A is not" are the basic forms of all judgments. "Is" must be a sign of approval, "is not" sign of rejection–A, sign for the judged What, and this must, however, be present to us in the judgment. Therefore, the A is likewise expression of a mere presentation underlying the judgment. If we verbally find so many other forms of expressions–besides existential ones–then that must be merely for empirically-grammatical reasons: Language is really not a merely logical creation suited to

logical requirements. Since, the Is and <the> Is-Not as signs of approval or rejection pertain to every judgment in the same way, the Is does not have a different meaning in the categorical statement and in the existential statement, and the concept of existence arises through reflection on the approving judgment. All that is systematically thought out. But, I do not need to say that it is definitely untenable and only very interesting as a sample of an extremely psychological <logic>. Only, we have to give *Brentano* a great deal of credit on behalf of formal logic for having directed general attention to existential judgments, completely shoved aside up until then, and having made them the center of the new judgment-theoretical investigations. Those who at some point have made clear to themselves the difference between judgment 353
as judgment-experience and judgment as meaning and who comply with the principle of scientific work never to let go of a directly beheld distinction–be it even in favor of ever so beautiful fashionable theories, as the psychological ones are–cannot initiate much with such theories. Of course, it is easy to criticize from my standpoint, from the standpoint of someone who has seen through the illusion of psychologism. Some decades ago matters stood otherwise. In those days, every, the main, philosophy–even idealistic philosophy–was subject to the psychological bias, and the differences were only in the greater or lesser clarity and consistency of the implementation. The paradox of the *Brentan*ian theory was manifest. Healthy human understanding resisted accepting it. Giving in to this, one's only choice though was to reject the *Brentan*ian theory, but then to be inconsistent and end up with vague half-truths, or to be consistently psychological and then give in to his theory, or at most to improve it along psychological lines. So, we then actually see that *Brentano*'s critics indeed otherwise know how to say many a good thing, but nothing that could neatly bring clarity to the matter in dispute.

Certainly, one can say that the essence of judging is to be defined as being convinced, and every instance of being convinced has a What about which one is convinced, a "matter". But, this What is precisely the whole judgment in the sense of logic, judgment in the second sense, in the sense of meaning. That holds for whatever content and in whatever form I judge, whether I judge

"God is spirit!" or "God exists!" Of what am I judgingly so convinced? Well: "God is spirit!" or "God exists!" And if I judge negatively "God is not a physical being", then the whole What that is stated in the words there is also that of which I am convinced. Therefore, belonging to this What, to the *Brentan*ian judgment-matter is both the Is and the Is-Not. The conviction is not stated in any words–I believe: means– only the What of the conviction is stated. And, were the conviction said, then, for instance, the word conviction would have to occur, or some other word in which the conviction is presented and is meant by means of its meaning. If I say, "I am convinced that God exists", then I am speaking about my conviction. I am making a statement about it, i.e. I am making a judgment about it. I then have a new judgment with a new meaning–instead of the judgment "God exists", the judgment "I am convinced that God exists", to which a new conviction corresponds that is expressed in this statement without any word. And so on *in infinitum*. What I am stating in my statement is not my acts of stating, but the What of my convictions, i.e. a judgment-meaning, a judgment in the logical sense, exclusively finds expression. Consequently, no reason at all, and no possibility at all, exists to identify the psychological moment of the conviction with the ideal meaning-moment of the Is or Is-Not, and in further consequence then to identify the Is of the categorical proposition and that of the existential proposition.

Similar fundamental errors run throughout the entire recent
354 theory of judgment. So it is, for example, in *Lotze* and *Bergmann*. (Lotze, *Logik*, 1874; the shortest and most concise explanation of *Bergmann's* theories is set forth in the *Grundproblemen der Logik* (1895)). *Brentano* had given in to the illusory thought that every judgment believes something and this "something" would however have to be presented, and that accordingly every judgment depicts an act of believing that is founded in a mere presentation of what is believed. Therefore, if we judge, "S is P"–*Bergmann* then says to himself–, then to begin with I have the mere presentation or, as he says, the mere predicating "S is p". And then, however, the judging is considering-true or considering-false, therefore, a "critical attitude" of declaring-valid is added to the "S is P" in the judgment "S is P!" and of declaring-not-valid in "S is not P". In other

words, according to *Bergmann*, the genuine meaning is, "That S is P is valid"; "That S is P is not valid". In a similar sense, *Lotze* had already earlier said that validity and non-validity are factual predicates that hold for the whole judgment-content. The latter is expressed in its own right without deciding in questions, and we would hence, according to *Lotze*, actually have to distinguish among three judgment-qualities: questioning, affirming, and negating.

Those are practically incomprehensible theories. If we judge, "S is P", then we can certainly also further judge, "it is valid that S is P", and again further, "It is valid that it is valid, that S is p", and so forth. It is, however, evident for this reason that the judgment state-of-affairs, i.e. the judgment in the logical sense, constantly changes. And, if we take as a principle that it belongs to the meaning of every categorical statement to be predication about validity, then that would also have to hold of this predication about validity, and infinite regress would, however, be inevitable and would be *widersinnig*.

So, here again, the source of this folly is the confusing of the conviction-quality that one grasps as an experience of being affirmatively or negatively convinced and what belongs to the statement's meaning-content. Generally, one calls judging a considering-valid or -as-true, or considering-false. In a certain sense naturally rightly so. And then, one seeks an "is valid" and <an> "is invalid" in the meaning-content of the statement and falls into falsely attributing the essentially new "that S is P is valid" to the logical judgment itself, to the "S is P". (I am passing over other errors that are made here and that are connected with the false interpretation of considering-valid.) –The theories for the different specific judgment-forms, for example, <for> the forms of universal and <of> particular judgments, also proceed in similar ways. *Bergmann*, for example, thinks that declaring a predication to be valid can either be related to the whole extension of the subject-presentation, or to a mere part of this extension. It is accordingly universal or particular. The meaning of the "Some or all S are P" is accordingly: The predication "the S that are P" holds for all S or holds for some S.

It will be easy for you to offer criticism here on the basis of my
earlier analyses. However, you see over and over, since with the
wrong approach with which the investigation is undertaken, com-
355 plex and reflexive forms of judgment are given for the simple basic
forms, the simple forms themselves are lost, and thereby the mean-
ing of the interpreted judgments changes everywhere, and perhaps
not even the equivalency is still maintained. However, enough for
criticism that ultimately would over and over lead back to the
same main points.

<c) Comprehensive Overview of the Main Differences in the Sphere of Propositionally Simple Judgments>

With our analysis of existential propositions and of the meaning differences established in relation to this which determine the level of being of the presentations underlying the judging and of the judgments themselves, the main differences in the sphere of the propositionally simple judgments are established, namely, as I stress, the differences that are determined by the pure forms in the strictest sense by aiming at the basic differences among primitive presentational nuclei. The forms naturally become complex with the complexity of the presentation-components that occur in them. So, for example, the form "An A really is, an A exists" owing to which the A itself is a complex presentation that for its part can then in turn include different components that partly do, and partly do not, include reality-positing, for example, "An inhabited country on the North Pole does not exist." A specific problem that has not yet been sufficiently tackled up to this point would go into more closely studying the meaning modifications generated in the series of forms I established through the changes in the level of being, and thereby successively separating off new and partly characteristic forms. Thus, I had, for example, originally only established the one form, "An A is something." However, it now breaks down into two forms having different meanings. In the one case, "an A" has a positing-quality, in the other not. And, in this sense, the individual forms would be in general gone through and it <would be> to be asked to what extent changes in the level of being are to be undertaken in them. The main considerations that have emerged as generating basic forms were: 1) the difference between singular and particular predication; 2) the difference between universality and non-universality. This was related to the fact that the predications break down into ones in which no indefinites occur, and those in which indefinites occur, and in the latter case, each of the indefinites could: either be of the nature of the *quoddam*, a certain something, something thought indeterminately, but a certain something; or of the nature of what is in general, that of generality. The fundamental distinction between universal and particular generality then concerns the latter. Judgments thus emerged having one or many universality or particularity places, and potentially both at the same time. For this reason, the distinctions go further that are contingent upon the standpoint of level of being–i.e., upon the differences between existential positing and ass<umptive> positing on the side of the presentations founding the judgment. And still occurring under this is yet a primitive form of its own, that of existential judgment as <judgment> predicating reality. 356

Intertwined with all these differences is the difference fleetingly touched upon in first position in our presentation and always set in the forefront by logicians, that of what is called judgment-quality, i.e. the difference that is generated through the operation of negation. It presented itself to us in a form associated with the basic form "S is P", namely, as the difference between "S is P and S is not P". Before I say anything more about negation, I still emphasize that all primitive judgment-forms that we found were of the categorical type, i.e., a subject and a predicate is to be distinguished (or on the side of meanings, a subject-presentation and a <predicate presentation>) in every primitive judgment, and the predicate is either affirmatively attributed to, or negatively denied, it. The *Brentan*ian theory of subjectless and predicateless judgments, namely, in the real sense of one-place judgments–which he interprets the existential judgment as being–was not tenable. In these judgments, existence is stated of the relevant, but merely assumptive, subject concerned. An outcome of our analyses is therefore so far: Every primitive propositionally simple judgment is categorical.

Appendix XII (to §40a): <Logically Incomplete Meanings>[1]

Logically incomplete meanings or meanings containing "gaps", namely, ones that are "logically", i.e. as thought-meanings, incomplete. The incompletion is extra-essential when, for example, with a "this", reference is made back to a foregoing complete meaning in the same thought context, for example "S is p–from this it follows".

Some terms of the incompletion. Every "this" that receives determinacy of meaning through an alogical presentation. Likewise: "a certain something". How do matters stand with empirical meanings, with individual proper names of things, etc.? An individual thing can be thought as existing in a specific period of time or at a specific instant of its duration, namely, then, as beginning, as stopping (starting- and end-point of its duration), or at a midpoint, in-between-point of its duration. It can, however, also be thought as lasting eternally and at a specific point of everlasting time, in which case no starting and stopping point is signalized. A definite thing, something thought by its own presentation, can then also be thought indefinitely with respect to its time, as simply existing–its duration and its whole temporal determination then remaining undetermined. It can also be thought as "now", i.e. the duration is not fully determined by thinking, but indicated deictically by the "now", otherwise indefinitely with respect to its extension, starting- and end-point. It can also be indicated as starting "now" or stopping now, as starting "earlier", etc., similarly with respect to determining place, spatial extension, etc. Finally, its space-filling and causal… qualities <are to be named>. When we
357 have a presentation of an individual thing, it must be presented as something lasting in time, as having spatial extension, qualities. However, the presentation cannot at all be fully determined in every respect. What then does the proper name express?

Many years ago, I already said that a proper name is actually something similar to a deictic meaning. A proper name does not conceptualize. It contains nothing of a conceptual determination. It names what is indefinitely identical in possible varied presenta-

[1] December 1910

tions, even thought-presentations of the same object. Missing is only the form of the reference. Just as the This, the PN lacks any thought-determination of content, i.e. any moment of the meaning that determines the nominally presented object in accordance with what it is. This identical thing is not presentable and not thinkable without determining and determinability. It therefore contains the presentation of what is identical in all the indeterminacies making up its necessary determinabilities. Proper names having a grammatical form, a type, letting it be seen that it is a matter of a person, a city, a river, are naturally not pure proper names, but are tantamount to saying "the person Fr", "the city", etc.

We must now distinguish between apophantic (analytic) indefinites (empty places) and indefinites in the apophantic stuff (synthetic indefinites, synthetic empty places). The former pertain to the analytic form, to the apophantic form, to the form of predication as such, the latter to the form of apophantic matter. Can one perhaps also say, to the form of the alogical presentations (presentation-meanings) underlying the *apophansis*? No. Because, for instance, the intuitive presentation that underlies proper names and is repeatedly different always has different moments of indeterminacy. Proper names, however, always have the same absolute indeterminacy of meaning, namely, factual speaking, <with> respect to what is determining the object. Apophantically, a PN is something full, namely, nothing containing a form and empty place. On the other hand, however, the meaning PN is indefinite inasmuch as it fixes what is identical without any determination.

In the theory of validity it is said: Just as a "this", a "certain thing", also a "something", is not called the same thing everywhere it reoccurs, neither is the PN. "This" does not validly exclude non-contradictory predicates, and PN just as little–namely owing to indeterminacy. Socrates is sick during time t. He is not sick during time t′. During his existence he is sometimes sick and sometimes not sick. The predicate "sick" is itself indefinite. It requires determination of the point in time to which it is to refer.

Logically incomplete meanings, namely, those which as logical meanings, as thought-meanings, are incomplete. We can then again distinguish between the incompletion that lies in the nucleus, and the incompletion that is to a certain degree incompletion of the

expression, inasmuch as not everything thought is expressed, for example, “The paper!” Namely, horrified, I see that it is beginning to burn, and I think, “The paper is burning”, but that is otherwise
358 not expressed, and the meaning “is burning”–and consequently the entire sentence–is not expressed. The thought-meaning is, however, not at all realized either. The unexpressed remainder is intuitively presentable–“The child!” (is falling), etc. Deciding is difficult in the case of “the” and <in the case of> this (“The paper is white”, namely, that there, in front of me). There the “the”, however, probably represents a proper name. It names the object without any mediation by general concepts, by determinations of any kind, and the object is not presented as having such. It is “directly” presented (perceived, remembered, clearly or unclearly). But, the form of the reference that is lacking in the case of the proper name is still there. The incompletion then belongs in the nucleus. Or, more precisely, the This is incomplete here, insofar precisely as it does not make the object itself present in thought by means of a proper-meaning, as is however thought “directly”. I think a number 4 and I say there, “This!” Then, the incompletion consists in the fact that I use the form of reference and do not call what was referred to as four by its proper name. The latter is fully complete. In contrast, were I to have had a proper name for “this paper”, then that would be in a different sense, namely, in the sense in which each direct presentation of an object is incomplete, uncompleted. Consequently, I must surely say that the isolated This is in each case a place of incompletion. We may call these places extra-essential places of incompletion when the missing meaning is either a proper-meaning, or another one with thought-meaning standing before me. So it is, for example, when the This functions in the “the same one” that is again identifyingly still placing something before me, and before me in thought, without repeating it in the explicit expression. It is likewise extra-essential incompletion when I say “and so forth” and thereby only indirectly refer to a completing, for instance, of a complex predicate, standing before me collectively. In contrast, I call essential incompletions those in which a part of what was thought is not grasped in thought and expressed. For example, if we say, “Here, today it is raining”, the “here” functions similarly to a proper name of a person and at the

same time a deixis is present. That it is a matter of a time is clear. By the same token, the here is precisely the here that is prevailing. A direct presentation presents it. If, however, I say, "it is raining," then I mean "here and now," but I am not thinking that and am not expressing it.

Empty places are in a certain sense gaps in meaning, but they are meant gaps. It is part of the meaning of the "this" to have a gap in its meaning, of the meaning of "a certain thing," of the meaning of every "something." But this having gaps is a property of certain meaning elements and potentially of the whole meanings (like the functions) containing them. It is different when the gap is actually filled up, filled, but not filled in thought. It makes a big difference whether a thought with indefinites is performed in which something indefinite–something with gaps is thought (here the filling up is more precise determination and leads beyond the given meaning to new meanings)–and whether, on the other hand, an incomplete thought is performed, namely, whether a unit of belief (*Meinung*) with a meaning is performed, while only a part of this 359
belief is grasped in thought, only a part is present as thought-meaning, while the other part is indeed believed (*gemeint*) (in a broader sense, therefore, meant), but is not logically believed in the form of a logical meaning. So it is, when we say, "There!," namely, there–lies the (long-sought) knife. We must thereby divorce an expression's having gaps in the external sense and an expression's and a meaning's having gaps in the internal sense, when, namely, the conceptual grasping is missing. Of course, one does not completely clearly transition into the other.

Appendix XIII (to §40 b): <Propositional Function and Proposition>[1]

The fact that A is b holds: a) occasionally in "one case," in several cases; a′) also holds in a certain case, holds in several certain cases, in a certain plurality of cases; and b) holds generally.

We have to distinguish 1) the "judgment-form," "An A is b," "A human is mortal"–the function, <the> propositional function. 2)

[1] Probably the end of 1910. (Editor's note)

The judgment and the propositional thought, "A human is mortal". The latter is necessarily either a universal or a particular <judgment> or a *quidam* judgment. The former is in general not a judgment, but precisely a function. And I can state of the function that it is "filled" in certain cases, either particularly <or> universally. The judgment "function" does not hold in the sense a judgment holds, but it holds formed, in the form of universality, etc. It has modal forms, but through different modal forms is identical throughout. The function can be a "mathematical" function or a non-mathematical one. In any case, the mathematical function must not contain any others as "mathematical" terms.

We would first have to distinguish between two levels of functions. Namely, if we consider the judgment-configurations of the pure theory of meaning-forms, then they are functions with A, b, etc. as arguments and, if we take forms of the kind "An A is b", then that can be thought as a configuration of judgment-functions (not of judgements), under which "A human is mortal" etc. falls, that are themselves already functions with the empty place "a, an". As a general configuration, it is itself a function with two different arguments, "an" A, b. The pure configurations of apophantic meanings and the configurations of meaning-functions, which we—as is to be seen—must differentiate, are both mathematical. Mathematical too are filled configurations in which mathematical concepts such as "theorem", "number", "nominal meaning", etc. occur. Consequently, a function like "A human is mortal" is not mathematical. A function can be *a priori* or *a posteriori*. Namely, it can contain pure and empirical concepts. "Any red is a color" is an a priori function. Every mathematical function is a priori. There are, however, also other a priori functions.

360 Appendix XIV (to §40 f): <The Nominalizing Conversion of Matter. "Being" as Predicate>[1]

Instead of making judgments about meanings and of stating, as *Bolzano* says, objectivity and objectlessness of them, we can then also make statements about the objects meant, but purely as

[1] Perhaps 1911. (Editor's note)

objects meant, i.e., we make judgments there about what is supposed as such, as something identical that can be the same both in the given meanings and in several different meanings. So, we make a judgment about the "object" regular decahedron, i.e. about what is supposed as being identical as such in the meaning "decahedron". We posit it as something meant, something supposed, and then state being or not-being of it. We can then conclusively express what is there in this way: If we make a judgment, then what has been judgingly supposed as such is stated, a certain state-of-affairs in its mode of being (propositional matter with its noematic quality). But, being is not thereby emphasized, let alone predicated ("Gold is yellow!", "This table is red!"). Something similar holds for every judgment-component that as such makes a particular assertion. It is unity of quality and matter, and the quality is thereby the being-quality. Every judgment–and every nominal-thetic judgment-component–can then undergo a specific shift in which a new judgment arises, in which what was previously quality, being-quality of the judgment or judgment-component, is turned into a predicate "being". Subject is then a nominalizing conversion of the matter coming to be emphasized, namely, it becomes the state-of-affairs in quotation marks that is meant in a certain meaning-content, posited as subject, and then imparts being upon it as a predicate. Every judgment–and every nominal judgment-component–admits of this conversion. Therefore, <an> existential judgment is actually embedded in every one–*implicitly*, however, not every judgment is an existential judgment.

Appendix XV (to §45 b): <Analytic and Synthetic Truths. Concept-Truths and Truths of Matters-of-Fact>[1]

We therefore ask: What kind of an essential line of demarcation within pure laws is this by which analytic laws set themselves apart from all other laws? There, a distinction with respect to the ultimate terms, the distinction between material (in the broadest sense) and analytic-categorial then gives us a clear-cut demarca-

[1] From the lecture course of WS 1908/09 (inserted into the lecture course of WS 1910/11). (Editor's note)

tion. In accordance with it, meanings, propositions, judgments divide into ones exclusively containing formal-categorial terms and ones that also contain material terms. So, purely conceptual presentations and propositions also divide up in the same way,
361 therefore, also pure concept-propositions and laws.

You will immediately understand what it is a matter of here if we consider some examples. If <we> take concepts, if we take some terms, and preferably primitive terms, equipped with full-nuclei, then an essential difference however exists between: the <concepts> before us in the examples "green", "sound", "lion", –more intensively also "psychical presentation lion", "presentation sound"–; and, on the other hand, 1) concepts like "object", "quality", "property", "genus", "multiplicity", "cardinal number", "magnitude", "state-of-affairs", "relation", "existence", and so on–concepts I include under the heading "formal-object categories", a) concepts under which every object of every imaginable sphere of objects can in principle be brought, formal-ontological ones, and b) formal-logical ones.

2) Likewise, though, also concepts such as "presentation", "proposition", "truth", "subject-presentation", "nominal presentation", and so forth–in short the pure meaning categories.

This logically, and for this reason also epistemologically, fundamental division of terms (to which the fundamentally essentially different species corresponds, since concepts of one and another species arise), therefore creates a difference of truths. Examples on the one side are: "Red is a species of color", "A sound of quality c is deeper than a sound of quality d", "If any sound a is qualitatively lower than sound b, then b is qualitatively higher than sound a", "If a is higher than b, b higher than c, then a is higher than c", and so forth. Those are plainly obvious truths of the law-type. Outwardly, the proposition "If sound a is higher than b, then it is not true that a is not higher than b" would then belong here. In any case, material concepts, factual ones, occur everywhere here, for example, "sound", "color", "red", and so forth. On the other hand, let us take examples such as, "Every object has some property", "Any two objects stand in some relationships", or "If it holds for any pair of terms S and α of the proposition 'S is α', then the proposition 'S is not α' does not hold", or expressed equivalently, "If any object S has any property α, then it is not true that the same object does not have the property α". So, all truths of the theory of forms of meanings belong here, for example, the fact that any coherent meaning remains a coherent meaning if a nominal presentation is in turn substituted for every nominal presentation occurring in it, and so in general for substitutions of any kind in which only the grammatical category of the presentations in question is preserved.

We can also characterize the distinction that emerges in this way: the class set forth in the second position is the class of truths falling within the bounds of pure logic. The one clear-cut concept of pure logic is thereby defined. It embraces meaning-logic and, in further consequence, the formal ontology corresponding to both groups of categories. All remaining truths overstep these bounds precisely because they contain material concepts. Precisely because of its "formal", i.e. purely categorial nature, formal logic need not take notice of them. Instead of pure logic, we can also say analytics or science of what is analytically knowable in general, science that establishes and systematically substantiates analytic 362
laws. Material truths, and especially material law-truths, then break down again into two groups: 1) extra-essential material truths, and 2) essential material truths. (With this distinction, I am of course appealing to the highest logical principle, the principle of substitution.) Every analytic-formal law changes into a valid truth howsoever we substitute a material term of universality of the corresponding category for any formal term of universality. For example, corresponding to the formal law, "If something is a, then it is not not-a" are real propositions such as, for example, "If something is red, then it is not not-red". That is an extra-essential material truth. It is not essentially bound to the concept "red" and in general to a factual genus. It holds in relationship to the property "red", because it holds for arbitrary properties of arbitrary objects. It holds, we can also say, by analytic necessity precisely as an individual case of a formal analytic law. Extra-essential material truths of this kind fit into a broader sphere of analytic necessities. So far as <we> bring the concepts "analytic" and "formal" to coincide and consequently understand analytic laws among categorial laws, we also then have <to> understand individuations and singularizations (individualizations) of analytic laws among analytic necessities. The individualizations differ from the particularizations due

to the fact that individual presentations, and hence existence-positings, are used in substitution. For example, if we compose, "If Socrates is human, then he is not not a human". Extra-essential material truths or analytic necessities are, as I can also say, characterized by the fact that they change into categorial law-truths when we replace their material terms with indefinite terms of the corresponding categories.[1] Every formal-logically valid inference as such is characterized in this manner, for example, "If all humans are mortal and Socrates is a human, then Socrates is mortal". This truth is extra-essentially material, i.e. it is not essentially bound to the material concepts figuring there. It is merely a particular case of the law that is a formal-logical law, "If All A are α, and x is an A, then x is also α". Every such inference is therefore an analytic necessity.

If we defined analytic laws as purely categorial laws, then to hold fast to the *Kant*ian analytic-synthetic terminological dichotomy, it seems we would have to call the remaining laws synthetic. On both sides, it is a matter of pure concept-propositions, therefore of a priori ones. We would therefore have analytic and synthetic a priori propositions, or rather truths. But the latter divide into analytic and non-analytic necessities, and since it is the intention of talk of synthetic laws to eliminate the analytic altogether, also as analytic necessity, then the classification is not satisfactory. Performs this elimination, one then arrives at the following distinction: Under the concept of analytic a priori propositions belong 363
both the purely categorial concept-truths (purely analytic propositions) and also those concept-truths that are the material individuations of such truths. On the other hand, synthetic a priori truths were those material concept-truths that are not be grasped through mere subsumption as individuations of purely analytic truths, which therefore when they are in general generalizable eventually lead to laws still having material content and then resist any generalization. In other words, counted among analytic a priori truths are both the purely analytic truths and also extra-essential material concept-truths. The essential synthetic truths were

[1] The a priori analytic then represents a narrower concept of the analytical, the higher mathesis, a broader one.

those a priori truths that are not graspable as mere purely conceptual individuations of purely categorial laws. (*Kant* was aiming at precisely such a finding with the original coining of the concept "synthetic".) Important too is the parallel transaction on the side of existential propositions–and more precisely of existential truths–of a posteriori truths. Through the existential positings, they contain a, so to speak, non-rational element. In this sense, they are "contingent". By contingent truths, or a posteriori truths in the concise sense, I understand, however, those that are not only contingent insofar as they imply existential positings, but also insofar as, apart from that, they contain nothing of essence-necessities, i.e. do not present themselves as singular individuations of any a priori truths. Belonging in there are all specifically natural scientific statements, all singular statements about experimental findings, all perception- and memory-judgments etc. But, also all natural laws. All remaining existential propositions, which therefore have in common the circumstance that they are singular individuations of a priori truths, break down into analytically necessary existential truths (namely, singular individuations of analytically a priori truths) and synthetically necessary existential truths.

Analytically necessary existential truths were all applications of purely logical laws to singular individual cases, for instance, in natural science. Synthetically necessary laws, however, were all applications, for example, of geometry or similar disciplines–I cite in addition pure kinematics–to empirical nature.

People commonly say that all analytic propositions are a priori and thereby also include analytic existential truths among analytic propositions. That is justified to the extent that in the case of an empirical statement, one does not want to make statements about the existence of the objects named in the existential terms, but rather about what is attributable to them, or what would be attributable to them under such and such conditions, or was attributable, and the like. Precisely that, however, is the side of the necessity, that which derives from the corresponding analytic law. On the other hand, anyone who attaches value to perspicuity in thinking and in distinctions cannot but adhere to the divisions I have made.

364 Appendix XVI (to §50): <Indirectness Pertains to the Essence of Theorems, Directness of Insight to the Essence of Principles>[1]

I am now broadening my reflections. I am leaving behind the reflection on the essence of inferences and the description of the general characteristics belonging to them. Before I turn, which is my next goal, to the apophantic theory of law, which also embraces the laws of apophantic inferences in systematic theory, let us cast yet another look at the essential characteristic of the domain of laws and ponder the question suggested to us by the distinction between direct and indirect inferences as to whether something does not also correspond to it on the side of laws of inference and of laws in general, and hence also <ponder> whether, as in the case of the hypothetical necessities, we would not also have to speak of inherently direct and indirect, inherently prior and ulterior ones in the case of the remaining necessities. These questions will surely have to be answered affirmatively.

I bring to mind the difference between inherently direct and indirect concept-truths. If we imagine a singular red, then we see with Evidenz that the truth, this red is a color, is a necessity, namely, a direct one. And likewise, <we see> that the pure concept-truth "red is a color" is a direct truth. I shall likewise speak of direct and pure concept-truths in the case of propositions like $2 + 1 = 1 + 2$ and, more generally, $a + 1 = 1 + a$—all this in contrast to necessities, or to pure concept-truths that are remote from such "axioms" in a system of deductive theory.

Truths like those of the theorems of *Euclidean* geometry, or the theorems in a system of arithmetic, are only intelligible to us, subjectively speaking, through proofs, therefore through a chain of inferences. They thereby acquire the character of deductive necessity, but, mere deductive necessity–i.e. the necessity that every proposition deduced in any manner as such has–does not. Rather, it is part of the essence of genuine theorems that they can only be indirectly understood, and part of the essence of principles, of axioms in the concise sense that they can be directly understood. That

[1] Probably WS 1908/09 (inserted into the lecture course of WS 1910/11). (Editor's note)

is, though, not something subjective and contingent, but something belonging to the <essence> of the truths concerned. Aside from all actual understanding or not-understanding, certain truths–law-truths here–are inherently direct and others inherently indirect. Being a principle and being a theorem therefore involve something essential, something permitting one to speak of an ordering of truths in themselves, at least within closed spheres of truth, namely, a priori ones.

Accordingly, the concepts argument and conclusion, and correlatively argument-truth and conclusion-truth acquire a concise 365
meaning entirely of their own, that of a not merely relative, but absolute difference. The concept of inherently dependent truth being necessary conclusion is indeed to be divorced from that of generally deduced, generally inferred<truth>.

A concise concept of deductive, of inferential argumentation, and in further consequence, a more rigorous concept of deductive theory is also thereby determined. We could use the *Aristotelian* word *apodeixis* here. An apodictic inference, an apodictic proof, proceeds from genuine apodictic arguments to apodictic conclusions. An arbitrary inference and proof (an arbitrary inference, for example, through mere subsumption of a particular proposition under the universal one known beforehand) is, however, not at all of this nature. Of course, in a system of rigorous concept-truths, as in mathematics, one can infer back from truths standing in a later position in the system to ones by their very nature standing closer to basic truths precisely when one formulates these later truths purely in their own right as premises and infers in accordance with formal logical laws, with the addition of some axioms.

What is characteristic of an apodictic theory consists in the fact that a group of genuine principles, of axioms that have essential unity through a relationship to a higher generic concept–such as cardinal number, set, spatial magnitude, and so on–provides the foundation for a systematic apodictic derivation of theorems. In this sense, every purely mathematical discipline is of the nature of an apodictic theory.

With such theories, I have, therefore, also in these questions of apodicticity, opposed the psychologism and relativism that believes it can trace all theorization and all ordering of truths in the form of systematic theories back to merely psychological or biological principles. People talk here of a thought economy, which in turn connects with the complex of facts of adaptation showing that humans have their adaptation to nature intellectually, just as in every other biological respect.

I need not embark here on criticism of the commonness of such theories. In any case, according to them, the principles in deductive theories–considered in themselves and *a priori*–would have no intrinsic merit. It would just merely be a matter of accidental human development that precisely propositions having such content, and not some far-flung theorems in the system instead, make sense to us directly. If, however, one delves into the meaning-content of argument-truths and of theorem-truths, one soon realizes that the distinction between indirectness and directness is purely intrinsically and essentially inherent in them themselves. Theorems in themselves refer back to indirect relationships and in themselves require going back to the same analyzing subsumption under principles and associated primitive inferences. The theorem is valid in itself because the principles hold, and this "because" has a status different from that of any random "because". If I already know, no matter how, for instance, because my teacher has told me so, that of A and B, only one is valid, and if incidentally, I know in
366 some other way that A is not valid, then I naturally then know that B is valid. But, although this propositional inference assures me of the validity of B, it does not substantiate the truth of B in the apodictic sense. I naturally also say here, "B is valid, because of A and B only one is valid etc."; but the "because" does give any genuine essence-argument, any apodictic argument here. "Deduced" is then no longer the same as "substantiated".

I must mention one more thing. If we consider the properties and relations pertaining *a priori*, therefore, purely conceptually, to the objects of a general concept G, therefore, encounter a series of direct concept-truths destined to act as axioms for a deductive discipline, then it is not excluded that, despite their directness, these direct axioms are not then deductively independent from one another. In other words, it can be there that in the series of direct axioms A_1, A_2,.. A_n, one or another of them can turn out to be a mere deductive consequence of the remaining ones, without however one's being able to say for that reason that it is apodictically substantiated it them. A technically perfect deductive theory is then constructed in such a way that one takes a system of principles deductively independent from one another as axioms as a basis and from there apodictically derives, therefore potentially also proves, some propositions that are in themselves of the nature of principles as good as the axioms given priority. Similarly, for further progress, one will not be able to say either that this must be unambiguous. Groups of other equivalent propositions in the stage of indirectness will be possible whose ap<odictic> derivation offers various possibilities in such a way that if one of them has been proven from the axioms, the others ensue as consequences, but also the other way around. So the system as a whole is on the whole apodictic while it cannot be said that every other course is not apodictic. Examples of this are provided by our mathematical sciences which, as is well known, fail to produce a construction that one could say is the absolutely necessary one. However, all that would require yet closer exploration.

Appendix XVII (Original Version of §§60–62): <The Problems That Make a Systematic Investigation of Entirely Different Regions of Possible Being and Possible Science Necessary>

Formal ontology–or, as we for good reasons also might call, the *mathesis universalis*–is the ultimate unfolding, in a different respect, however, also purifying, of the Idea of traditional formal logic. It is logic in a finalized and no longer extendable sense. It <is> analytics inasmuch as it brings a prominent tendency of *Aristotelian* analytics to pure development. It is absolutely complete in itself, entirely divorced from all other sciences in a clear-cut way and stands in the same relationship to them all, in a relationship that we could call science-theoretical. Its immense meaning for all of human knowledge in general is based precisely 367
on this.

If we again take up[1] the Idea of the theory of science guiding us from the beginning, then, as the formal theory of meaning and ontology described, logic is the first manifestation of this Idea.

[1] This paragraph corresponds to the first paragraph of Chap. 12 above. (Editor's note)

Knowing (*Wissen*) in the sense of science (*Wissenschaft*) is thinking or thought-state-of-mind that refers back to thinking. Corresponding to thinking is something thought, and so corresponding to every science is a system of judgments in my meaning-theoretical sense, a system of postulated truths and probabilities, and these refer to objects and states-of-affairs. The science of meanings in general, of truths, possibilities, probabilities in general, of objects in general in absolutely pure, formal universality, yields a system of absolute truths to which every science is obviously bound, and which are prior in terms of validity to every science in general—as already to every judgment in general.

Naturally, if it is itself thereby prior in this sense, it refers back to itself. This reflexivity characterizes it precisely as genuine theory of science and obviously does not involve any vicious circle. One would initially like to find a problem here. Formal logic sets forth the truths that are conditions of the validity of every particular truth in general. But, it is itself nevertheless a science. It establishes particular truths, deduces particular truths from particular truths, constructs theories, and so forth. Does it not thereby presuppose the truths that it very first establishes? Now, obviously, it would cancel itself out if it presupposed those truths as premises. But, it is precisely—and no less obviously—not a question of that. Formal logic begins with the fact that it establishes axioms as direct givens having absolute validity, and if it draws consequences and deduces, for example, derived laws of inference, precisely every step itself falls directly under axioms to be fixed. Formal logic's reflexivity does not therefore harbor within it any puzzle, although it is otherwise a striking characteristic of formal logic as compared to other factual sciences.

I[1] now ask whether we cannot broaden the Idea of the theory of science, whether we cannot characterize new fields of investigation, and perhaps whole disciplines that–even if in a different respect–are prior to all particular sciences and consequently do not presuppose any of them in terms of their content.

---

[1]The following paragraphs correspond to the text of §§60–62 of the lecture above that starts here and ends with the words "in the usual sense of physical nature". (Editor's and translator's note)

I add an explanatory remark here. We can initially call generally science-theoretical the exploration of everything concerning the Idea, the essence of a science as such in unconditional generality. That initially refers to knowledge that concerns what is constitutive of the Idea of a science, i.e. concerns essence-moments that necessarily construct the Idea of science. The Idea of meaning, 368
more precisely of judgment, is such a constructive moment, and so the whole of apophantics is theory of science in a magnificent sense—and likewise the whole of formal ontology, since the Idea of object in general, along with all related Ideas, such as property, relation, and so on, is constitutive of the Idea of science. Science is indeed science of any objectivity that is called its domain.

We can then, however, surely stretch the concept of science-theoretical to such an extent that <it> includes everything that, according to its essence, no preexisting science presupposes or that belongs within the sphere of any particular science and is otherwise of such a nature that every particular science could in principle make free use of it, without for that reason forfeiting its own domain, whether it actually finds reason for that then or not. In this sense, the formal theory of theories is already science-theoretical. It is not part of the constitutive essence of every science to be theoretical science in the precise sense I have described, therefore, to derive theoretical deductive systems out of basic laws as a priori or empirically obtained foundations and, thereby, to have to engage in theoretical explanation. That does not however change anything about the fact that the theory of theories is science-theoretical, because the Idea of deduction, of proof, of theory, or the Idea of manifold is by essence prior to any science, and it is thereby a matter of higher essence-configurations that further form constitutive concepts of science in pure universality.

We shall from now on hold fast to this most universal concept of "science-theoretical" as theory of science and reflect together about what may emerge *a priori* as science-theoretical fields of investigation. It is clear that from now on, we are changing outlook, that we must forgo abstraction from the nature of the nuclei. Analytics is the scientific field of pure form. We now have to take into account what is called the matter of knowledge.

Let us take up the Idea of mathematics. According to my analyses, there was a difference between analytic-formal and non-analytic mathematics. Everything that pertains to analytic categories in mathematical disciplines figures then on one side, therefore, the arithmetic of cardinal numbers, the arithmetic of ordinal numbers, and so on, as well as the formal theory of manifolds, not, however, geometry as theory of space, pure chronology as theory of time, and so forth.

Are these disciplines then not science-theoretical? They are obviously not so in the most universal sense, because not every science deals with spatiotemporal objectivities. That is certainly not true, for example, of the *mathesis universalis*, or of analytics itself with all its disciplines. On the other hand, one will nevertheless be able to point to the fact that, within the vast complexes of
369 sciences of nature referred to as non-analytic (therefore synthetic), the mathematical disciplines play a role similar to the analytic mathematical disciplines, therefore, that within the bounds of nature and natural science, they are science-theoretical in nature.

If we survey the host of natural sciences, then each one of them theoretically explores some domain of the totality of nature. In theory, though, each one of them can make just as free use of all the synthetic-mathematical disciplines as of the formal-mathematical ones. The reason is obvious. Just as the objectivities falling under the heading of "nature" are precisely objectivities–subjects of properties, relation terms of relations, components of conjunctions, parts of wholes–, in short, have an analytic-ontological form, on account of which they come under analytics, so it as nature has a form. *Kant* spoke of *natura formaliter spectata*. Everything worthy of the name thing has a bodily figure that takes its place in space, that has the formal properties of a three-dimensional *Euclidean* manifold, and consequently requires the familiar geometry as unfolding of its essence. Furthermore, any concrete being exists in time and moves in space, and is consequently subject–something which concerns the temporal-figures and movement-figures–to chronology and the theory of motion (kinematics). Furthermore, every thing has its real properties and is a temporal unit amid change in these real properties. In this respect, we call it a substantial unit. And substance again stands in a close relationship to causality. The thing changes preserving its real identity in the changing, but it changes in such a manner that its changes are dependent on those of other things, namely, in accordance with fixed causal laws. In other words, every change has its causes. Each is necessary. Each points to laws of change that unify and govern the functional dependencies of the changes in different concrete things. No thing is isolated. Every thing is a component of a causal-coherent nature. Similar characteristics belong to the essence of thing and nature.

Specific natural sciences explore specific things, specific real properties, real changes, specific real laws as causal laws of the changes. Every <natural science> accepts things experienced simply as things and only asks what holds of them more specifically. In all its ways of being experienced, in all its statements about things, each is, however, bound to what the Idea of thing and nature prescribed *a priori*, or in *Kant*ian terms, to the form of nature. The knowledge proper to the form, to the spatiotemporal form, as well as, *formaliter*, to the store of real spatiotemporal plenitude owing to which the thing is a natural reality, all that is prior to the specific natural sciences, is a common base of scientific knowledge of which each specific natural science can make free use and which does not belong in any of them. That also holds for the most universal, what are called abstract, natural sciences. They concern specifically given nature, though with the greatest universality. They already presuppose that nature exists as a system of things that is 370
subject to the form of nature as an a priori, as an *eidos* that is prior to all specific nature to be given empirically.

All that is beyond doubt, and so in the realm of natural knowledge, analytic-formal ontology comes before synthetic-formal ontology, namely, the ontology of nature. It would break down into several separate disciplines, <into> the disciplines that explore the a priori of space and time, therefore geometry, chronology, and kinematics, on the other hand, <into> those disciplines that would correspond to *Kant*ian "pure" natural science, therefore, explore the a priori of spatiotemporal reality (matter)–except for pure space- and time-form. One then observes that the analytic categories, the categories of objectivity in general in the natural sphere would come before the categories of nature. We could say: the categories in the original *Aristotelian* sense, where we would, however, have to exclude all "psychical nature", to which *Aristotle* surely thought his categories also referred. We could again say the *Kant*ian categories, since in his theory of categories, *Kant* was essentially guided by nature in the usual sense of physical nature.

All the same, ontology of nature is, however, science-theoretical only in a limited sense, insofar as it only precisely deals with an area of science, with that defined by means of the Idea of nature. Besides, there may be other sciences and areas of science concerned with objects whose unity is due to a different Idea binding them. And, accordingly, we would also have ontologies that would develop the ideal essence of such objectivity and that, with regard to the sciences of objects of this nature, would have to be science-theoretical in nature in a similar way–but precisely again limitedly science-theoretical. The full universality of the concept of the theory of science would then be abandoned.

Different would be an investigation setting itself the task of looking for the entirely different regions of possible being and possible science, and within each region, the chief basic concepts that had earned them the name "categories" and would represent the defining concepts of the ontology proper to any such domain. Those highest concepts that define the regions would then differ from those concepts in which the essence of the relevant primitive concept, of the highest of the domain, unfolds from the point of view of objects. I could say that the regional concept theoretically fixing a scientific region must differ from the genuine categories.

We know two regions in advance, the analytic <region>, whose regional concept is "something in general" or "object in general", and the natural scientific region, whose regional concept is "thing of nature". One also at the same time observes, notes in passing, that the *Aristotelian* table of categories begins with οὐσία and proceeds to quality, quantity, place, time, relation, and so forth, that the first component has an entirely different status and differs in
371 nature from the remaining ones, which alone are real predicables, while substance actually only denotes the identical thing, insofar as it is the bearer of such predicables. At one point in the *Prior Analytics*, *Aristotle* himself contrasted οὐσία with all other categories and called the latter συμβεβηκότα.

Were one to succeed in breaking down the realm of possible knowledge into regions in essential and not haphazard ways, then this would result, for one thing, in a fundamentally essential classification of sciences–and along with that at the same time a theory of categories, a theory of the regional and specifically categorial concepts.

Belonging to every regional <concept> would at the same time be an a priori ontology, which would have to act as a theory of science specific to the corresponding scientific area.

All these ontologies would be formal in the same sense, just as geometry and pure natural science are called "formal"; namely, <as> sciences of the "form" of possible nature-thingness as such or of the possible natural world in general.

Taken together, these formal disciplines would lend knowledge a formal scaffolding to which every possible individually actually existing objectivity and conjunction of objects is bound as a priori schema. All such investigations are then obviously science-theoretical in my sense. They really do not arrive at any knowledge within the particular sciences and do not presuppose any such <science>. They do not state anything at all about the fact that there really are such kinds of objectivity and what really holds for them. They are only concerned with the Idea of possible objectivities in general–thoroughly *a priori*–and inquire precisely into the regions and categories into which possible objectivities must fall.

In the last lecture, we reflected upon which general science-theoretical investigations may possibly present themselves above and beyond those of apophantics and ontology, and in initially continuing to use the ontic approach, we came upon the characteristic features of the nuclei and the characteristic features of the scientific fields corresponding to them. Analytic-formal ontology abstracts from all such characteristic features. It explores the a priori pertaining to meanings and objectivities in general when we abandon all "matter" of knowledge in indefinite generality. It is precisely universal mathesis. It mathematizes all nucleus-contents within their nucleus-categories, and for that reason its universality is precisely all-encompassing, because every meaning–let us say, every judgment, and likewise correlatively every objectivity–first and foremost has a mathematical form, an analytic form that can

be considered in ideal universality and then prescribes laws for possible meaning-validity and possible existing objectivity in general. If we then also take into consideration what is called the matter of knowledge, we do not mathematize the ultimate concept-contents that we called "nuclei" (and correlatively the actual store of the objectivities), then it becomes apparent that, for this reason, we must not yet be in the particular individual sci-
372 ences–provided namely we can still maintain our investigation at a high level of generality–that it may deserve to be called formal, and even science-theoretical. I illustrated that using *Kant*ian talk of a form of nature. We recognized that there is a formal ontology of nature that, along with the disciplines of geometry, chronology, kinematics, and the rest of pure natural science, plays an analogous role for the series of diversely defined natural sciences, just as analytic-formal ontology does for all sciences in general. Corresponding here to the analytic *Widersinn* springing from the violation of analytic-formal laws is the natural scientific *Widersinn* springing from an infringement of the laws pertaining to the a priori form of nature, and this *Widersinn* is still prior to every empirical natural science.

Once again, I said that the analytic-formal categories of being in general in the sphere of nature have a counterpart in the categories of natural-material being.

If, then, a formal ontology like that of nature is really not science-theoretical in the most encompassing sense of all, then we however readily see that in this most encompassing sense, an investigation would have to be called science-theoretical that would survey the entire field of possible particular sciences in general and go in quest of the essential demarcations that break it down radically into regions of being. Therefore, such an investigation is not thought of here as a merely comparative reflection on factual sciences, therefore presupposing that there already are such sciences and that particular objectivities exist that they explore. On the contrary, an investigation is possible that starts from the Idea of knowable objectivities in general and then asks into which domains they would have to split *a priori*, which categories would have to pertain to them as forms by virtue of the regional concepts defining the domains and, furthermore, which formal ontologies would then be to be constituted, defined by precisely those concepts, more precisely by the regional <concepts>. That does not preclude looking at the given factual sciences and at the fields under consideration in them as factually obtaining spheres of objectivity, only that it is just not a matter of facticity and the investigation is an a priori one. Similarly, in the formal theory of meaning and formal ontology, we really start from examples that may be taken from the sphere of actual judging and actual objectivities. But, the existential positings remain excluded where it is a matter of pure investigation of essence.

The problem indicated is now an exceedingly difficult one, and its solution outstrips our capacities. The fact that it has not yet been solved up until now, indeed, not actually properly grasped in its full generality, is partly due to certain prejudices stemming from natural habitual ways of thinking, partly to the fact that the method of such investigations is more idiosyncratic and differently oriented than one would at first like to assume.

As far as the first is concerned, the prejudices, those come into consideration here that grow out of the habits of the natural atti- 373
tude that initially incline us to see only nature, nature in the narrower and in an extended sense. Indeed, in certain ways, the circle is still closing in owing to the fact that since the rise of the magnificent sciences of physical nature, there has been a growing inclination to extend unjustifiably concepts and methods of physical natural science, on the other hand, however, to explain away what is actually given and the corresponding chief differences of objectivity.

The realm of the spirit, which has again and again been wrongly naturalized in the psychology of the last centuries, is already afflicted by this.

As for what secondly concerns the method, let it just be indicated to start with that an avoiding of all errors of the kind indicated earlier and a true grasping of the essential demarcation between the regions are impossible without a phenomenological analysis also resorting to the consciousness of givenness itself.

Before I now enter into more detailed explanations, I realize that it is natural to begin with the difference between individual objectivities and the eidetic-ideal objectivities—the difference that

I have also called in the most general way that of the a priori and a posteriori. In a certain sense, all essence has some universality. An extension corresponds to it as long as the ability to refer to an undetermined multiplicity of particulars corresponding to it is part of the essence. So, for example, the essence "number" has its extension in an undetermined and unrestricted generality of possible sets that are denumerable by means of it. The Idea (the essence) of a specific red has its extension in just such a multiplicity of individually singular red-moments. It is otherwise to be noted that not every essence is a genus. The difference between genera and species, as it obtains between color and red, is not to be confused with the difference between Idea (or essence) and extension of possible particulars. The generic universal color is differentiated as far down as red, as far as the totally determined shade of red. In the gradation of generic differentiation, the last one is ideal, what is called the final difference. However, this itself in turn has an extension, which however is not now an extension of species, but of singular particulars. The relationship of this final difference to the particulars is thereby essentially different from that of the genus to the difference. The genus and every essence have an extension of individual particulars. For example, the genus color <has> the individual moments, which are individual singularizations of colors. And, it is thereby part of the essence of these situations that this singularization is only possible because some least difference between colors is singularized at the same time. The individual color moment is a singularization of a specific shade of color and only for this reason of color in general.

Assuming this,[1] it is clear that the eidetic sciences will not sepa-
374 rate off in accordance with individual essences, but in accordance with higher genera, under which all related series of other genera and species fall, where whether the disciplines have an intrinsic relationship and intertwine or not will depend on the store of essence relationships of the governing genus-essences. In the first case, precisely higher generic concepts will create a unity. This is what happens, for instance, with the theory of cardinal numbers,

[1] Compare the "Textkritischen Anhang", p. 515 f. (to p. 281, line 4) of Hua XXX to the following text up to the end of the Appendix. (Editor's note)

set theory, and so in general <with> the disciplines of formal ontology, and also <with> disciplines such as geometry and kinematics within the ideal science of the essence of nature.

It is furthermore clear that by virtue of the a priori relationship that obtains between <the> Idea and its individual extension, all eidetic knowledge makes unconditional general statements without further ado, or by turning things around implies statements about possible individual particulars that correspond to the Ideas. If we therefore compare eidetic sciences or sciences of essence to sciences of individuals, then we are obviously thinking of sciences in which individual existence is posited and in which new knowledge is determined by the givenness of actual existence, knowledge that is not merely valid as analytic or synthetic necessities in relationship to corresponding knowledge in analytic or synthetic essence-sciences. Essence-knowledge is, however, naturally applied to what is individual, enters into the sciences of individual existence, and plays its role there.

If we then look around at what can be judged about in the sphere of the individual existence of the a posteriori, then we first encounter the region of nature as physical nature. One sees or, so to speak, feels that[1]

Appendix XVIII (to §64): <The Problem of Reason>[2]

<a) The Ideal Structure of the Realm of the Spirit: Individual Experience and the Thought-Acts Logically Processing These, on the One Hand, the Experienced Objects and Thought Truths, on the Other Hand>

In the last lectures, I spoke of the essential demarcations in the realm of the sciences. I spoke of the demarcation that concerns all empirical sciences of nature. Nature is a basic regional concept, and pertaining to it is the ontology of nature that describes and brings to theoretic-scientific elaboration an a priori prior to all

[1] Continuation in the second sentence of the third paragraph of §62 above (Editor's note)

[2] From the lecture course of WS 1910/11. (Editor's note)

possible natural sciences. Nature founds the realm of the spirit. A new regional concept is "Animal" <as> bodily-spiritual being, and pertaining to it is the concept "mind"; initially as individual mind. I
375 also then further discussed the unity of animal-human society and the corresponding pure intellectuality, namely, the communal spirit, with its different correlates. Pertaining to that were the psychological disciplines and those of the humanities and social sciences, as well as, by virtue of the relationship between mind and nature, the necessary psychophysics, inasmuch as, besides the inner mental relationships, the functional dependencies of the bodily and mental ones must be explored. Pertaining to the new regional concepts are new a priori ontologies that, on the one hand, have to explore the immanent a priori of individual and social consciousness as well as, on the other hand, the a priori that may pertain to the form of psychophysical unity.

Have I thereby situated all sciences, all empirical existential sciences on one side, all ontologies and all a priori sciences in general on the other?

Let us reflect once more on the scope of the realm of the spirit and what belongs in it. The individual mind, let us say, the individual human being, is an experiencing being, and its experiencing is constantly I-experiencing. We all know what that means. Every one of us says, "I experience, I think, I feel, I desire and will"—and thereby the difference that *Hume* had in mind as the difference between impressions and Ideas is everywhere. For example, I experience, namely, I perceive, I remember, I expect, and I perform such "acts" with varying degrees of clarity up to the point of full <clarity> and intuitiveness. For example, when experiencing clear perception, something tangible itself figures presently existing before me. And that "figuring before me in that way" is something I can attend to. I can be entirely given to the object perceptively, or also be attentively directed toward the perceiving as consciousness in reflection. In contrast, we also engage in act modifications. We speak of fantasy. If we fantasize, something self-presently existing does not actually figure before us, but, as it were, just imaginarily. And similarly, when we engage in memory-fantasy, expectation-fantasy. We see that standing opposite the originary acts in which something is "actually" conscious are everywhere certain modifications in which something in principle the same or similar is not actually conscious to us, but is so in a modified way, in a "quasi" mode–in mere fantasy, or whatever we might otherwise call it. Let us therefore limit ourselves to the unmodified consciousness, the originary acts, the act impressions, to use the *Hume*an expression.

There, we therefore initially have experiencing acts, such as perceptions, memories and the ones related to them in which direct existence is conscious. The ego is directly conscious of a present, past, or future existence, and that concerns either tangible existence, existence of nature–then, the acts are called external experiences–or it concerns existence of its own acts of any kind, insofar as the ego reflectively directs its gaze toward its experiences. We then speak of inner experience. Likewise, in the direct reciprocal understanding in which the ego enters into a communicative relationship with another ego, <we> naturally also <speak of> experience of someone else's inner life, experience of human beings as such, of their experiences, and so forth.

In so doing, express reference is however still to be made to the 376
correlates to be grasped with these ego experiences. While I am perceiving and perceiving precisely this thing there, I can also take heed of the fact that it is given to me now in this, and then in that, way of appearing. I can take heed of the perspectival adumbrations, namely, of the adumbration of tangible forms, colors, and so forth, precisely as they are given to me. I can not only direct reflection to the perception-consciousness, but also to the ways of appearing of the object, toward which I am not turned when I look at the object and toward which I nevertheless can turn at any time.

The ego does not, however, only have experiences, whether primary and reflectionless or reflective experiences. It also thinks, and in thinking makes judgments. That means that the plain experiences form the foundation for new, explicating, relating, concatenating acts, thereby, the acts of considering-true and <as-> probable, and above them, yet another fully new layer, the comprehending, the thinking layer in the specific sense. It is the layer associated with words coming to "expression" in stating. Or better yet, full stating is thinking basing itself upon plain experiencing and explicating acts and understanding them in themselves.

Therefore, people judge. They judge individually. They come to an understanding regarding their judgments. They influence one another's convictions, and in this reciprocal coming to an understanding and reciprocal influencing, they make ever new judgments that are built upon one another in different forms. In other words, people prove, theorize, practice sciences, and in such a way that the accomplishments of individual ones–reciprocally determining other ones–enter into the cultural construct "science" having its intersubjective, collective-mental objectivity. Thereby, however, both in relationship to individual human beings and in relationship to the human community, the fleeting store of simple acts of judgment, acts of inferring, of proving, of theorizing differs from what figures in these acts as true and probably valid, as directly grasped truth or as indirectly proven truth, or even as proof of this truth, as recognized theory and the discovered, demonstrated end result of the theory, and so forth. Each of us makes these distinctions whenever we but become aware of them. Claimed truth figures as an objectivity, as do claimed theory and science. And for us, everything is *de facto* science, a term for a sphere of intersubjectively supposed or valid truths that ought to hold as it holds, whether we or anyone else is concerned judgingly with them in momentary, actual experiences or not.

<b) The Idea of Reason in Relationship to the Correlation Between Consciousness and What One Is Conscious of as the Subject of a New Science. Ontology as Science Directed Straight to That Itself Needs Reason-Theoretical Elucidation>

Accordingly, within the ideal framework that the sciences of the individual mind and <of the> communal spirit explore, we
377 therefore find, on the one hand, all experiences occurring in individual actual minds and also all experiences of logically processing thought-acts, on the other hand, however, their correlates, therefore the experienced objectivities, the thought truths and probabilities that refer to such objectivities, and highest of all, the empirical sciences. Within the framework of the sciences of human civilization belong, for example, also all natural sciences, namely, insofar

as we take these sciences to be cultural phenomena in the different stages of their development.

But, we thereby immediately become aware of the constraint. Each empirical act and thought-act, all inferring, proving, and theorizing belongs in the individual field of the humanities and social sciences, or in psychology. Precisely for this reason, each thing thought as such, believed as such, deduced, proven, <each thing> theorized as such, belongs in them. And, precisely the same likewise belongs within the framework of the history of the sciences of the communal spirit in general if we take human beings as a component of the human scheme of things, for instance in their individuality, as object of the historical life of the spirit. Nothing at all though is thereby stated about the truth of what has been thought, about the truth of the theory, of the science that is humanly thought, explored, conceptualized.

Psychology, history, empirical humanities and social science in general deal with facts. Experiencing, experience with respect to judging, i.e. being convinced or presuming are psychological and social facts. Implied there is that the people involved are convinced by one fact or another. Substantiating as proving and as theorizing are likewise facts. But in this respect, alchemy is just as much a fact as chemistry, astrology as much as astronomy. Alchemists had their convictions, had their arguments, their theories, just as much as chemists have them today. Therefore, it is clear that the exploration of thinking as fact and the exploration of science as cultural phenomenon are something different from the exploration of the rightness of such thinking, or the answering of the question as to whether the thought, supposed truth, is in fact truth. It is clear that the historical findings about the content of Greek or Arabic natural science only state what was considered to be scientific truth in the psychological forms of knowing and substantiating thinking of the Greeks and Arabs, <which> though still does not say anything about whether <and> to what extent this supposed science is science in the sense of Idea.

Through this contrasting of supposed science and "actual" science, we then immediately become conscious of the Idea of reason and with it of a new science. The outcome is therefore the following: In multiple acts of experiencing, something experienced

supposedly stands directly before us as something existing (whether it is something present, something past or something in the future). In thinking it is thought, is determined as being in such and such a way. At the same time, something like an empirical truth figures before our mind's eye in judging empirical thinking, in like manner, an empirical theory, an empirical science. But, this figuring before us amounts to no more than a supposing figuring
378 before us. Experience and empirical thinking can, people also say, be rational and non-rational, and only when it is rational, can it be legitimately said that the objectivity may in reality be so, one theory or another may in truth be valid on account of that, this or that natural law, and so forth. We therefore have an encounter with the Idea of reason, more specifically here: experiencing reason relative to the remarkable correlation between consciousness and what one is conscious of, between experience and what is experienced, empirical thinking and what is empirically thought. All such thinking thinks–as can be stated evidently in virtue of its essence–something, an empirical objectivity with such and such properties. And yet, this thinking need not be either rational, or legitimate. How does thinking show its legitimacy, its rationality? How does it show that there are things, properties of things, laws about things, and the like positing with such and such content, and that something is thereby posited that actually is? And likewise in the empirical sphere in the broadest sense, which also includes along with it psychical being and existence in the communal spirit.

Now, one will at first perhaps presume that we are only reverting back to formal ontology and the ontologies of the realm of the spirit. For what is existing is indeed something objective. Formal ontology explores what pertains to what is objective as such. If what is objective is nature, then we in turn know that nature explores ontology–what pertains in natural-formal generality to precisely all existence of things as such in conformity with the Idea of thingness. And some such thing <holds> for the realm of the spirit.

However, the first thing to be said there is that all ontology is for us though the correlate of ontological investigation. Investigating ontologically, we recognize that such and such formal or real ontological laws hold. We recognize this. Namely, we perform such and such presentations and such and such thought-acts.

These can, however, be valid or invalid, rational and non-rational. How is this difference clarified? How do we come in this way to distinguish evaluatively, so to speak, and ontologically to accord certain thought-acts the value of valid and rational <thought-acts>, to others the <value> of non-rational <thought-acts>? If our ontology is really a system of truths, then truth must be able to show itself in our actual thinking. What does this self-showing itself look like? How is its legitimacy understood as opposed to the illegitimacy of all of the possible errors occurring for us in these spheres?

Second, it is to be noted that ontology exclusively states propositions precisely about ontological systems and laws. In stating these propositions and thinking them, we are turned, for instance, toward numbers and number relationships, spatial figures, number systems, the Ideas "substantiality" and "causality", and so forth, and all that from an a priori perspective. However, we are not at all turned to the consciousness of counting, to the consciousness of intuiting space and thinking about space, or to the spatial phenomena pertaining to them with their remarkable sequences of adumbrations. In the exploration of the essence of substantiality and causality, we are not turned toward the consciousness of the unity 379
of multiple phenomena in which substance comes to givenness and toward the consciousness of necessity whose correlate is causality. We must, however, expressly turn toward all of them, just as we pose questions about reason, so we want to have it understood what ontological knowledge of this kind may justify, how it–which, however, occurs in the factual order as an empirical fact in human consciousness–may come to accord the status of an unconditional law independent of human existence to what is putatively known in it. Investigation of reason, therefore, aims at consciousness, if it is to be valid, and precisely thereby at consciousness in relation to its meaning- and object-correlates. Ontological investigation aims, however, at objects, more specifically, at objective essence and, if need be, at the same time, at meanings.

Third, it finally needs to be said that ontology only generally, and in formal universality, says what in general holds for what is existing, for instance, for what is existing in general in formal universality, or for what is existing in nature, what is existing in the mind. In empirical science, however, we do not judge formally and we ultimately establish details down to their individual specifications. The question of reason concerns all these findings. They are all realized in human experiencing and thinking, and all such experiencing and thinking is subject to the question of reason, the question of legitimacy. The general ontological-noetic question was: How is it comprehensible that only thinking corresponding to the ontological laws can be objectively valid, rational? The present question reads: If thinking is then already bound to ontological laws, what guarantees its objective truth, its truth for the specific case going beyond that? Or, expressed in *Kant*ian terms, the one question reads: How is pure natural scientific knowledge possible, how is pure geometrical, pure kinematic knowledge possible? Following upon that is however also the further questions: How is empirically-natural scientific knowledge possible so far as it is not predetermined by ontological knowledge? How is the difference between rational, objectively valid, and non-rational experience, or empirical scientific knowledge to be understood?

Before I delve into the unique nature of noetic problems, I wish to draw yet further lines of thought.

I have intentionally only limited myself to looking at experiencing as the complex of acts positing existence and thinking only insofar as it presents itself as empirical thinking. Human beings do not, however, merely experience. They adopt other position-takings. They evaluate things as beautiful and ugly, where we understand beauty in the broadest sense that concerns the esthetic only as a special case. Human beings are, however, also striving human beings. They act and create. And in thinking, human beings are also active in these axiological and practical realms. Naturally, all these position-takings and the accomplishments of the individual and social mind proceeding from them as facts belong in the domains of the relevant sciences of existence, of psychology and psychophysics, history, sociology, and so forth. To the extent that deeds implying physical existence thereby come into consider-
380 ation, <they belong> also in the realm of the physical natural sciences. But, this again does not suffice, and it is of the greatest importance to grasp the parallels pertaining to my foregoing series of observations.

<c) Recapitulation: The Question of Reason Is to Be Posed in Relation to the Possibility of All Kinds of Knowledge>

In the last lecture, we have made the need for a theory of reason clear to ourselves, namely, to begin with, in the empirical realm. In experience and in empirical thinking, every kind of existence ought to be thought, known, and ultimately scientifically known. How is the fact to be understood that among all the acts in which experiencing and empirical thinking is performed, certain <acts> do and can spontaneously claim to be rational, legitimate, objectively valid? All empirical judging is supposing to have objective truth and with that to state something that holds for actual existence. The fact that anything exists is only true for us because it is experienced, it is potentially substantiated, and conceived in thought. On the other hand, being, for example, the being of nature, is not the same as consciousness of it, the experiencing it, thinking it, substantiating-its-being. How is one to understand the fact that consciousness, while being consciousness of, is capable of spontaneously positing something existing as something being in itself that, however, ought to be what it is, whether it is something someone has been conscious of or not?

If one makes a distinction between rational and non-rational experiencing and empirical thinking, and if the rational ought to involve authority for determining what is intrinsically being, then precisely this distinction must be explored and be made understood. If the empirical science of the individual mind and communal spirit explores consciousness as something existing within the context of the life of the spirit–for example, judging at the moment, theorizing at the moment–as a fact of human thought-life, then obviously needed is then a science of judging reason, of the relationship among judgment, truth, objectivity being and being so–a science that deals with these relationships, not as relationships of psychical facts and facts of the social sciences and humanities, but from the viewpoint of validity or <of> rationality. And, more specifically, needed is a science that not only establishes the conditions of legitimacy or rationality, but procures for us clarity about the essence or meaning of such claims to legitimacy and proofs of legitimacy on ultimate grounds. Two layers of problems emerge

here with respect to the existential sphere. On the one hand, with regard to the ontology of the existing order, <we> have the problem as to how it may be "possible", i.e. how the possibility of ontological knowledge may be understood, the possibility of synthetic a priori knowledge, knowledge in the existential sphere, therefore <the possibility> of geometrical, chronological, and kinematic kinds of knowing in general, how the claim to legitimacy of such knowledge <may be> understood–as a claim to prescribe laws for
381 all possible natural existence, although it nevertheless is what it is, even if it is never ever known. Added to that, however, is the same problem for the empirical sciences. How is reason or the authority of a posteriori natural knowledge, the possibility of empirical knowledge in all non-ontological natural sciences elucidated? *Kant* did not raise the latter question. In his thought, the first one comes up in the questions about the possibility of synthetic a priori knowledge of mathematics and pure natural science.

Of course, we have to bring up precisely both these sets of problems for the ontology of spirits and the empirical humanities and social sciences. At the same time, a more general question stands out for us that was completely overlooked by *Kant*. It concerns the possibility, not <of> synthetic, and more precisely, <of> existential knowledge, but <of> analytic <knowledge>. How is, were we to have to formulate it, analytic knowledge a priori possible? How is analytic reason to be understood and to be elucidated? For all natural knowledge is first, in general, knowledge. Analytics, analytic formal ontology, stands above ontology of nature and above all ontologies in general, and to be asked in relationship to it is also the question of reason. The latter is, so to speak, in a position of its own, and if yet other object domains and ontologies should emerge than those discussed up until now, then it naturally lies just as high above them as above the ontologies and sciences of the existential sphere.

Before we then delve further into the particularities of noetic problems, we must then draw new lines of demarcation, present new domains and ontologies, and precisely thereby new noetic clusters of problems. It is indeed already becoming clear that we have to have just as many theories of reason as domains and ontologies.

INDEX

E. Husserl, *Logic and General Theory of Science*,
Husserliana: Edmund Husserl – Collected Works 15,
https://doi.org/10.1007/978-3-030-14529-3

GPSR Compliance

The European Union's (EU) General Product Safety Regulation (GPSR) is a set of rules that requires consumer products to be safe and our obligations to ensure this.

If you have any concerns about our products, you can contact us on ProductSafety@springernature.com

In case Publisher is established outside the EU, the EU authorized representative is:

Springer Nature Customer Service Center GmbH
Europaplatz 3
69115 Heidelberg, Germany

Batch number: 10740146

Printed by Printforce, the Netherlands